Heavy Water
and the Wartime Race for Nuclear Energy

Also by Per F Dahl

Ludvig Colding and the Conservation of Energy Principle:
Experimental and Philosophical Contributions
(1972)

Superconductivity: its Historical Roots and Development
from Mercury to the Ceramic Oxides
(1992) American Institute of Physics

Flash of the Cathode Rays:
A History of J J Thomson's Electron
(1997) Institute of Physics Publishing

Related Titles Published by
Institute of Physics Publishing

Cockcroft and the Atom
G Hartcup and T E Allibone (out of print)

Operation Epsilon: The Farm Hall Transcripts
Edited by Sir Charles Frank

Radar Days
E G Bowen

Echoes of War: The Story of H_2S Radar
Sir Bernard Lovell

Boffin: A Personal Story of the Early Days of
Radar and Radio Astronomy and Quantum Optics
R Hanbury Brown

To be published Autumn 1999

Technical and Military Imperatives:
A Radar History of World War II
L Brown

Heavy Water

and the Wartime Race for Nuclear Energy

Per F Dahl

Lawrence Berkeley National Laboratory,
CA, USA

Institute of Physics Publishing
Bristol and Philadelphia

British Library Cataloguing-in-Publication Data

A catalogue record for this book is available from the British Library.

ISBN   0 7503 0633 5

Library of Congress Cataloging-in-Publication Data are available

The author has attempted to trace the copyright holders of all the figures reproduced in this publication and apologizes to them if permission to publish in this form has not been obtained.

Published by Institute of Physics Publishing, wholly owned by The Institute of Physics, London

Institute of Physics Publishing, Dirac House, Temple Back, Bristol BS1 6BE, UK

US Office: Institute of Physics Publishing, Suite 1035, The Public Ledger Building, 150 South Independence Mall West, Philadelphia, PA 19106, USA

Printed and bound in the UK by Bookcraft (Bath) Ltd

Contents

Preface   ix

Acknowledgments   xiii

List of illustrations   xv

1   **PROLOGUE**   1
    1.1   Fornebu airport, 12 March 1940   1

2   **MANCHESTER AND PARIS, 1919**   3
    2.1   Manchester: how it began   3
    2.2   Paris, and the Joliot-Curies   7

3   **THE NEUTRON**   11
    3.1   Prelude to 1932: Chadwick and Bothe on the hunt   11
    3.2   A discovery narrowly missed, and the neutron at last   17

4   **HEAVY WATER**   22
    4.1   Deuterium: a comedy of errors   22
    4.2   Gilbert Lewis and Leif Tronstad: the promise of deuterium   27
    4.3   Birkeland and his gun; Eyde and Birkeland   33
    4.4   Tronstad and Norsk Hydro: an auspicious union   41

5   **ARTIFICIAL RADIOACTIVITY**   49
    5.1   Another French miss, and triumph at last   49
    5.2   Rome: another discovery, and a discovery missed   57
    5.3   Meanwhile, back in Paris   63
    5.4   Escape, in the nick of time   66

6   **NUCLEAR FISSION**   73
    6.1   Berlin: December 1938   73
    6.2   More neutrons?   78
    6.3   Prospects for a chain reaction on the eve of war   89
    6.4   A moderator of choice   99

7   **HEAVY WATER REVISITED**   104
   7.1   The Allier mission   104
   7.2   Attack on Norway   110
   7.3   The battle for Rjukan; Tronstad goes into action   113

8   **THE BRITISH INITIATIVE**   118
   8.1   MAUD   118
   8.2   Broompark   123

9   **GERMAN ARMY ORDNANCE TAKES CHARGE**   130
   9.1   The Uranium Club; a tritium episode   130
   9.2   A serious error   138
   9.3   Joliot's guests   143

10 **HEAVY WATER TAKES CENTER STAGE**   147
   10.1 Pressure on Norsk Hydro mounts   147
   10.2 SIS, SOE, and the Galtesund affair   156
   10.3 Mild sabotage; frank talk   158
   10.4 Exit Jomar Brun   165

11 **AMERICA JOINS THE QUEST**   170
   11.1 Stirrings in the new world   170
   11.2 'Graphite versus deuterium' once more   173
   11.3 North American heavy water in abundance   178

12 **ACTION VEMORK**   186
   12.1 Germany's uranium machines: off to a promising start 186
   12.2 Freshman: an unqualified disaster   192
   12.3 Gunnerside: a qualified success   197

13 **NEUTRONS DESPITE BOMBS**   206
   13.1 Aftermath at Vemork; neutrons in Berlin-Gottow   206
   13.2 The Americans strike   211

14 **WAVERING OUTLOOK FOR HEAVY WATER**   219
   14.1 Penultimate pile experiments   219
   14.2 The ferry   224
   14.3 Prospects for heavy-water production in Germany   233

15 **CANADA ENTERS THE RACE**   239
   15.1 ZEEP   239

16 **FEARS AND FACTS ON THE CONTINENT**   247
   16.1 Alsos   247
   16.2 In the Haigerloch cave   253

17 **SWABIAN JURA AND UPPER TELEMARK: FINAL EVENTS** 259
  17.1 The rush for Haigerloch 259
  17.2 Interrupted Sunshine 264

18 **HIROSHIMA REVEALED; FURTHER CONTESTANTS FOR NUCLEAR ENERGY** 271
  18.1 Farm Hall: Operation Epsilon 271
  18.2 Belated entries: Russia and Japan 275

19 **EPILOGUE** 282
  19.1 Whither heavy water; what if? 282
  19.2 A few of the personalities 287

Appendix A: SOME PROPERTIES OF HEAVY WATER (D_2O) COMPARED TO WATER (H_2O) 293

Appendix B: A CHRONOLOGY OF HEAVY WATER 294

Abbreviations 300

Notes 303

Select bibliography 359

Name index 365

Subject index 385

Preface

IMI, P-9, SH-200, X, Z, Juice, Glucose, Lurgan, Polymer. As befitting the many code names for deuterium oxide, commonly known as heavy water, the substance played a sinister role in the race for nuclear energy during the Second World War. About 120 people—German soldiers, Allied commandos, and Norwegian civilians—died in its cause. It was a key factor in Germany's wartime bid to harness atomic energy, primarily as a source of electric power; its acute shortage was a factor in Japan's decision not to seriously pursue nuclear weaponry; its very existence was a nagging thorn in the side of the Allied powers. Books and films have dwelt on the Allied efforts to deny the Germans heavy water by military means; however, a history of heavy water in itself remains to be written. This book attempts to fill portions of that gap, concentrating on the circumstances whereby Norway became the pre-eminent producer of heavy water, and on the scientific role the rare isotope of hydrogen played in the wartime efforts by the Axis and Allied powers alike. Instead of a purely technical treatise on heavy water, the book might better be described as a social history of the subject.

More specifically, the book covers the haphazard discovery of heavy water by Urey and company, and its early promise in the hands of Gilbert Lewis and his Berkeley colleagues. In Europe, the isotope caused equal stir, principally in Britain and in Germany. Nor did these developments go unnoticed in Norway, where conditions were opportune for its large-scale production. The book traces the background for Norsk Hydro's ascendancy as the sole industrial producer of heavy water, from the amazing Kristian Birkeland to the 1930s when the prodding of several German scientists caught the attention of Jomar Brun and Leif Tronstad—the latter a recurring personality in the ensuing developments. The chronicle of heavy water runs concurrently with the rise of nuclear physics and chemistry, from the neutron in Cambridge to slow neutrons in Rome, and from tell-tale radio-chemistry experiments in Paris to uranium fission in Berlin on the eve of World War II. The French–German contest for Norwegian heavy-water stocks in 1940 dramatically underscores the real or perceived importance of the isotope for the newly warring powers. German pile experiments were ultimately doomed as the Allies mounted a ceaseless campaign to eliminate their Norwegian key to success. Long before 'action Vemork' was a *fait accompli*, the Allies' own multi-faceted nuclear effort was steadily gaining momentum. Heavy water loomed large, albeit late in the behemoth atomic undertaking in the Americas, but in a curiously secondary role: as a massive back-up production effort within the Manhattan Project, and as the central element of a tottering Canadian–British–French heavy-water reactor effort that came to fruition as the war ended.

My primary sources include the wealth of material in the Norsk Hydro archives at Oslo, Rjukan, and Notodden, the archives of Norsk Industriarbeidermuseum, Vemork, and of Riksarkivet in Oslo. In Germany I had access to the archives of the Max-Planck Society and of the Max-Planck Institute for the History of Science, both in Berlin, and of the holdings in the Deutsches Museum, Munich. In the United States I relied first and foremost on the extensive holdings of the Niels Bohr Library, American Institute of Physics (particularly oral history interviews), helped by their excellent *Guide to the Archival Collections*. The Niels Bohr Library holdings include a copy of the Archives for the History of Quantum Physics, which I also accessed at the Office for History of Science and Technology of the University of California at Berkeley. At Berkeley I have also availed myself of the Gilbert Lewis papers and the E O Lawrence papers, both in the Bancroft Library, University of California. The Lawrence papers are important in the present context, as portions of the correspondence deal with the American wartime heavy-water project, and particularly with Urey's role in it. In the archives of Rensselaer Polytechnic Institute, the staff gave me speedy access to the Paul Harteck papers, spanning the years 1927–1945 and to the end of Harteck's life.

Among the secondary sources pertinent to the subject at hand, I have profited particularly from David Irving's venerable *The German Atomic Bomb*, Mark Walker's *German National Socialism and the Quest for Nuclear Power*, and Spencer Weart's *Scientists in Power*. Until the volumes commemorating the 100th anniversary of Norsk Hydro in 2005 appear, the standard work on the history of Norsk Hydro will remain Kr Anker Olsen's 1955 tome, *Norsk Hydro Gjennom 50 År*. Kr Anker Olsen's volume has recently been supplemented by the chronology by Ketil Gjølme Andersen and Gunnar Yttri of Hydro's research projects, *Et Forsøk Verdt*, though heavy water *per se* receives scant attention in that work. A definitive insider's account of wartime events surrounding Norsk Hydro, and heavy water in particular, is Jomar Brun's *Brennpunkt Vemork 1940–1945*. There is no dearth of books on the purely military 'battle for heavy water'. The earliest is Per Bøhn's *IMI: Norsk Innsats i Kampen om Atomkraften* (1946), if we overlook Colonel Wilson's retrospective and privately circulated memoir of 1945. On the average two books on the military heavy-water operations have appeared each decade since the end of World War II.

In as much as wartime German heavy-water pile research is a focus of interest here, not the uranium project as a whole, volume 14 of the FIAT *Review of German Science (Nuclear Physics and Cosmic Rays)* sufficed for the most part as a record of those developments, supplemented by other sources (including the G-reports) enumerated in the notes and bibliography.

I am indebted to many individuals and institutions for archival assistance. In Norway, they included Kari Kolstad (Norsk Hydro, Oslo), Søren Sem (Norsk Hydro, Porsgrunn), Torleif Lindtveit (Norsk Teknisk Museum, Oslo), Frode Sæland (Norges Industriarbeidermuseum, Vemork), Alv Egeland (Fysisk Institutt, University of Oslo), Anne-Karin Sønstebu (Norges Hjemmefrontmuseum, Oslo), and Odd Halvorsen (Riksarkivet, Oslo). In Germany, I relied primarily on Dieter Hoffmann (Max-Planck-Institut für Wissenschaftsgeschichte, Berlin), Marion Kazemi, and Andreas K Walter (Archiv zur Geschichte der Max-Planck-Gesellschaft, Berlin), and Wilhelm Füßl (Deutsches Museum archives, Munich). In the United States, I had the benefit of Joseph Andersen and his staff (Niels Bohr Library, American Institute of Physics, College Park, MD), John Dojka and Tammy Gobert (Archives and Special Collections, Folsom Library, Rensselaer Polytechnic Institute, Troy, NY), Rita Labrie (Research Library, Lawrence Berkeley National Laboratory, Berkeley, CA), and the staff of the Bancroft Library (University of California, Berkeley, CA).

I thank the referees for their careful and insightful suggestions. I thank Frode Sæland, as well, for his careful reading of a draft extract of the manuscript, and for numerous corrections and suggestions. Above all, I am indebted to Per Pynten (Norsk Hydro, Rjukan), for assistance, stimulating discussions, and marvelous hospitality during my stay at Rjukan.

Along the way, I have been helped directly and indirectly by numerous other individuals, including Marc Friedman, John Heilbron, Bjørn Landmark, Shoaib Karin Naeer, Thomas Northrop, Jens Anton Poulsson, Helmut Rechenberg, Ronald Scanlan, Ruth Sime, Asbjørn Søreide, Marilyn Taylor, and Leif Trondstad Jr. Claus Helberg rates particular mention; he took the time to share with me insightful glimpses of his fellow saboteurs, in the appropriate setting of Oslo's Resistance Museum, and rendered other assistance. Special thanks go to Professor Roger Hahn and the staff of the Office for the History of Science and Technology, Berkeley; Hahn for my Research Associateship, and the whole crew—staff and students alike—for hospitality and good company while the present manuscript took form.

I thank my publisher, Jim Revill, and his editorial staff at Institute of Physics Publishing, and Al Troyano of the production department, for their splendid support and cooperation in all stages of manuscript processing and production of the volume.

Finally, the book depended, as always, on my personal editor, companion, and wife. Eleanor and I have shared many experiences over the years. The 'field work' for this volume, and its preparation, rank high among the 'ups and downs' that constitute an interesting life.

Per F Dahl

Acknowledgments

The author gratefully acknowledges permission to quote excerpts from the following works:

Goldsmith M 1976 *Frédéric Joliot Curie: A Biography* (London: Lawrence & Wishart). Copyright © 1976 Lawrence & Wishart. Reprinted with permission from Lawrence & Wishart.

Sime, Ruth Lewin 1996 *Lise Meitner: A Life in Physics* (Berkeley: University of California Press). Copyright © 1996 The Regents of the University of California. Reprinted with permission from the University of California Press.

Segrè, Emilio 1970 *Enrico Fermi: Physicist* (Chicago: The University of Chicago Press). Copyright © 1970 The University of Chicago. Reprinted with permission from the University of Chicago Press.

Weart, Spencer R and Szilard, Gertrud Weiss 1978 *Leo Szilard: His Version of the Facts* (Cambridge, MA: MIT Press). Copyright © 1978 Gertrud Weiss Szilard and MIT Press. Reprinted with permission from MIT Press.

Brun, Jomar 1985 *Brennpunkt Vemork: 1940–1945* (Oslo: Universitetsforlaget AS). Copyright © 1985 Scandinavian University Press/Universitetsforlaget. Reprinted with permission from Universitetsforlaget.

Peierls, Rudolf 1985 *Bird of Passage: Recollections of a Physicist* (Princeton: Princeton University Press). Copyright © 1985 Princeton University Press. Reprinted with permission from Princeton University Press.

Anker Olsen Kr 1955 *Norsk Hydro Gjennom 50 År: Et Eventyr fra Realitetens Verden* (Oslo: Norsk Hydro-Elektrisk Kvælstofaktieselskab). Copyright © 1955 Norsk Hydro-Elektrisk Kvælstofaktieselskab. Reprinted with permission from Norsk Hydro-Elektrisk Kvælstofaktieselskab.

Groves, Leslie R 1962 *Now It Can Be Told* (New York: Harper & Row). Copyright © 1962 Lt Gen R H Groves. Reprinted with permission from Lt Gen R H Groves (Retired).

Weart, Spencer R 1979 *Scientists in Power* (Cambridge, MA: Harvard University Press). Copyright © 1979 Spencer R Weart. Reprinted with permission from Spencer R Weart.

Chadwick J 1964 Some personal notes on the search for the neutron *Proceedings of the 10th International Congress on the History of Science (Ithaca, 1962)* (Paris: Hermann). Copyright © 1962 Hermann editeurs des Sciences et des Arts. Reprinted with permission from Hermann.

Bernstein J 1996 *Hitler's Uranium Club: The Secret Recordings at Farm Hall*. Reprinted by permission of J Bernstein.

Bernstein J and Cassidy D 1995 Bomb apologetics: Farm Hall, August 1945 *Physics Today* **48** 32–6. Reprinted with permission from American Institute of Physics.

Tronstad, Leif, *Diaries 1941–1945*, National Archives of Norway, Oslo. Portions reprinted with permission from Leif Tronstad Jr and Riksarkivet.

Archives, Norsk Industriarbeidermuseum, Vemork. Portions reprinted with permission from Norsk Industriarbeidermuseum.

Norsk Hydro Archives, Oslo, Rjukan, and Notodden. Portions reprinted with permission from Norsk Hydro-Elektrisk Kvælstofaktieselskab.

List of illustrations

Figure 2.1.   Rutherford's apparatus for observing nuclear disintegration.
Figure 3.1.   Bothe and Becker's apparatus which revealed penetrating γ-radiation from beryllium bombarded with α-particles.
Figure 3.2.   Apparatus used by the Joliot-Curies for revealing recoil protons caused by neutral radiation from Po–Be.
Figure 3.3.   Chadwick's neutron apparatus.
Figure 4.1.   Protons versus 'nuclear electrons', showing the pattern that led to Urey's discovery of a heavy isotope of hydrogen.
Figure 4.2.   Harold Urey at work.
Figure 4.3.   Birkeland and his Terrella.
Figure 4.4.   Southeastern Norway, including Telemark district.
Figure 4.5.   Harnessing the Telemark waterfalls.
Figure 4.6.   The original high-concentration cells at Vemork.
Figure 5.1.   Frédéric Joliot and Irène Curie.
Figure 6.1.   Kowarski, Halban, and Joliot manning a cloud chamber.
Figure 6.2.   Calculated neutron density-distribution curve.
Figure 6.3.   Experimental neutron density-distribution curves.
Figure 9.1.   Centers of the German uranium effort.
Figure 9.2.   Paul Harteck.
Figure 9.3.   Bothe's apparatus for measuring neutron absorption in graphite.
Figure 10.1. German plans for producing heavy water at Vemork.
Figure 10.2. Telemark water-course, with power stations indicated.
Figure 10.3. Flow scheme for Vemork plant.
Figure 10.4. A heavy-water cell and supporting apparatus at Vemork.
Figure 10.5. A heavy-water cell in greater detail.
Figure 12.1. L-IV, the promising Leipzig uranium pile.
Figure 12.2. Ready for action: a few of the Norwegian saboteurs in England.
Figure 12.3. Vemork from afar.
Figure 12.4. The Vemork plant, closer up.
Figure 12.5. The electrolysis building at Vemork, viewed from the east.

Figure 13.1. Vemork electrolyzers after sabotage.
Figure 13.2. Rjukan in flames.
Figure 14.1. The train-ferry *Hydro*.
Figure 14.2. The dual-temperature water–hydrogen exchange process for producing heavy water.
Figure 16.1. B-VIII, the Haigerloch pile.
Figure 17.1. Leif Tronstad.
Figure 19.1. Vemork today.

Chapter 1

PROLOGUE

1.1. Fornebu airport, 12 March 1940

Late in the second week of March 1940, a Junkers airliner prepared for boarding in Scandinavia, bound for Amsterdam on a regularly scheduled flight. Among the passengers were two Frenchmen, both officers in civilian dress, accompanying ten grayish metal canisters, each the size of a household bucket. The place was Oslo's Fornebu airport on the Snarøya peninsula jutting into the Oslofjord, three miles south of the center of town. The timing for the junket was delicate. A state of war had existed between Germany and France and England since Hitler's invasion of Poland in September of the previous year, and the French couple and their sponsors had good reasons for believing the Germans viewed this particular flight, or rather its cargo, with special interest. There could be no doubt that preparations for the flight had been closely monitored by German agents.

The uneasy calm prevailing in the West just then was not shared by Eastern Europe. No sooner had Germany and the Soviet Union carved up Poland than the Soviets annexed the Baltic States without further ado, and launched their winter attack on Finland on 30 November. Despite heroic resistance, Finland sued for peace in our month of focus, March 1940. In the West, however, Hitler's desire for a fall offensive through Belgium into France had been frustrated by bad weather and hesitation among his Generals. All the same, the six-month 'Phony War' was about to come to an abrupt end—within a month of the Fornebu departure, and within the very approaches to the Oslofjord glittering in the haze below the Junkers aircraft once airborne. Only days before, Hitler had issued to his three service headquarters, Army (under Brauchitsch), Navy (under Raeder), and Air Force (under Goering), a formal directive for the occupation of Norway and Denmark. To be sure, the directive was issued over the objections of his commanders, who now viewed Operation *Weserübung* as a diversion from a strike to the west. But Hitler was not to be dissuaded, other than demoting General of Infantry Nikolaus

1

von Falkenhorst from overall commander of the operation, to appease Goering who was livid with rage, complaining that Hitler had kept him in the dark. Even as the little airliner from Fornebu headed out over the Skagerrak, the arm of the North Sea between Norway and Denmark, the German naval staff was canceling all naval operations so that ships might be ready for the onset of *Weserübung*. Weser Day was designated for 9 April, the start of the April new moon period [1-1].

The French apprehension proved justified not long after airborne from Fornebu. Intercepted over the North Sea by two of Goering's *Luftwaffe* fighters, the plane was forced down in Hamburg and the cargo compartment forced open. Alas, the German authorities found not what they expected: merely a load of crushed Norwegian granite carefully stowed in a wooden box. What they were looking for, in fact, was heavy water, alias deuterium oxide—a much more precious Norwegian commodity. Known to mankind for less than a decade, the rare substance was of vital importance in an embryonic scientific bomb development effort then under way in laboratories of the contending powers: Britain, France, and Germany. An old ruse had been played on prying German agents, in the waning days of unoccupied Norway. The incident, one arguably of the greatest importance in the political affairs of the Western World, will be described more fully in due course. For the reason for all the fuss, and events leading to the incident on 12 March, we must backtrack fully two decades. Our chronicle begins shortly after the end of the First World War, in the laboratory of Ernest Rutherford at Manchester University, UK.

Chapter 2

MANCHESTER AND PARIS, 1919

2.1. Manchester: how it began

The year 1919 was an auspicious one for mankind, beginning with the Treaty of Versailles, that marked the end of the Great War. The year marked, as well, a small but important milestone in the intellectual history of mankind, with the completion of an experiment in England by the physicist Ernest Rutherford, then nearing 50 years of age. It would, in fact, be the last year Rutherford found time for experiments with his own hands; from now on, administrative duties and other affairs would increasingly claim his attention.

Rutherford's experiment that year was an outgrowth of a prewar experiment under his guidance by Hans Geiger and Ernest Marsden at Manchester University—on the scattering of α-particles, the nuclei of helium atoms, off thin metallic foils. Rutherford's pondering, during the winter of 1910–1911, over the results of that classic experiment, had culminated in the announcement of his nuclear atom in the spring of 1911. Featuring a central, positively charged nucleus with discrete electrons forming the periphery of an almost empty atom, it contrasted sharply with the distributed 'pudding' of positive electricity embedded with negative charges characterizing the atomic model of his peer, J J Thomson, whose professorship Rutherford inherited at Cambridge late in our year of focus, 1919.

In the experiment of Geiger and Marsden, collisions had been observed between α-particles emitted from radium and atomic nuclei massive enough that recoil of the target nuclei, typically gold, could be ignored. The question naturally arose as to what might happen in a close encounter between an α-particle and a much lighter nucleus, say one of hydrogen with one quarter the mass of the incident α-particle. Charles G Darwin, grandson of the great Charles Darwin and a young mathematical physicist on Rutherford's team, had given the matter some thought. He concluded that head-on collisions in hydrogen should give rise to 'H' particles recoiling with nearly twice the velocity and four times the range of the α-projectiles [2-1].

Marsden took up the problem again experimentally on the eve of World War I, with preliminary results largely as predicted for collisions in hydrogen. However, under certain conditions, among them collisions in air, something was evidently wrong: now and then H-particles of a range too great to be accounted for as recoils showed up. Marsden's results were hastily written up as he dashed off to war. The paper ended on a tantalizing note, expressing 'a strong suspicion that the H-particles are emitted from the radioactive atoms themselves' [2-2].

Rutherford was intrigued by the possibility of some new particles (besides α-, β-, and γ-rays, already known) emitted from the radioactive atoms of the source. He at once set about extending the experiment, making sure he was not stepping on Marsden's turf without the latter's consent. Getting an experiment going again was sooner said than done, however, with the war on the Western Front soon in full swing and bringing the normal routine to a halt at Manchester and most everywhere else. About the only one left at Manchester was William Kay, Rutherford's trusted laboratory steward. Moreover, Rutherford himself was soon in the thick of the war effort. Thus it is doubly surprising that he found time for this, his *experimentum crucis*, that was carried out largely by himself alone over the next four years, when he could find 'an odd half-day' for it [2-3].

The experimental arrangement, similar to Marsden's, is shown in figure 2.1. The RaC (polonium-83) source of α-particles, in the form of a tapered disk D, was mounted near the center of a rectangular brass vessel AA that could be evacuated or filled with the gas to be studied. One end of the vessel was closed with a brass plate E with a small rectangular opening covered with absorbing foils of various thicknesses, presenting different stopping powers for α-particles when the vessel was under vacuum. The foils were located far enough from the source to be beyond the range of α-particles in the gas under study. A zinc sulfide screen F was fixed close to the opening, just outside the vessel. Scintillation flashes from particles striking the screen could be adjusted at will by varying the gas quantity or foil thickness. The whole apparatus was placed in a strong magnetic field to keep lighter β-particles (swift electrons) out of the way.

With dry oxygen or carbon dioxide admitted, the number of scintillations, from what appeared to be hydrogen-like particles identical in mass and range to the H-particles in Marsden's experiment, decreased with increasing gas concentration. This was as expected from the stopping power of the column of gas between source and screen. However, admitting dry air instead brought a definite surprise: the number of scintillations actually *increased*. Having shown that oxygen

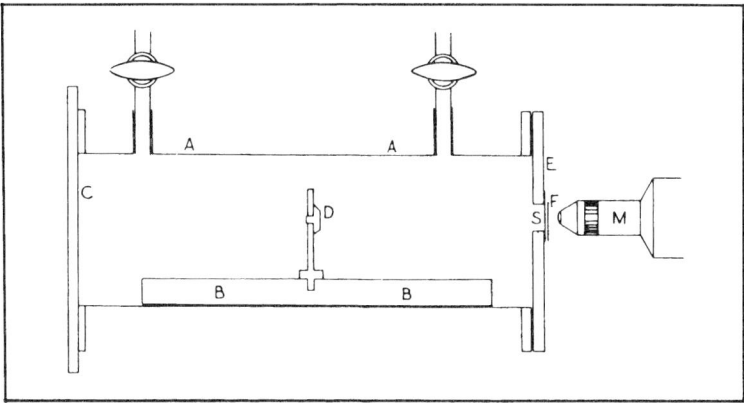

Figure 2.1. *Apparatus used by Rutherford in observing the first nuclear disintegration. Rutherford E, Chadwick J and Ellis C D 1931 Radiations from Radioactive Substances (Cambridge: Cambridge University Press).*

alone decreased the count, the increase could only be due to the nitrogen in the dry air. This, in turn, implied that the α-particles had collided with nitrogen atoms and knocked hydrogen ions out of them. There was no longer any doubt. A new phenomenon was at hand.

The significance of this, Rutherford's crowning experimental discovery, was, in fact, twofold. First, it provided the first example of artificially induced transformation of chemical elements (as distinct from the naturally occurring, spontaneous transformations of radioactivity), namely nitrogen into oxygen. Second, it established that among the constituents of atomic nuclei are hydrogen ions or *protons* [2-4]. It was not so much a matter of *disintegrating* the nitrogen atoms, as Rutherford first penned it in his Manchester notebook [2-5], as shuffling of an overstuffed nucleus. Crashing through the electrical barrier surrounding the nitrogen nucleus of atomic weight 14, the α-particle, weight 4, was absorbed, forming an overweight compound nucleus which promptly spat out a proton of weight unity and settled into the oxygen isotope of mass 17 [2-6].

The path-breaking wartime experiments of Rutherford were completed by mid-spring of 1919, and published in four back-to-back papers in the *Philosophical Magazine* for June of that year. Part IV, the centerpiece of the quartet, ends as follows:

> The results as a whole suggests that, if α-particles—or similar projectiles—of still greater energy were available for experiment, we might expect to break down the nucleus structure of many of the lighter atoms. [2-7]

Rutherford himself harbored no doubt as to the importance of the experiments just completed. While the war was still going on, he is said to have arrived late one day for a War Research Committee meeting, explaining, 'I have been engaged in experiments which suggest that the atom can be artificially disintegrated. If it is true, it is of far greater importance than a war!' [2-8]. Their importance for him was, to be sure, not so much the key to practical applications—say as a new source of power, or the production of rare elements—as purely scientific, opening up the possibility of 'a search into one of the deepest secrets of nature' [2-9]. Moreover, though the energy released in a nuclear encounter exceeded that of the incident α-particle, only one in a million bullets hit their mark, resulting in no net gain in energy, only in fundamental knowledge. Higher bombarding energy would help, as would be demonstrated by younger colleagues of Rutherford in 1932, but the electrical barrier between incident and target nucleus loomed as the ultimate arbiter in the quest for nuclear energy. As late as 1933, with the uncharged neutron in hand, thanks again to the discovery of a younger disciple the year before, and almost five million electron volts of accelerating potential available for atom smashing, Rutherford declared, in an address before the British Association, that 'any one who expects a source of power from the transformation of these atoms is talking moonshine!' [2-10].

Not so, opined the British author H G Wells in a prophetic novel making the rounds on the eve of the Great War, inspired by the habitual broodings of Rutherford's old colleague at McGill University in Canada, Frederick Soddy [2-11]. Soddy had, in 1903, elected to join William Ramsay at University College in London, to pursue in greater detail his own and Rutherford's prediction that helium was continuously produced by the radioactive decay of radium. A great optimist then, he had spoken confidently of atomic energy from radioactivity as the salvation from industrial strangulation which depleted coal reserves. He even saw atomic energy as 'virtually providing anyone . . . with a private sun of his own' [2-12].

Wells was sufficiently awed over Soddy's pronouncements that he dashed off to Switzerland with a batch of scientific reprints in hand. There he churned out in frenzied haste his novel *The World Set Free*, that foretold of atomic energy leading to 'atomic bombs' and a world in rubble by 1956. Though dismissed by the *Times Literary Supplement* as 'porridge', his novel would haunt scientists in the years to come, including Werner Heisenberg and Leo Szilard—both of whom will loom large in our forthcoming tale. Among the fans of Wells was the influential French physicist Jean-Baptiste Perrin, who himself viewed Rutherford's disintegration experiments as a possible stepping stone towards controlled nuclear energy, and whom we shall meet as well. 'Next to that discovery', he mused, 'the discovery of Fire would be a small matter in the history of Humanity' [2-13].

2.2. Paris, and the Joliot-Curies

The discovery of radioactivity by Henri Becquerel at the Musée d'Historie Naturelle in Paris, on a dreary day in 1896, has been often told [2-14], and needs no retelling here. Curiously, its discovery at first went largely unnoticed in the scientific community. The dearth in interest can be attributed to the nearly universal preoccupation with x-rays discovered by Wilhelm Conrad Röntgen in Würzburg barely four months earlier, and their startling images compared to the feebleness of the photographic shadows from Becquerel's rays. Only in 1898 was interest in radioactivity rekindled, when Marie Sklodowska Curie, casting about for a research topic for her doctoral dissertation at the Sorbonne, began a systematic search for other emitters besides uranium. Thorium, next to uranium in the periodic table, seemed a good starting candidate. To her chagrin, she soon learned that Gerhard Schmidt of Erlangen had anticipated her in the radioactivity of thorium by a narrow margin. Schmidt soon called it quits, but her husband of three years, Pierre Curie, was sufficiently intrigued to drop his own magnetic researches and join her in the hunt. Thus began their legendary quest.

Marie had already found that pitchblende, an ore of uranium, was more 'active' than the activity attributed to the uranium it contained. Evidently small amounts of an unknown, very active element was present in the source—a surmise based on Becquerel's original conclusion that the invisible radiation was intrinsic to the element, and on Marie's after him that radioactivity (her term) was an atomic property. The rest is history, as the saying goes. Obtaining hundreds of kilograms of pitchblende gratis from the Joachimstahl mine in Bohemia courtesy of the Austrians, they isolated, by a herculean task, a substance linked to bismuth (a component of pitchblende), and one tied to barium (ditto). Substances many thousand times more active than the purest uranium or thorium, they named the first substance polonium in honor of Marie's native land, and the other radium.

Amidst the backbreaking effort in a ramshackle shed in the yard of the École de Physique et de Chimie Industriélle de la ville de Paris (EPCI) where Pierre taught, Marie had to put off defending her doctoral dissertation. Her examining committee accepted her thesis with enthusiastic approval in June 1903, a year after the Curies met with success. By happenstance, Rutherford and his wife, Mary, showed up at their lodgings the very day she received her degree. The celebration, with Paul Langevin and Jean Perrin and his wife joining in, ended near midnight in the garden, recalls Rutherford, when Pierre drew from his pocket a vial with radium in solution. 'The luminosity was brilliant in the darkness and it was a splendid finale to an unforgettable day' [2-15].

That fall Marie and Pierre Curie shared the Nobel Prize in physics with Henri Becquerel, though both were too worn out to make it to Stockholm for the ceremony. Two years later, Pierre, long denied a chair in physics at the Sorbonne, had one created for him. Not long afterwards, on 9 April 1906, struggling to open his umbrella on a rainswept Paris street, he fell to his death under a heavy horse-drawn van. Marie succeeded to Pierre's chair, while doggedly continuing the work they had begun together. In 1911, she was awarded an unprecedented second Nobel Prize, this time in chemistry, for her discovery of radium [2-16]. Such recognition called for some tangible act of appreciation from French academic circles. Marie had, in fact, been talking with representatives of the University of Paris about prospects for establishing a physics research center patterned after the Pasteur Institute. Thus, the Institut du Radium was born in July of 1914, on the eve of World War I. It was completed on the rue Pierre Curie at a cost barely more than the price at the time of 1 gram of radium. The war, however, prevented staffing of the Institute until 1919.

The war over, Marie was joined by her elder daughter Irène, as her personal assistant. Born in 1897, Irène was educated privately and at the Collège Sévigné, was a budding radiochemist, and became in due course the wife of Frédéric Joliot [2-17]. Joliot himself first set foot in the Radium Institute in 1924, and must now claim our attention. It was on his initiative that the world's stock of heavy water turned up at Fornebu airport in March of 1940, as we shall relate in due course.

Frédéric Joliot, born in 1900 to a well-to-do Parisian tradesman, had been prepared from an early age for admission to EPCI [2-18]. He passed the entrance examination to the prestigious school in 1919, the year Marie Curie's Radium Institute opened for business, as noted earlier. At the EPCI, Paul Langevin, its director of studies as well as successor to Pierre and his former pupil, saw to it that Joliot received a solid grounding in scientific research, with stress on laboratory practice. He also instilled in Joliot the streak of pacifism and humanism that eventually steered him to socialism. Opting for physics over chemistry, Joliot graduated first in his class in 1923. After some delay, he accepted an assistant's stipend at the Radium Institute, where he showed up for an interview with Marie Curie on a winter day in 1924. She had, in fact, already decided, on the recommendation of Langevin, to appoint him her *préparateur particulier* (personal demonstrator).

Thus began Joliot under Marie's guidance, supported under a grant from the Rothschild Foundation. The pay was not much, however, and, as was common practice, he augmented his wages by teaching a course in electrical measurements at a private school. That, in turn, required him to obtain a second *Baccalaureat* and his *licence* (Master's degree) in short order, amidst learning the rudiments of radioactivity. By the time he actually began teaching at the École d'Electricité Industrielle Charlat, he had even more need for some extra money. Irène Curie had become his wife.

Though they were married in 1926, Frédéric and Irène continued working on separate problems for the time being. Joliot took up electrochemical studies of polonium, the results of which passed as his doctoral thesis in 1930. Irène, while also personal assistant to her mother, began research on the range of α-particles in matter, which qualified as her own doctoral thesis in 1925. (To avoid upstaging her, Marie Curie did not attend the public occasion when Irène defended her thesis in the auditorium at the Sorbonne.) Thus, Joliot the physicist obtained his doctorate based on a chemical investigation; Irène, ostensibly a chemist, chose a problem in pure physics! At any rate, only from 1928 onwards do their names appear jointly in papers to the Académie des Sciences (*Comptes Rendus*). At first they were signed Irène Curie and Frédéric Joliot, though popular articles written jointly were usually signed Irène and Frédéric Joliot-Curie. It seems that Joliot was rather sensitive to criticism about the addition of Curie to his name [2-19].

As a couple, the two differed in many ways. Irène, according to her younger sister Eve, 'was as calm and serene as he was impulsive'.

> By nature very reserved, she found it difficult to make friends, while he was able to make human contacts with everyone. She took little interest in her appearance and dress, while he was good-looking, elegant, and always a great success with the opposite sex. In argument, Irène was incapable of the least deceit or artifice, or of making the smallest concession. With a hard obstinacy she would present her case, meeting her opponent head on, even if he occupied a high social position. Frédéric, on the contrary, without yielding on anything basic, knew magnificently how to use his intuitive understanding to put his opponent in a condition to accept his arguments. [2-20]

As for Marie Curie, she started out as a rather difficult mother-in-law, skeptical as to how long the union could last. She was in the habit of introducing Joliot as 'the young man who married Irène'. In time, however, she warmed to Joliot. The birth of a daughter in 1927, Hélène, and a son Pierre in 1932, helped win her over and undoubtedly had a stabilizing influence on what was viewed by many as a casual marriage [2-21]. Gradually the two came to accept each other's characters as different but complementary. One thing they had in common was enthusiasm for outdoor activities—skiing at Mégève in the winter, sailing and swimming at l'Arcouest in the summer, and a fondness for long walks in the forest of Fontainebleau, but the essential bond was their developing scientific teamwork.

Their first collaboration effort was in keeping with the principal business of the Radium Institute and the labor of the Curies that led to its foundation in the first place. Availing themselves of the nearly two grams of radium D acquired by Marie Curie in her work and through a gift from the women of America (to the tune of one hundred thousand dollars), Frédéric and Irène set about isolating a supply of polonium as a powerful source of α-particles. Radium emits both α-particles and γ-rays, whereas polonium, a radioactive byproduct of radium, is a very pure source of α-particles. Its preparation, variously by electrolysis and evaporation, was both a laborious and hazardous undertaking due to its extreme toxicity; if it enters the body, it lodges in dangerous concentrations in the lungs, liver, and spleen. By the end of 1931, after three years of effort, they had 200 millicuries in the form of highly active surface deposits.

Equally important were methods for measuring the ionizing radiations from the interaction of fast α-particles with atomic nuclei. The Wilson cloud chamber was to remain Joliot's instrument of choice. In 1932, he built his own chamber capable of operating from sub-atmospheric pressures to several atmospheres, in a magnetic field of 1500 gauss. He augmented his cloud chamber technique with thin-walled Geiger–Müller counters and ionization chambers connected to a highly sensitive electrometer. Thus armed with a polonium source second to none and state-of-the-art detection apparatus, Frédéric and Irène were well positioned to tackle forefront problems at the start of 1932. Their preparations were in the nick of time, as it turned out.

Chapter 3

THE NEUTRON

3.1. Prelude to 1932: Chadwick and Bothe on the hunt

Rutherford's alchemy of 1919 had driven home the doctrine that matter consists solely of two fundamental building blocks, electrons and protons. On that supposition, the helium nucleus—or α-particle—consisted of four protons and two electrons. In general, a neutral atom of mass number A and atomic number Z was supposed to contain A protons, all in the nucleus, and A negative electrons, $A - Z$ in the nucleus and the rest making up the external electron shells. This supposition held sway for over a decade, though it was increasingly viewed with distrust with the advent of quantum mechanics in the mid-1920s [3-1].

The two-particle orthodoxy suggested several other interesting ramifications to Rutherford, who had succeeded his old professor, J J Thomson, as the Cavendish Professor at Cambridge in late 1919. In his Bakerian Lecture to the Royal Society on 3 June 1920, he observed that, within experimental limits, the masses of all isotopes hitherto studied were represented by whole numbers, when tabulated on a scale in which oxygen is taken as 16. The only exception appeared to be hydrogen, with a mass of 1.008. This anomaly suggested to Rutherford that 'it may be possible for an electron to combine much more closely with the H nucleus [than in the ordinary hydrogen atom], forming a kind of neutral doublet' [3-2]; that is, an atom of mass unity and zero nuclear charge.

Such an atom would have very novel properties. Its external field would be practically zero, except very close to the nucleus, and in consequence it should be able to move freely through matter. Its presence would probably be difficult to detect by spectroscope, and it may be impossible to contain it in a sealed vessel. On the other hand, it should enter readily the structure of atoms, and may either unite with the nucleus or be disintegrated by its intense field, resulting possibly in the escape of a charged H atom or an electron or both. [3-3]

If an electron can bind with one H nucleus, why not with two? This would imply 'the possible existence of an atom of mass nearly two carrying one charge, which is to be regarded as an isotope of hydrogen' [3-4]. Thus were born, in what is normally considered a retrospective lecture embodying conservative views, the concepts of the neutron *and* of the deuteron. Both would be experimentally verified within a month of each other, some twelve years hence [3-5].

Among those in Rutherford's Royal Society audience that evening in Crane Court was James Chadwick, whom in fact Rutherford acknowledged at the end of his lecture 'for his assistance in counting scintillations'—a drudge for which Rutherford had little patience. Chadwick had followed Rutherford from Manchester to Cambridge the previous year. As of late, he had not had an easy time.

James Chadwick was born in Bollington near Manchester, UK in 1891 [3-6]. When his father left for Manchester to establish a laundry business, young Chadwick stayed on with his grandmother, and received his primary education at the local school. After attending the Manchester Municipal Secondary School, he sat for two scholarships at the tender age of 16, with a view to entering Manchester University. He was successful in both, accepted one, and showed up at the university. He had intended to take a degree course in mathematics, but by a quirk of circumstances got into physics instead. All beginning students were required to attend an interview before the beginning of term. Chadwick landed in the wrong line in the crowded hall, and realized soon that he was being interviewed as a potential physics student, not for mathematics. Being too shy to speak up about the mistake, and impressed by what the physics lecturer had to say, he decided to read for physics instead.

During his first year he regretted his decision; the physics classes were large and noisy. The second year went better. The classes were much smaller. More important was his first contact with Rutherford, whose lectures on his early experiments on electromagnetism in New Zealand enchanted young Chadwick no end. The third year course in physics involved the execution of a minor research project. Rutherford assigned him the task of developing a method for assaying radium. Chadwick noticed a snag in the proposed procedure but, shy as usual, did not point it out to Rutherford, whom he found enthusiastic but quite overwhelming. Rutherford, for his part, was disappointed his student had not caught the snag. Nevertheless, good progress was made thereafter, and Chadwick was awarded a first-class degree in 1911.

Upon graduating, Chadwick was accepted by Rutherford as a research student for the MSc degree at Manchester. The department of physics at Manchester was then at its height. Besides Rutherford, there were Geiger and Marsden, Darwin, Henry G J Moseley, and Gyorgy Hevesy, among a host of others. Niels Bohr, too, hung about for five months in 1912, and the Rutherford and Bohr atoms, respectively, date from this period.

In 1913 Chadwick obtained his master's degree, and Rutherford recommended him for an Exhibition of 1851 Senior Research Studentship. A condition of the scholarship being that its recipient carry out research in a laboratory removed from where he had earned it, and Chadwick wishing to carry on in radioactivity, he cast about but found no other British laboratory suitable for his purpose. As it happened, however, Hans Geiger had shortly before taken up the newly established position as director of the laboratory for radium research at the Physikalisch-Technische Reichsanstalt (PTR) in Berlin, and there Chadwick went. He chose the PTR over joining Curie's laboratory, which had too much of a chemical flavor for his liking, though he himself acquired a reputation among physicists for his competence in that field.

The roster of scientific luminaries at the PTR was fully on par with the staffing of Rutherford's laboratory during Chadwick's first year in Berlin. Geiger, for all practical purposes his supervisor, went out of his way to ensure that Chadwick met leading German scientists, including Albert Einstein, Otto Hahn, and Lise Meitner, as well as his own assistants, among them Walther Bothe who joined Geiger just then. Bothe, too, will figure in our chronicle before long. Meanwhile, Chadwick also got back to work, studying the magnetic spectrum of β-rays with the aid of the newly developed Geiger point counter, and in so doing made a notable discovery, which we must, however, pass over in the present context [3-7]. At any rate, working conditions at the PTR were rather different from those he had experienced at Manchester; for instance, all the technical activities took place amidst a large workshop organization, which made little room for improvization.

Alas, the good life was brought to a sudden stop when World War I broke out in 1914. Even before war was declared, Geiger himself was called up as a reserve officer; however, before leaving he saw to it that Chadwick had some cash in case of need. Chadwick received conflicting advice on what to do from his various associates at the PTR. Finally, he applied to the local Cook's Tours office for a ticket home by way of Holland, but Cook's urged him to travel via Switzerland instead. His German friends advised him to lie low in Berlin until troop movements were over. No sooner was war declared, however, than he was arrested in the middle of Berlin, ostensibly for making subversive remarks (that is, speaking English!), and detained in a prison cell until sprung on the intervention of his PTR hosts. Before long, though, the authorities decided to intern all Englishmen, and he found himself quartered with his countryman Charles D Ellis in a camp for civilian prisoners, actually stables of a race course, at Ruhleben near Spandau, for the duration of the war.

In the event, he was treated rather decently, and even managed some token research, 'making the best of it' thanks to the kindness of Heinrich

Rubens, Walther Nernst, and Emil Warburg at the Reichsanstalt. He got a home-made electroscope going, as well as a magnet. With the electroscope he found that thorium oxide, used by some for toothpaste, was radioactive. With the magnet, assembled from iron and copper supplied by his guards, he set about investigating whether order can be produced in liquids through a magnetic field. In the company of Ellis, he embarked on a study of photochemistry. Their dabbling in molecular physics in this unlikely setting inspired Ellis to give up his plans for a career as artillery officer, and take up physics instead when the opportunity allowed it. He would, in fact, become a first-rate colleague of Rutherford and Chadwick in the years to come.

As for Geiger, he wound up serving as artillery officer on the Western front, across the lines from his former colleague Marsden, and keeping in occasional touch with Chadwick and even Rutherford through neutral parties.

In 1918, Chadwick rejoined Rutherford in Manchester, in reasonably good shape except for a digestion in permanent disarray from his inadequate wartime diet. In 1921, settled in Cambridge, he was elected to a research fellowship at Gonville and Caius College, and the next year was appointed assistant director of research under Rutherford at the Cavendish Laboratory; in effect, he became Rutherford's lieutenant in day-to-day charge of all in-house atomic and nuclear investigations.

Chadwick's administrative duties exempted him from any teaching load, and did not prevent him from undertaking research on his own—to be sure, under direction of his mentor, and sometimes in the company of Ellis, his Ruhleben companion. Ellis, by the way, had secured a place at Trinity College, Cambridge, in 1919, and soon became a student under Rutherford. During his subsequent tenure at the Cavendish, where he stayed until 1936, he became the leading in-house expert on γ- and β-radioactivity. His work would be pivotal in unscrambling the mystery of the continuous β-ray spectrum found by Chadwick in Berlin. At any rate, Chadwick's own variegated researches at the Cavendish need not concern us here; it suffices to note that among them was his and Rutherford's program on demonstrating the transmutation of other light nuclei under alpha bombardment besides nitrogen, namely, sodium, boron, aluminum, phosphorus, and fluorine, just as Rutherford had anticipated in 1919. In so doing, they got into a protracted argument with the Vienna group, which is another story but interesting in its own right [3-8].

In 1923, Chadwick returned to the question of the neutron. He and Rutherford had discussed it early on, whiling away time getting dark-adapted before counting scintillations. The name for the hypothetical particle seems to have cropped up even earlier than 1919, when W D Harkins at the University of Chicago introduced it in his independent study of the constitution of the nuclei [3-9]. If, indeed, a close combination of a proton and electron were possible, the very process of combining ought to give rise to γ-radiation, according to the rules of electrodynamics. Chadwick searched for spontaneous γ-rays from hydrogen, with an ionization chamber and a rather primitive point counter, but without success. In the meanwhile, his old colleague Hans Geiger, by now ensconced again at the Reichsanstalt, had become equally depressed over the limitations of the scintillation counting technique. For one thing, it was insensitive to β-rays, and was slow and unreliable; it may well have contributed to the spurious results coming out of Vienna. Geiger set about improving the electrical point counter he himself, prodded by Rutherford, had devised at Manchester in 1908, despite the latter's inherent distrust of complicated instrumentation. In 1928, assisted by his student Walther Müller, Geiger upgraded his counter into what is universally known as a Geiger counter, one very sensitive to β- and γ-rays. That same year, fortified with his own version of the new counter, Chadwick looked again for spontaneous radiation from hydrogen, but still found nothing. Might the Geiger counter not yet be up to the task?

For the next development in the saga of the neutron, we again switch back to the Reichsanstalt. Walther Bothe, 'a real physicist's physicist' in the opinion of Emilio Segrè [3-10], had, we recall, joined Geiger in Berlin in 1914. What happened to Bothe next is very reminiscent of Chadwick's wartime experience. Son of Fritz Bothe, a merchant, Walther was born in Oraninenburg near Berlin in 1891, and became one of the select pupils of Max Planck, under whom he obtained his doctorate on radiation theory in 1914. No sooner had he joined Geiger than the war broke out, and he too was called up. He was taken prisoner by the Russians in 1915, and sent to Siberia where he was interned until 1920. As resilient as Chadwick, he honed his mathematical skills and studied theoretical physics under the worst of conditions. On his return to Germany, he accepted Geiger's invitation to return to the laboratory for radium research at the Reichsanstalt. There, among other things, Bothe came up with a modification of the Geiger counter, the counting speed of which was still not much greater than the scintillation method. Bothe introduced the so-called coincidence method, correlating counts between pairs of counters in a technique that proved invaluable in, among other areas, the young field of cosmic-ray research. As far as we are concerned, it was also the first reliable method for measuring the energy of weak γ-rays.

In 1928, just as Chadwick searched again without success for spontaneous γ-rays in hydrogen, Bothe and his student Herbert Becker began a systematic study of γ-radiation excited by α-bombardment of light nuclei. Their apparatus is shown schematically in figure 3.1. Attempting to minimize the natural γ-background, they used polonium for the same reason as the Joliot-Curies; it is a very pure alpha emitter.

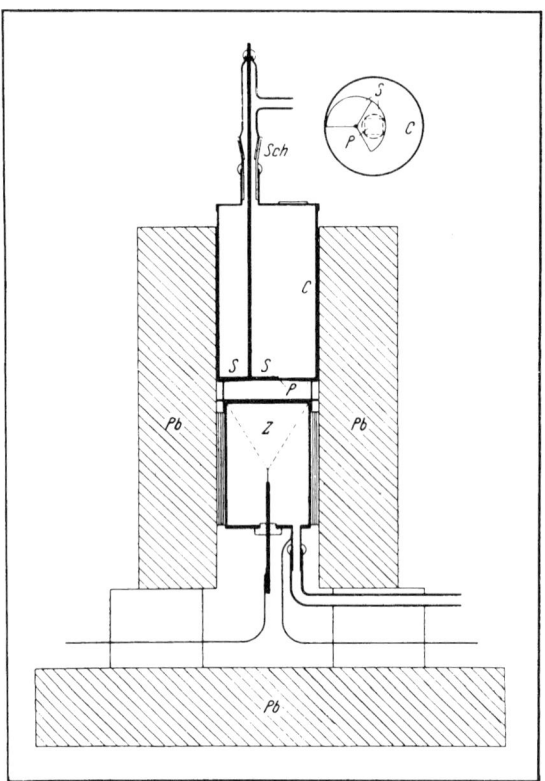

Figure 3.1. *Apparatus with which Bothe and Becker revealed penetrating γ-radiation emitted from beryllium bombarded with polonium α-particles.   P is the polonium preparation, S is the substance under study, and Z is the Geiger counter.   Bothe W and Becker H 1930 ZP* **66** *290.*

In retrospect, Chadwick, who had obtained some polonium from Lise Meitner at the Kaiser Wilhelm Institutes in Berlin-Dahlem, albeit not enough to challenge Bothe's work, puzzled how Bothe obtained his source [3-11]. Polonium was, for all practical purposes, only available in Paris. In fact, Bothe obtained it the hard way: he prepared his own source by a process of chemical extraction [3-12].

At any rate, among the light elements, beryllium occupied a special place.   Named glucinium until 1957, it is a highly toxic, difficult-to-handle, steelish gray and lightweight metal, discovered in 1798. Beryllium did not disintegrate under α-bombardment, and hence was not a proton emitter. To Bothe and Becker's everlasting surprise, the metal nevertheless gave off a highly penetrating radiation. The radiation was nearly ten times as intense as observed with any other element, and more energetic than the incident alphas, but it was definitely not protons. Bothe concluded that they were observing exceptionally hard

γ-radiation, and so reported to the German Physical Society in 1930 [3-13]. Presumably the radiation came from nuclear disintegrations despite the absence of protons. The same was soon observed in lithium and boron.

We will rejoin Bothe in due course. Someone of great authoritative stature in German scientific circles, his name, by an unfortunate laboratory error, would figure prominently in the background for Germany's important decision to predicate its wartime nuclear effort on scarce Norwegian heavy water.

3.2. A discovery narrowly missed, and the neutron at last

News of the 'beryllium radiation' reached Cambridge and Paris almost at once, and Chadwick and the Joliot-Curies lost no time in getting experiments going. It was late in the year 1931. The French had the distinct advantage of a powerful α-source at hand. Frédéric and Irène, the latter pregnant with their second child, placed one of their polonium preparations against a beryllium plate. Sure enough, they soon confirmed that no protons were emitted from beryllium under α-bombardment, but copious quantities of energetic γ-rays emerged from the steelish metal. To make doubly sure they were dealing with a very penetrating radiation, they passed it through 15 mm of lead, before it entered an ionization chamber (figure 3.2). Next, on the supposition that the penetrating radiation, in traversing matter, produced secondary, less penetrating radiation that could be more readily detected in the ionization chamber, they made the radiation enter the chamber through an aluminum window only 5 μm (five thousandths of a millimeter) thick. Various sheets of absorbers were also inserted in front of the aluminum window. In this way, they hoped to trap secondary rays of very low penetrating power. The current through the chamber should have lessened. In fact, it hardly diminished at all.

An even greater surprise was in store. When they interposed a block of paraffin wax (a mixture of hydrogen and carbon atoms) or similar substances rich in hydrogen (cellophane, or simply water), numerous high-speed protons shot into the ionization or cloud chamber. It appeared that high-speed γ-rays collided head-on with hydrogen nuclei, knocking protons out of the way. If so, the γ-ray energy had to have been at least 50 MeV, ten times as energetic as any previously known γ-rays. Energetic electrons were also knocked out, again emphasizing the high-energy γ-component of the beryllium radiation.

Joliot and his wife promptly reported their discovery in a note to the Paris Academy on 18 January 1932 [3-14]. It had been tiring work, and they availed themselves of several opportunities to escape to Marie Curie's summer place in l'Arcouest in Brittany. On 2 April, Joliot wrote from there to his friend Dmitry Skobel'tzyn in Moscow that they had a son, born on 12 March, with Irène resting comfortably in Paris.

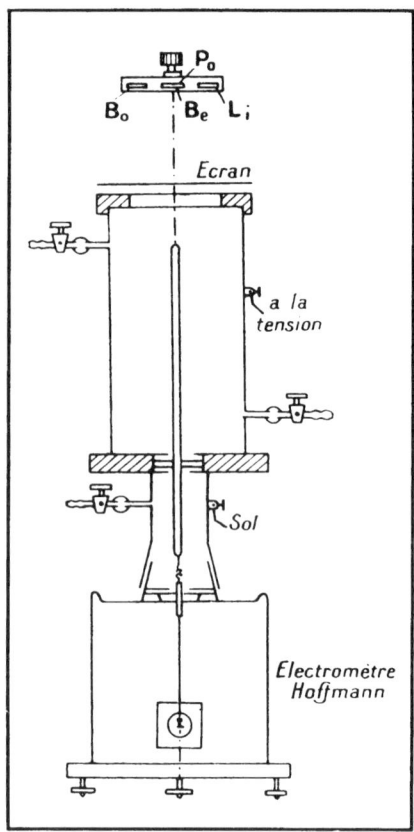

Figure 3.2. *Apparatus used by the Joliot-Curies for revealing recoil protons caused by head-on collisions between hydrogen nuclei and high-speed γ-rays emitted from the polonium–beryllium source (on top). In the center is an ionization chamber connected to an electrometer (bottom). Curie I and Joliot F 1932 CR* **194** *273.*

We have been working hard during the last few months [he added] and I was very tired before I left for Brittany. We had to speed up the pace of our experiments, for it is annoying to be overtaken by other laboratories which immediately take up one's experiments. In Paris this was done straight away by M. Maurice de Broglie with Thiabaud and two other colleagues. In Cambridge, Chadwick did not wait long to do so either. [3-15]

Indeed, Chadwick too had been hot on the trail of the mysterious beryllium radiation, from the moment he received word of the Bothe–Becker results. All along he harbored a suspicion that the radiation actually consisted of neutral particles—quite possibly Rutherford's elusive neutron. Then, one morning in February of 1932, he was taken aback by the startling communication of the Joliot-Curies in the latest issue of *Comptes Rendus*. What happened next, Chadwick would not soon forget.

> Not many minutes afterwards [Norman] Feather came to my room to tell me about this report, as astonished as I was. A little later that morning I told Rutherford. It was a custom of long standing that I should visit him about 11 a.m., to tell him any news of interest and to discuss the work in progress in the laboratory. As I told him about the Curie-Joliot observation and their views on it, I saw his growing amazement; and finally he burst out, 'I don't believe it'. Such an impatient remark was utterly out of character, and in all my association with him I recall no similar occasion. I mention it to emphasize the electrifying effect of the Curie-Joliot report. Of course, Rutherford agreed that one must believe the observations; the explanation was quite another matter. [3-16]

By chance, Chadwick was just then ready to begin a new round of experiments with a polonium source of his own. Soon after the German announcement, he had set H C Webster, an Australian research student, to work examining the γ-radiation from beryllium, using an ionization chamber in conjunction with the new linear amplifier (vacuum tube, or valve) then coming into use at the Cavendish and abroad. Chadwick had been struck by the value of polonium as an α-emitter while in Vienna, sorting out the controversy between the Cavendish team and Hans Pettersson and Gerhard Kirsch at Stefan Meyer's Radium Institute. However, what polonium he was able to lay his hands on back in Cambridge was too weak, in part due to the parsimonious attitude of Rutherford, a source of considerable friction between the two.

By a happy coincidence, Norman Feather had just returned from a sabbatical year at Johns Hopkins University in Baltimore. There, at the Kelly Hospital, he chanced across a generous supply of old radon ampoules, no longer in use. Thanks to the passage of time, and the radioactive decay of the radium emanation, the tubes now contained a considerable quantity of polonium—a decay product of radium. The hospital gladly donated them to Chadwick, and Feather had them shipped back to the Cavendish. Performing some risky chemical separations, taking a cue from Bothe and chemistry being second nature to himself as well, Chadwick was soon able to deploy them in an α-source on a par with those in Paris or Vienna.

Using the new source (figure 3.3), Webster found before long that the radiation from beryllium emitted in the forward direction was more penetrating than that emitted backward. This made no sense, assuming the laws of conservation of energy and momentum held, if the radiation consisted of γ-rays—that is, of massless quanta of electromagnetic radiation. With Webster off to the University of Bristol, Chadwick took over. The correct explanation was not long in coming.

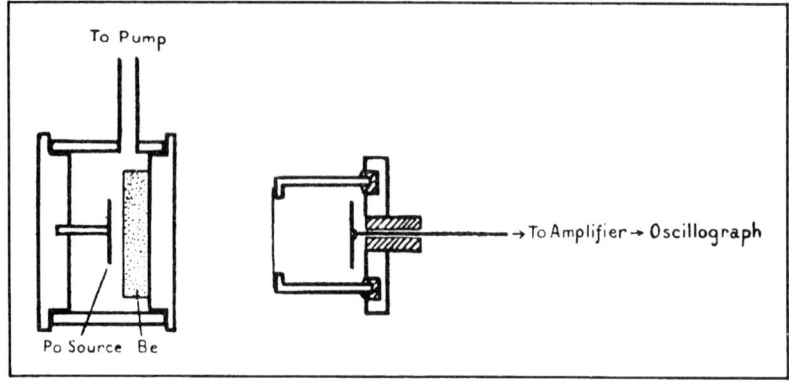

Figure 3.3. *Apparatus with which Chadwick discovered the neutron. On the left, the Po–Be source; on the right, the ionization chamber connected to an amplifier and oscillograph. Chadwick J 1932 PRS* **136** *692.*

> A few days of strenuous work were sufficient to show that these strange effects were due to a neutral particle and to enable me to measure its mass: the neutron postulated by Rutherford in 1920 had at last revealed itself. [3-17]

Cautious as always, Chadwick submitted a letter, 'Possible existence of a neutron', to *Nature* on 17 February [3-18]. On the 24th, he sent proofs of the letter, which appeared in print on 27 February, to Niels Bohr in Copenhagen [3-19]. Bohr responded in kind, inviting him to report on the discovery and its aftermath at a small informal conference at Bohr's Institute planned for the second week of April—an annual affair at the Institute [3-20]. Meanwhile, on 2 April Joliot continued his aforesaid letter to Skobel'tzyn as follows:

> Chadwick has, by the way, published the very attractive hypothesis that the penetrating radiation from Po(α)Be is composed of neutrons. I tell you this because you are in touch via *C.R.* (*Comptes Rendus*) and *Nature* with these experiments, concerning the projection of atomic nuclei. We have recently been carrying out new experiments on the Po(α)Be radiation and the results will be published on Monday in the *C.R.* Here is a summary. Po(α)Be radiation is composed of at least two parts: one part is gamma rays of energy between 5 and 11 MeV and is scattered by the Compton effect. The other part is radiation of *enormous* penetrating power—about half is absorbed in 16 cm of Pb following collisions with nuclei. This radiation is probably composed of neutrons. [3-21]

In the event, Chadwick was unable to attend the Copenhagen meeting in April, a meeting that by happenstance marked the tenth anniversary of the founding of Bohr's Institute of Theoretical Physics at Blegdamsveien 17. The neutron figured heavily in the technical discussions, and also in the script of a play, a parody on Goethe's *Faust*, written and performed by students of Bohr during the meeting. The finale, 'Apotheosis of the True Neutron', delivered by Wagner (standing in for Chadwick), goes, in part, as follows:

> The *Neutron* has come to be.
> Loaded with Mass is he.
> Of Charge, forever free.
> Pauli, do you agree? [3-22]

The question remains: how did the Joliot-Curies miss the epochal discovery? Looking back, years later, Frédéric admitted that he and Irène had not remembered the speculation about an uncharged particle making the rounds in the 1920s, much less read Rutherford's prediction in his Bakerian Lecture of 1920 [3-23]. This curious lapse on the part of the Joliot-Curies must remain among the inexplicable trivia in the annals of 20th century physics.

On a purely technical note, it has been pointed out that the point counter detector used by Bothe and Becker was almost exclusively sensitive to γ-rays, whereas Chadwick's ionization chamber responded to neutrons. However, the current ionization chamber of the Joliot-Curies was sensitive to both neutrons *and* gamma-rays. In the opinion of someone close to Joliot, 'Such was the unconventional bonus which the otherwise enviable possession of the strongest polonium source in the world at that time conferred on these single-minded investigators in the autumn of 1931' [3-24].

Chapter 4

HEAVY WATER

4.1. Deuterium: a comedy of errors

Just as Chadwick was in the habit of scanning scientific journals as they arrived in his morning mail, so was the American physical chemist Harold C Urey at Columbia University. What caught his attention one warm summer day in 1931 was a letter to the editor in *The Physical Review* by Raymond T Birge, a professor of physics at the University of California, Berkeley, and Donald H Menzel, professor of astrophysics at Lick Observatory on nearby Mount Hamilton [4-1]. Birge and Menzel suggested that isotopes of hydrogen might explain an apparent discrepancy in the atomic weight of that element. The atomic weight of hydrogen had been a sore point of contention among scientists for a long time. Its value led Rutherford to propose, some ten years earlier, not only his hypothetical neutron, but also one or more isotopes of hydrogen. Experimental isotopic research had made considerable advances in the decade since Rutherford's Bakerian Lecture in 1920.

In Urey's time, there were actually two systems of atomic weights in use, one by chemists and one by physicists. The two came about as a result of the discovery of two isotopes of oxygen by William Giauque and Herrick Johnston in 1931: one of weight 17 and one of weight 18 [4-2]. The chemists chose the value 16 as the atomic weight of the naturally occurring mixture of O^{16}, O^{17}, and O^{18}, while the physicists adopted 16 for the relative mass of O^{16} alone. As a result, the atomic weights of a single isotope or element should differ on the two scales. This was indeed the case, except for hydrogen for which the atomic weights agreed within the claimed experimental errors: the chemists quoted $1.007\,77 \pm 0.000\,02$; Francis Aston claimed $1.007\,78 \pm 0.000\,15$ on the physical scale with his celebrated mass spectrograph at the Cavendish Laboratory. The near coincidence of the two values suggested to Birge and Menzel that one part in 4500 of the common variety of hydrogen contains a heavy isotope, to which they assigned the symbol H^2.

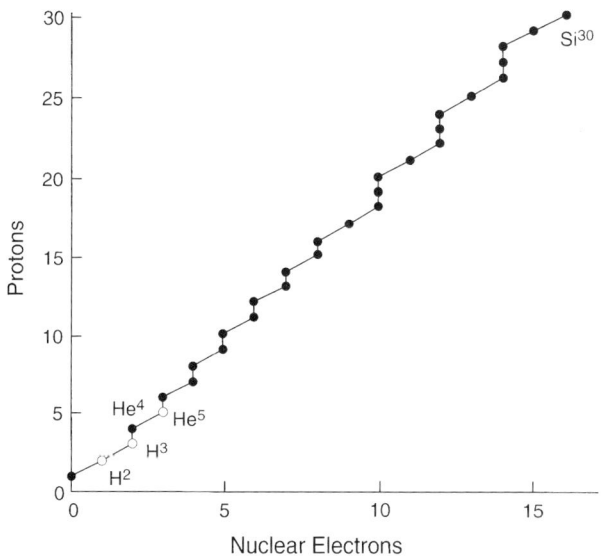

Figure 4.1. *Nuclear protons plotted against 'nuclear electrons' for known atomic nuclei (filled circles), with missing nuclei (open circles) for H^2, H^3, and He^5 (Brickwedde F G 1982 PT* **34** *210).*

Their proposal was just what was needed to help complete a chart which hung on Urey's laboratory wall at Columbia (figure 4.1). In keeping with the orthodoxy then still prevailing (though not for long!) that atomic nuclei consisted of protons and electrons, the chart shows the number of nuclear protons (presumably the mass number A) as ordinates and nuclear electrons ($A - Z$, where Z is the atomic number) as abscissae. Filled circles represented nuclei from H^1 to Si^{30} that were known before 1931; open circles are missing nuclei, namely H^2, H^3, He^5. (Also missing is a point for He^7, and Be^8 was later shown to be unstable.) Thus the H^2 of Birge and Menzel would fill in a prominent blank spot in the apparent regularity of stepped segments exhibited by the chart [4-3].

Urey's broad interest in physics and chemistry alike is epitomized by his classic textbook coauthored with Arthur E Ruark the previous year. *Atoms, Molecules and Quanta* was the first comprehensive text on atomic structure in English [4-4]. Urey's scientific mentor was the chemical thermodynamicist Gilbert N Lewis at Berkeley, a scant decade earlier. He was born on a farm in Indiana in 1893, the son of a school teacher–farmer and minister [4-5]. His father passed away when the boy was six. After graduating from high school, he earned some money teaching in public schools in Indiana and Montana for three years, then entered Montana State University. There he majored in zoology with a minor in chemistry. 'Money was tight for him as a college student', recalls Ferdinand G Brickwedde, who would soon cross his path. 'During the academic year he slept and studied in a tent. During his summers he worked on a road gang laying railroad track in the Northwest' [4-6].

Urey graduated with a BSc degree in 1917, just as the United States entered World War I. Though leaning towards pacifism as a result of his Brethren church background, he went to work with a chemical industrial laboratory in Philadelphia specializing in munitions. That experience soured him on industrial chemistry and steered him into academia when the war ended. For want of adequate income, he taught chemistry at Montana State, and in 1921 began graduate work in chemistry, with a minor in physics, at Berkeley under Lewis—probably the only physical chemist in the US on a par with Irving Langmuir at General Electric. His doctoral work centered on the rotational contributions to the heat capacities and entropies of gases. He received his PhD in 1923, and spent the next year as an American–Scandinavian Fellow at the Bohr Institute in Copenhagen, sharpening his expertise in atomic physics. 'Bohr thought I was a physicist', he quipped years later. 'If he had known I was a chemist, he probably would not have let me in' [4-7]. He then joined Johns Hopkins University as an instructor in chemistry, where he coauthored the aforesaid text and stayed until 1929 when he accepted a professorship at Columbia University.

Urey would remain at Columbia until the end of World War II, when he joined the Fermi Institute for Nuclear Studies at the University of Chicago. University of Chicago president Robert Hutchins later recalled that he had tried to recruit Urey before the war. 'It seems to me, as I look back on it, that I spent most of my time at the University of Chicago trying to persuade Harold Urey to join the faculty. I think it is too bad that I owe my eventual success in this high endeavor to General Groves' [4-8].

A hint that something was amiss with the composition of water came to light as early as 1913, when Arthur B Lamb and Richard E Leen at New York University were unable to come up with a definite value for the density of pure water. Their highly accurate results varied by as much as 8×10^7 g cm^{-3}, depending on the sample. Theirs was apparently the first reported experiment in which an isotopic difference in properties was clearly in evidence [4-9]. The existence of isotopes was, in fact, proposed that same year by Frederick Soddy in England and independently by Kasimir Fajans in Germany.

Urey lost no time in following up on the letter by Birge and Menzel. Teaming up with his colleague George Murphy at Columbia, he soon identified hydrogen and its isotope spectroscopically observing the Balmer atomic spectrum with a 21 foot grating spectrograph (designed by himself and Murphy) and Wood's-type discharge tube. Modest exposures sufficing with gas from a commercial tank of hydrogen, the experiment seemed too simple! Urey decided not to rush into print, but make doubly sure that the new spectral lines were not impurity lines or

Figure 4.2. *Harold Urey at work. Courtesy AIP, Emilio Segrè Visual Archives.*

'ghosts' (caused by irregularities in the ruling of the Columbia grating). 'If we had been brave', muses Murphy, 'we would have stopped work on that first day and written our paper announcing the discovery of heavy hydrogen' [4-10].

To proceed further, a method of isotopic concentration was needed and fractional distillation of liquid hydrogen seemed a good bet. (The normal atoms should boil off faster than those of mass two, yielding a residue rich in the heavy isotope.) Urey got in touch with F G Brickwedde, a former student of his then at the National Bureau of Standards in Washington, DC, and already an authority on low-temperature studies. Brickwedde gladly agreed to lend a hand, and ship some samples via Railway Express. Unfortunately, the Washington liquefier was down for repairs, causing some anxiety at Columbia; others might also be looking for the predicted isotope. But a sample was soon on the way, consisting of a few cubic centimeters of residual liquid evaporated from six liters of liquid hydrogen. To Urey's consternation, the intensity of the spectral lines attributed to heavy hydrogen was too feeble to be sure. More waiting. At last, more samples arrived, evaporated at a lower temperature where the rate at which heavy hydrogen is concentrated should be more rapid. Success! The new lines showed a six- or sevenfold increase in intensity of the Balmer lines of deuterium. (However, a line representing H^3, another blank spot on Urey's wall chart, was not seen.) To clinch the matter, on Thanksgiving Day, 1931, they found that the photographic image on red-sensitive plates of the most intense line, the $H\alpha$ Balmer line, was actually a doublet, as theory said it was.

The first public announcement of the discovery of deuterium was scheduled for the Christmas meeting of the American Physical Society at Tulane University. Urey almost did not make it for lack of travel funds. At the last minute, Columbia University President Nicholas Murray Butler saved the day with some travel money from his own account, and Urey could read his 10 minute abstract in New Orleans [4-11]. A letter to the editor of *The Physical Review* soon followed, and on 16 February, the full-length paper was mailed to the editor of the same journal, the day before Chadwick dispatched his own account to *Nature* announcing the neutron [4-12].

Soon after, Urey gave a colloquium on heavy hydrogen at Johns Hopkins, and the indubitable Robert Wood, a sharp spectroscopist if there ever was one, allowed that 'I rather think you are right about it; I think it's real' [4-13]. More assuring was Walker Bleakney at Princeton, who had a mass spectrometer in working order. In no time, he corroborated the isotopic abundance of their sample.

The story had a curious sequel. In 1935, it turned out that the atomic weight of hydrogen on the physical scale was slightly in error, having been revised upward by Aston, enough to reveal a real difference in the two hydrogen values after all. Nor does the story end there. The chemical atomic weight of hydrogen proved to be in error as well, as discovered when it was realized that the isotopes of hydrogen can be fractionated by electrolysis as well as by distillation. Thus, the discovery of heavy hydrogen came about quite fortuitously, based on no less than two incorrect values for the atomic weight of hydrogen. The comedy of errors does not, of course, distract from the accomplishment of Urey. As he himself observed in an addendum to the proofs of his Nobel lecture in 1935 (when he received word of Aston's correction), the prediction of Birge and Menzel—spurious or real—did, in fact, set in motion a search for the isotope, which might otherwise have remained undetected for quite some time. Such are the vagrancies of science!

Equally curious, Frederick Soddy, who in 1921 received the Nobel Prize in chemistry for his discovery of isotopy, refused to go along with the idea that deuterium is an isotope of hydrogen. Soddy dealt with isotopes of the naturally occurring radioactive elements, whose atomic weights are large and whose relative isotopic mass differences are correspondingly small. These isotopes showed no observable differences in chemical properties, leading Soddy to conclude that cases of chemical inseparability are actually chemical *identities*. His definition of isotopy, linking chemical identities of elements of different atomic weights, held sway until Chadwick's discovery of the neutron in 1932 [4-14]. From that time on, the isotopes of an element were viewed, not as inseparable bodies, but as consisting of atoms which have the same nuclear charge (number of protons) but different masses (number of neutrons).

4.2. Gilbert Lewis and Leif Tronstad: the promise of deuterium

In the short run, the fruits of Urey's discovery of deuterium would be reaped by his mentor, Gilbert Lewis, in Berkeley. Lewis was not the only one to be caught up in the flurry of excitement over deuterium, however. The subject sparked a special session at the general June meeting of the American Physical Society in Chicago in 1933, attended by Urey and Ernest O Lawrence, among others [4-15]. The ensuing discussion on the naming of the new isotope 'threatened to become acrimonious', noted Aston, a guest speaker on the occasion, who figured so prominently in the elucidation of deuterium and was someone not given to overstatements. The argumentation had to do with whether to retain the name 'hydrogen' for the new isotope. 'I had a talk with Professor Bohr at the time', continued Aston, 'and he certainly at first insisted that the new isotope should be called hydrogen; he said it was not a new element; it had the atomic number 1, and therefore it was hydrogen. I admit that is true, but it is quite impossible to go on with the chemistry of this substance without having a new word for it...' [4-16].

Before very long, the need for larger quantities of heavy water was felt in physical, chemical, and biological circles alike, as evidenced by an in-depth and more civil discussion of the same topic in a special meeting of the Royal Society in 1933, chaired by Ernest Rutherford; the meeting occupied a prominent column in *The Times* the following day, 15 December [4-17]. For one thing, it was to be expected that, on account of its greater mass, the rate of diffusion and the rate of chemical reactions will differ when H^2 is substituted for H^1, and compounds formed with the new isotope might in some cases exhibit different properties from normal hydrogen compounds. Similarly, the new discovery opened up interesting questions on the effect of heavy water in altering the normal physical and chemical processes in animal and plant life. Indeed, this soon proved to be the case, as we learn from Lewis, who deserves our attention at this point. As indicated in the previous section, Lewis influenced physical and organic chemistry at the University of California at Berkeley like no one else during the period 1912–1946 [4-18]. The preeminence of the Department of Chemistry at Berkeley was largely due to Lewis's direction for 29 years. His school of chemistry spread far and wide; among his students and those who served under him at Berkeley, the following received the Nobel Prize in Chemistry: Harald C Urey (1933), William F Giauque (1949), Glenn T Seaborg (1951), William F Libby (1960), and Melvin Calvin (1961).

Gilbert Newton Lewis was born near Boston in 1875, but his parents migrated to Lincoln, Nebraska, while he was still a child. His formal education up to the age of 13 was solely in the hands of his parents. At that age, he entered the preparatory school for the University of Nebraska, and starting at 16 he spent two years at the University. From there he went to Harvard, obtaining his BA in 1896. After graduation, he taught for a year at Phillips Academy in Andover, then returned to Harvard; there he received his PhD in 1899 under T W Richards on the electro- and thermochemistry of zinc and cadmium amalgams. Subsequently he was an instructor at Harvard from 1901 to 1903, and held a post as Superintendent of Weights and Measures at Manila during 1904–1905; his life-long passion for Philippine cigars stemmed from that experience. He then passed through the ranks from assistant to full professor at the Massachusetts Institute of Technology. In 1912 he was called to Berkeley, where he was appointed Dean of the College of Chemistry and Chairman of the Department as well; he held both positions until he was 65. His extraordinary tenure at Berkeley was interrupted for all of two years by World War I, when he served as Chief of the Defense Division of the Chemical Warfare Service. That short stint earned him the Distinguished Service Medal from his countrymen and the Cross of the Legion of Honor from France.

It would take too long to even list his variegated scientific researches. Two-thirds of his published works deal with thermodynamics and free energy of substances [4-19], atomic and molecular structure and the chemical bond, isotopes (especially deuterium), and the interaction of light with matter.

Confining ourselves to his work on deuterium, it was inspired by Urey's discovery of the isotope in 1931. Lewis and his research assistant Ronald T Macdonald were engaged in the separation of oxygen isotopes just then. When Lewis received word of a heavy isotope of hydrogen, he turned at once to it: the difference in properties between the two isotopes of hydrogen, he realized, should be far greater than between any other pair of isotopes. By the same token, the two hydrogen species should be the easiest ones to separate, notwithstanding the evidently small abundance of heavy hydrogen. Moreover, from the discovery of E W Washburn and Urey that the water in an old electrolytic cell contained an appreciably larger amount of H^2 than does ordinary water [4-20], Lewis knew that he had the starting material only doors away in Gilman Hall: in a large electrolytic cell that William Giauque had used for many years to liquefy hydrogen.

On 8 April 1933 Lewis took time out to acknowledge to Urey that a progress report was overdue.

> I have been so much occupied in studying the properties of your hydrogen isotope that I have not found time to write up the procedure by which I obtained it, but I am going to do this within

the next day and if you should care to have it, I will send you the paper for the Journal of Chemical Physics. [4-21]

Urey, as it happened, was managing editor of the brand new *Journal of Chemical Physics*. The manuscript barely reached him on 15 April, the editorial deadline for the next issue. Its receipt, moreover, proved slightly awkward for the editor.

> I find... that my editorial position is a little embarrassing, for we are working on methods of separating hydrogen isotopes ourselves. Up until the time your letter appeared, we had not worked on it very vigorously because I learned of the electrolytic method in Dr. Washburn's laboratory, and did not feel free to start work on the field. Now that you have broken in so gloriously, I also feel perfectly free to go ahead with my own ideas. Hence, receiving a paper from someone else in the field some two months before it will appear in print is a little embarrassing. [4-22]

The paper by Lewis and Macdonald appeared in the June issue of Urey's journal. Starting with 20 liters of residual water from Giauque's old cell, he and Macdonald had reduced an alkaline solution in stages with nickel electrodes; soon they had about 0.5 ml of concentrated water with about 70% deuterium, and by July a small amount of essentially pure (99.99%) material [4-23].

In the ensuing correspondence between Lewis and himself, Urey expressed surprise over the high current density utilized in the Berkeley electrolyzer, yielding correspondingly higher concentrations than achieved at Columbia, depending somewhat on assumptions made for the relative abundance of H^1 and H^2 in natural hydrogen [4-24]. The rest of the correspondence, during May and June of 1933, bogged down over nomenclature for H^2 and its nucleus. Lewis favored 'dygen' for the hydrogen isotope 'as an element' and 'dyon' for the nucleus of H^2— 'the word that we are just now using informally in [E O Lawrence's] laboratory' [4-25]. Urey disagreed, offering the unhelpful 'pycnogen' for the heavy isotope. For its nucleus, he suggested 'hydrogen two' might be good enough [4-26].

> I do hope the California group doesn't get so impatient about a name for the hydrogen two atom that they will not allow Dr. Murphy, Dr. Brickwedde and myself to name it, as is customary in such matters. We are working on the choice of a suitable name as rapidly as we can... If the California group will not give us a little time and be a little patient, they will force us into publishing a name for this isotope which may not be satisfactory for some reason or another. I would suggest, therefore, that the California group use the name hydrogen two provisionally and allow the discoverers of this element to name it when they see fit. [4-27]

Indeed, Urey had the last word, as he was entitled to, tentatively settling on 'deuterium' for the heavy hydrogen isotope, and 'deuteron' over 'deuton' (Berkeley's latest choice) for the H^2 nucleus—both 'suggested by one of the professors in the Department of Greek' [4-28].

Nomenclature notwithstanding, the two years 1933 and 1934 Lewis devoted himself to a study of the chemical and physical properties of heavy water, authoring or coauthoring some 26 papers on this subject. After that, he never touched the subject again in print. However, in the last of this torrent of articles, in *Science* for 16 February 1934, he summarized 'sporadic' attempts by Macdonald and himself to observe effects of water containing heavy hydrogen upon living organisms, albeit mostly small organisms. Thus, the germination of tobacco seeds was completely retarded in pure heavy water, and yeast cultures failed to grow in a nutrient medium dissolved in heavy water of similarly high concentration. In a more expensive, if less definitive, experiment involving Lewis's entire initial stock of heavy water, a mouse was fed doses of pure heavy water totalling 0.66 gram, which would be equivalent to a consumption of 4–5 liters of heavy water by an adult human being. The mouse survived, but showed 'marked signs of intoxication' [4-29].

There was, in addition, one question of particular interest to Rutherford and his colleagues at the Cavendish and to his competitors E O Lawrence and Charles C Lauritzen and their teams respectively at Berkeley and Pasadena: namely, the use of H^2 nuclei as swift projectiles for the transmutation of elements [4-30]. Carrying no more repelling charge than protons, but having twice their mass, they might well prove twice as effective. So they did, as we will relate in a subsequent connection. Rutherford himself, we might add, had his own opinion about nomenclature, and decreed—unsuccessfully, for once—that the projectiles be called 'diplons'. He thereby wished to avoid confusion, during casual spoken conversation, of 'deuteron' or 'deuton' with 'neutron'. The difficulty, Rutherford added, was accentuated by the recent discovery that neutrons are often liberated from elements bombarded by deuterons. Nevertheless, after fairly protracted correspondence with both Lewis and Urey and their camps on the matter [4-31], he did not press the point, noting simply that 'both sides have, in a sense, dug in' [4-32].

Alas, in his closing remarks at the Royal Society discussion, Rutherford allowed that

> ... at the moment, only a few scientific men in the country have even a cubic centimetre of heavy water for experimental purposes. I hope this difficulty will soon be removed and that our industrial friends will help us to obtain supplies of the new water at a reasonable price. The cost of making the new water in places like

Cambridge where the cost of power is high, is almost prohibitive. [4-33]

What little heavy water Rutherford could lay his hands on was initially courtesy of Lewis, who the previous May had shipped from Berkeley to Cambridge a vial of nearly pure concentrate; it was put to good use by Rutherford and Oliphant in their low-voltage transmutation experiments. Unfortunately, in August what was left of the material was badly contaminated by accident [4-34]. Compounding the situation, a subsequent drought in the Bay area of Berkeley prevented speedy replacement of fresh material from that end, though Lewis did his best by mailing off some older stock [4-35]. By that time, however, Paul Harteck, with Rutherford on a postdoctoral fellowship just then, had a small electrolyzer going at the Cavendish, based on Lewis's procedure and helpful advice. No sooner was Harteck in business than he, too, suffered from incessant demands on his scarce liquid, as we learn from R H Fowler's sympathetic reply to Lewis's complaint that the Berkeley stock was grossly overestimated.

> The fact is [wrote Lewis] that Dr. Macdonald and I have kept our first little apparatus running, when we could find time to attend to it, but the minute amounts produced have been used or given away so that I never have had more than three cubic centimeters at any one time. [4-36]

Replied Fowler,

> I am sorry you have not got as much heavy water as we believed, but the capacity of everyone for absorbing heavy water appears to be very much the same in Berkeley as in Cambridge. Harteck, who is running our small generating plant here very successfully on your lines, suffers in just the way you describe from the importunate demands from everybody who has a nice little experiment to perform. [4-37]

As for 'industrial friends', Rutherford was no doubt referring to Imperial Chemical Industries—the only potential bulk supplier of heavy water in Britain. The trouble was, according to Fowler, that as of February 1934 ICI had 'succeeded rather surprisingly in electrolyzing the water without any separation at all!' [4-38]. Unbeknown to Rutherford and his group, another such potential friend was Leif Tronstad of Norway, even then in Cambridge working under the physical chemist Eric Rideal—himself an expert on hydrogen and an active participant in the Royal Society discussion [4-39]. Before long Tronstad, whose scientific career was launched the year of Urey's great discovery, 1931, would spearhead the first production of heavy water on an industrial scale, in a most unlikely locale. In so doing, he set off the scientific intrigue of international proportions that culminated that spring day in

Oslo, 1940. Tronstad's name, more than any other, is inexorably linked with the heavy-water coup of 1940 and the subsequent, dramatic events in the rugged mountain district of south-central Norway during World War II.

Leif Hans Larsen Tronstad was born in Bærum, a suburb of Oslo, in 1903. He was the son of Hans Larsen Tronstad, farmer and coach-master, who died the year he was born [4-40]. Despite the ensuing hardship, his mother, by singular dedication, provided a good home and excellent education for the four children. In 1922, Leif graduated with a major in chemistry and high marks from Kristiania Tekniske Mellemskole (Oslo Technical School). The next year, with scarcely a year of additional study under his belt, he passed *examen artium* (the Gymnasium matriculation examination), and enrolled as a student of chemistry at Norges Tekniske Høyskole (Norwegian Technical University, or NTH for short) in Trondheim. There he completed a course of studies in inorganic chemistry during 1924–1927, specializing in electrochemistry, and graduated with top honors—an academic distinction high enough that word of it went automatically to the Royal Palace. At the same time, he underwent officer's training attached to the joint technical branch of the Norwegian army and was commissioned second lieutenant in the reserves—a commission that he would fulfill with distinction in active service a decade later. All the while he taught evening classes on the side to help support his mother and sister.

Upon graduating, Tronstad stayed on briefly as scientific assistant in qualitative analysis at NTH, mainly teaching freshmen students in the 'loft' laboratory, as it was called. Then followed a stint as research chemist at the nickel refinery works in Kristansand, where he was first exposed to the budding Norwegian electrochemical industry. Next, he held stipends variously back at NTH, at the Kaiser Wilhelm Institut für physikalische Chemie und Elektrochemie at Berlin-Dahlem, and at Metallografiska Institutet in Stockholm. During his stay in Germany he married Edla Obel, also from his home town outside Oslo. Later, when he pursued postgraduate studies at Trondheim she acted as his secretary and collaborator. They had two children, a son and a daughter. He received his doctorate (*doctor technicae*) at Trondheim in the spring of 1931.

Tronstad's earliest research of significance was carried out under the guidance of Professor Herbert Freundlich at Berlin-Dahlem during 1928–1929. There he extended qualitative methods for detecting thin oxide films on metals by optical polarization means, devised by Freundlich and others, on the model established by Paul Drude around 1890, into quantitative procedures for ascertaining the thickness of thin films and their nature generally. He continued his investigation of thin films, particularly those derived from electrochemical treatments, under Professor Carl Benedicks at Stockholm—work which he elaborated into his doctoral dissertation at NTH on 'Optische Untersuchungen zur Frage der Passivität des Eisens und Stahls'. With his doctorate secured, he was appointed Dozent in inorganic chemistry at Trondheim, and with a Christian Michelsen traveling stipend in hand proceeded at once to Cambridge University for a 10 month stay. His researches there, extending his optical polarization method to the study of monolayers of fatty acids and related areas, earned his respect in British circles—respect which would stand him and his native land in good stead in wartime years to come.

In the fall of 1934, Tronstad was elected to a temporary professorship in technical inorganic chemistry at NTH, succeeding the late Thv Lindemann; this chair became permanent in 1936, making him one of the youngest professors in Norway. He proved highly popular with students and colleagues alike, and acquired the reputation as a first-class teacher. He was a keen sportsman as well. As a student, he was a member of the relay team which set a Norwegian record for the 4×400 meter dash. His splendid health and robust physique would serve him well in the gruelling commando training in Britain during the forthcoming war, not to speak of his parachuting into the mountains of occupied Norway.

Tronstad's professional interests were by no means confined to his study of thin films. His lasting fame is primarily tied to his leading role in the organization of the Rjukan-Vemork hydrogen plant of Norsk Hydro for the large-scale separation of heavy water and, for that matter, for the subsequent destruction of the same plant while under German control in wartime Norway. How Tronstad, and Norway, became involved in heavy water is a story in its own right, and takes us back to the turn of the century, 1903 to be precise. It also affords us the opportunity to meet one of the most interesting characters in the burgeoning cathode-ray studies at that time: the Norwegian physicist Kristian Birkeland.

4.3. Birkeland and his gun; Eyde and Birkeland

One fair evening in February of 1903, prominent citizens of Kristiania, as Norway's capital was then known, were treated to a rare spectacle in the university's old banquet hall, namely the demonstration of a giant electromagnetic gun. Supposed to fire a projectile into a stout wooden target, the gun operated on the principle of flinging pieces of iron in the magnetic field of a number of aligned solenoidal electromagnets. According to its creator, not a sound would be heard other than the projectile hitting the target. Instead, the gun went off with a deafening roar, and the flash from the muzzle bathed the startled audience in a bluish glow. Though the demonstration was technically a fiasco, and the gun failed to bring about any revolution in weaponry, the spectacle indirectly played a role in a revolution nevertheless; not in armaments, but in agriculture. The man behind the gun was Kristian Olaf Bernhard Birkeland, professor of physics at Kristiania University.

Birkeland was unusual in many respects. Not only was he a pioneer in what we call today space physics—auroral phenomena and cosmic geophysics—but he fashioned himself as an entrepreneur on behalf of science. He unceasingly sought patrons, engaged in self-promotion and grandiose schemes and claims, and took out patents for the purpose of generating funds for research [4-41]. He was born in Kristiania in 1867, where his aptitude in mathematics and physical science was encouraged by his mathematics teacher in Aars and Voss Gymnasium [4-42]. Matriculating at the University in 1885, he studied theoretical physics and graduated with a *Candidatus Realium* degree in 1890. In 1892 he travelled abroad on a scholarship, to further his training in the manner of his countryman and role model Vilhelm Bjerknes. Following closely in Bjerknes's footsteps, he studied with Henri Poincaré in Paris, De la Rue and Edward Sarasin in Geneva, and Heinrich Hertz in Bonn. Among the fruits of his studies abroad are his general expression for the Poynting vector for the propagation of energy in an electromagnetic field, still valid, and the first general solution of Maxwell's equations. In 1898, only 31 years old, Birkeland was appointed full professor at Kristiania University, a chair he retained until he died prematurely at the age of 50. He died in Tokyo, attempting to reach Norway from Egypt during World War I. He had retired in Egypt under rather unsettling circumstances, in his private laboratory on a lavish estate in Helouan near Cairo.

Physically Birkeland was rather small in stature, slightly stooped, and partly bald from middle age onward. He sported a generous walrus moustache, and wore a fez on nearly all occasions, in and out of the laboratory. His health in later years was frail, by his own opinion from exposure to electromagnetic radiation; possibly also from chronic mercury poisoning. He maintained a large cadre of assistants who became prominent in Norwegian physics generally and in space physics in particular, and was a candidate for the Nobel Prize in physics at the time of his death.

Birkeland is perhaps best known for his researches on the mechanism of the aurora borealis, and for his efforts in establishing a network of geophysical observation stations in the Arctic. Struck by the resemblance between cathode-ray phenomena and the aurora borealis, he first suggested in 1896 that the aurora and magnetic storms are the result of charged particle rays emitted by the Sun and trapped in the Earth's magnetic field near the poles. In 1897, he embarked on his first northern light expedition to Finnmark in northern Norway to make a reconnaissance for establishing a mountaintop observatory. The same year he devised the original form of his experimental model of the Earth and a cathode-ray-emitting Sun: an anode in the form of a hollow metal sphere ('Terrella') covered with a phosphorescent coating, enclosing an

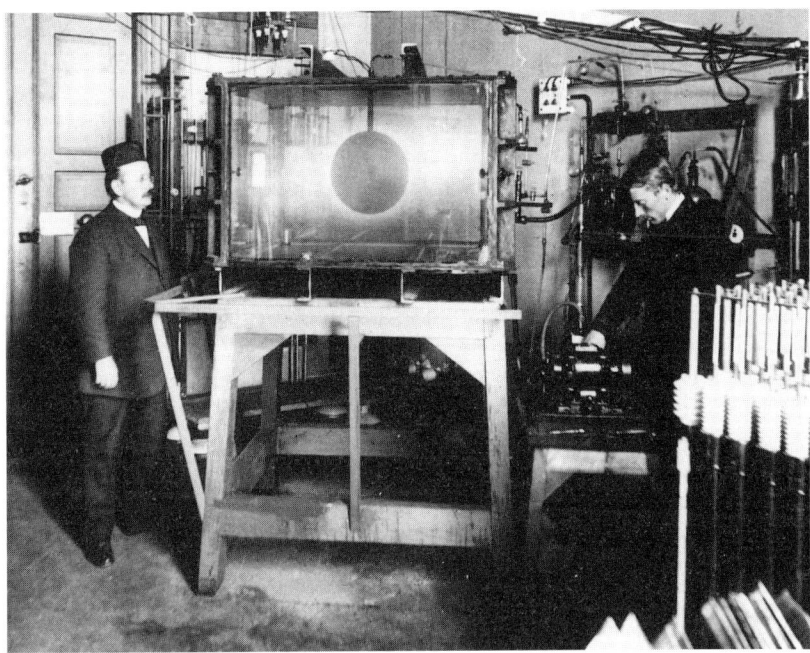

Figure 4.3. *Birkeland (left) and his assistant, Karl Devik, conducting an experiment with the 'world room'. The Earth, represented by the Terrella in the center of the vacuum box, is bathed in 'northern lights'. Courtesy Alv Egeland and Fysisk Institutt, University of Oslo.*

electromagnet, along with a cathode in an evacuated chamber. As the air became increasingly rarefied, electrons emitted from the 'solar' filament were sucked into the 'Earth's' magnetic field, producing 'auroral' rings around the polar areas. Together with an assistant, he built ever larger models in the cellars of the old university buildings, culminating in a 1000 liter 'world room'—the world's largest discharge tube, driven by a giant Swiss high-tension, direct-current generator capable of 300 mA at 15 000 volts.

A version of Birkeland's Terrella is on view at Norsk Teknisk Museum in Oslo, where his archives, including laboratory notebooks, correspondence, and personal transactions, are also deposited [4-43].

The mountaintop observatory was in due course erected 900 meters above sea level at Haldde near Alta in Finnmark, thanks to a generous grant from the Norwegian authorities and partly on the recommendation of Vilhelm Bjerknes, the father of modern meteorology. Bjerknes was himself also interested in possible links between electrical and meteorological phenomena in the upper atmosphere. Birkeland eventually established magnetic stations in northern Iceland, Svalbard, and Novaya Zemlya [4-44]. Analysis of simultaneous magnetic data from these far-flung stations led him to conclude that large currents flowed along the geomagnetic field lines during aurorae. He also established observatories in Egypt and Rhodesia to investigate correlations between the geomagnetic field and the zodiacal light.

To finance all these undertakings, not the least the analysis of the voluminous trove of geophysical data collected by himself and his assistants, Birkeland needed more funding than could be relied on from government and private sources. Not to worry! Anticipating the problem, he initiated his entrepreneurial activities based on technological research and applied physics. His first technical patent involved the aforesaid electromagnetic gun. The idea for such a device came to him from watching the escalating naval arms race between Germany and Britain at the turn of the century. The advantages of a gun of this type, he claimed, were twofold: it was silent and virtually unseen, and could strike targets at distances far greater than could conventional weapons. Its potential reach stretched from Kristiania to St Petersburg, the King of Norway was informed by the engineer Gunnar Knudsen, a backer of Birkeland and future Prime Minister. A first model, if not up to Knudsen's standards, was successfully demonstrated before influential industrial and military people in 1901, and the same year he launched a joint-stock company called 'Birkeland's Firearms' with Knudsen the first stock-holder. Recognizing the need for still more economic support, in 1903 Birkeland arranged his demonstration of a larger prototype before potential backers, including representatives from the giant German Krupp armament concern. The prototype had performed well in preliminary trials, so Birkeland had high expectations for the evening of of 6 February.

'Ladies and Gentlemen', Birkeland intoned before the hushed audience in the university's Festival Hall, 'you may calmly be seated. When I pull the switch, you will neither see or hear anything except the bang of the projectile against the target' [4-45]. The resultant explosion from 3000 amperes short-circuited not only blew the wits out of the frightened audience, but blew his venture capital to smithereens as well, with his stock plunging to nothing.

Three days later, unfazed, Birkeland attended a dinner party at the Knudsens, in the company of a gentleman fully as astute in business matters, and as visionary, as himself. His companion was the civil engineer 'Sam' Eyde [4-46]. Eyde had become preoccupied with a particular problem ever since listening to the opening address before the British Association meeting in 1898 by the colorful Sir William Crookes, savant of the British scientific establishment. Crookes had lamented the imminent depletion of Chilean guano, which provided fertilizer for most of the world's agriculture. Production of fertilizer from atmospheric nitrogen had been under consideration in various quarters, but would require enormous blocks of electric power. No industrial need on a massive scale had been convincingly identified for the mighty Norwegian waterfalls. Hydroelectric power, Crookes had stressed, might be the key to nitrogen fertilizers. Already at America's Niagara Falls, an experimental plant was being built by the American Products Company to explore utilizing the method of Bradley and Lovejoy for oxidizing atmospheric nitrogen. (The million-dollar operation did not prove a financial success, and was discontinued in 1904.) Eyde set about systematically buying up rights to Norwegian waterfalls which could still be had for modest sums. The farmers who owned the falls needed money for investing in their farms or emigrating to America, but had little appreciation for the value of the falls. One half of the mighty Rjukan falls went for all of 600 Norwegian crowns as late as the 1890s [4-47]. As a result, there ensued a veritable run on the waterfalls; many were snapped up by foreign companies. The matter came to a head in the Norwegian Storting, with the upshot stringent state controls on the harnessing and developing hydroelectric sites. This probably put a damper on the rate of growth of the economy, but at the same time ensured Norwegian control over the country's natural resources [4-48].

Over cigars that evening at the Knudsens, Eyde explained to Birkeland that what was really needed was the world's biggest electric arc. Behold! 'That's no problem', replied Birkeland. He had just the thing—the high-amperage arc from his electromagnetic gun.

Thus was born a partnership of unprecedented importance for the future economic well-being of a country still smarting from its troubled union with Sweden and second fiddle to its industrial neighbors to the south. Soon the two had experiments going with model furnaces for producing saltpeter (nitrate) from atmospheric gases. The arc process, originally an English invention, consisted in passing hot air through an electric arc and cooling it quickly to inhibit the decomposition of the nitric oxide which is formed. The Birkeland–Eyde approach took advantage of the well known fact that an electric arc is deflected by a magnetic field at right angles to it; furnaces employing this effect promised to be sufficiently rugged and large that commercial-scale production ought to be possible. In order to develop the production process they had managed to interest two influential Swedish brothers, bank directors Marcus and Knut Wallenberg, and in 1904 formed a corporation with Eyde as director and three million Norwegian crowns as starting capital [4-49]. The corporation, headquartered in Norway and subject to Norwegian laws, acquired not only Eyde's existing waterfall rights, but those of its various stockholders; it thus owned rights for exploiting the Birkeland–Eyde process at most of the hydroelectric sources in central Norway, stretching from Notodden and the adjacent Svælgfos falls, 60 miles southwest of Kristiania, to Rjukan in its majestic setting in the Vestfjord valley, 30 miles further to the northwest. The town of Notodden straddles the estuary of the Tinnelva river flowing into Lake

Heddal, at the northern end of the water course from Skien, 30 miles to the southeast and the administrative center of the county of Telemark.

Even before the corporation was a *fait accompli*, experiments were off to a fast start in the summer of 1903. A makeshift oven in the university cellars gave way to a succession of larger arc furnaces in a Kristiania warehouse [4-50]. The furnace power rose steadily: 3 kW, 30 kW, finally 60 kW with electric arcs 50 cm in diameter and 1 m in length flashing in the winter nights of 1903–1904. Next came a 100 kW test station at Ankerløkken coupled directly to the Kristiania power grid, followed by a true 500 kW prototype at Vassmoen adjacent to a power station. The Notodden plant, sized initially for 2000 kW, though eventually growing to 22 000 kW divided among several ovens, was inaugurated in May 1905 with a festive dinner in the town's former hotel 'Victoria'. In 1908, the project had advanced as far as it could at Notodden, and operation shifted to Rjukan.

It was soon apparent that Norwegian and Swedish capital would not suffice. In the event, the Wallenbergs were able to tap into French funds, principally Banque de Paris et des Pays-Bas, one of the great French banking houses. The French connection would prove pivotal in the events foreshadowing Norway's occupation in 1940. Late in 1905 was, at any rate, born Norsk Hydro-Elektrisk Kvælstofaktieselskap. The first phase of the giant concern, lasting some 24 years, would be dedicated to the Birkeland–Eyde electric arc method for producing nitrate from nitrogen in the air. In 1929 would begin a second phase of Norsk Hydro, when the concern introduced the so-called Haber–Bosch method for producing artificial fertilizer from ammonia (NH_3).

The town of Rjukan was, and is still in some ways, a rather gloomy locale in the district of Telemark, about 70 miles due west of Oslo as the crow flies (figure 4.4). The Sun is rarely on view, due to the surrounding mountains, dominated by Mount Gausta, 5600 feet in elevation, and the habitually unpleasant weather in those parts. Situated at the bottom of the narrow Vestfjord valley, Rjukan is blocked from 1 October to 15 March, though it is said that on a rare, good day one-sixth of Norway can be seen from the top of Mount Gausta. The original hamlet was even more gloomy in its utter isolation. Like other communities of Telemark in bygone days, it was impenetrable except on skis. Nor were visitors welcomed by the locals, who in 1580 were described as 'shameless bodies of the Devil whose chief delight is to kill bishops, priests, bailiffs and superiors—and who possess a large share of all original sin' [4-51]. Rjukan is named after the majestic 'smoking waterfall' dominating the scene at one time, 144 meters high, said to have first been 'discovered' in 1811 by the Prince of Hessen, though the industrial taming of the falls eventually eliminated them as a premier tourist attraction. Only rarely are the falls nowadays seen in their original thundering splendor, when the water load from the Hardanger Plateau exceeds the capacity of

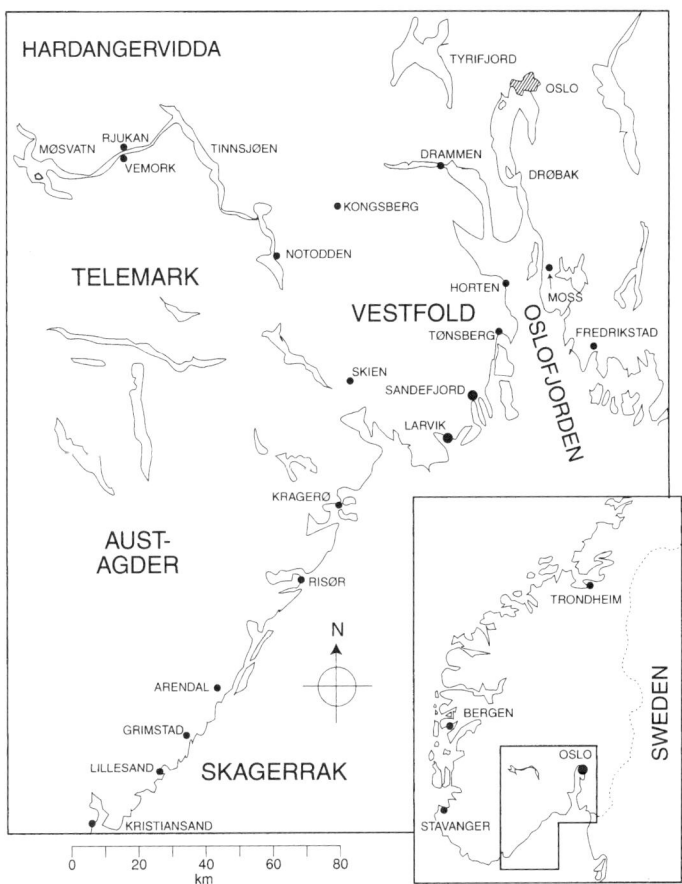

Figure 4.4. *The Telemark region, in relation to the Oslofjord and portions of southeastern Norway.*

the penstocks at Vemork and is sent cascading down the mountainside nearby. Still, their very harnessing inspired many artists over the years, none the more poignantly than the Norwegian painter Theodor Kittelsen in his Rjukan series of paintings and drawings executed between 1900 and 1910. Kittelsen has explained one of his pictures symbolizing Eyde's accomplishment thus: 'We behold the power station, glowing bright in the side of the mountain. The monster in the falls writhes, snorts in anger at having toiled with the machinery' [4-52]. At any rate, today Rjukan's fame is largely due to the heavy-water plant of nearby Vemork and the events associated with it—nowadays an industrial museum recalling the heyday of Norsk Hydro here and the drama forever linked with the area.

During the next decade, Norsk Hydro grew by leaps and bounds into Norway's largest industrial venture. The Birkeland–Eyde arc process

Figure 4.5. *'Taming the Trolls'—i.e., the forces of nature. Sketch from Kittelsen's Rjukan series, 1908–1909. The sketch depicts the Svælgfos power station, north of Notodden, that began operation in 1907. Courtesy Anker Olsen Kr 1955 Norsk Hydro Gjennom 50 År (Oslo: Norsk Hydro) p 68.*

[4-53] demanded unheard of quantities of electric power, transforming the modest villages of Telemark in south-central Norway into thriving residential and commercial centers. Thus Rjukan grew from a scant hamlet in 1907 to a town of 2200 in 1910 and 8350 ten years later [4-54]. In the early years, the scientific and technical solutions were largely due to Birkeland, whereas it took Eyde's organizational talent to put them into

practice. Before very long, Birkeland's impulsive and arrogant nature drove the two apart. Eyde rose steadily in the ranks of Hydro, and became at the same time a leading light in the organization of new, allied electrochemical and other concerns. He ended up a member of parliament and diplomat in Norwegian service abroad. Birkeland went his own way, with an enviable financial independence as a reward for his Hydro venture. He never spared himself in his unflagging pursuit of research, and perhaps wore himself out at an early age.

The memorial stone marking Birkeland's grave in Oslo, erected by the University, reads as follows: 'He fixed atmospheric nitrogen in the electromagnetic arc. He elucidated the nature of the northern lights, electric radiation from the Sun, and the geomagnetic field.'

4.4. Tronstad and Norsk Hydro: an auspicious union

We now rejoin Leif Tronstad on the eve of his academic tenure at NTH in Trondheim. The year after the demonstration of deuterium by Harold Urey and colleagues, Lewis and Macdonald succeeded in producing relatively pure heavy water by prolonged electrolysis of ordinary water, as noted earlier. That is, the method relies on the fact that hydrogen given off during the electrolysis of water contains five or six times less heavy hydrogen (deuterium) than the residual water left in the electrolytic cells. This method was, however, both difficult and expensive, since in ordinary water only one part in 4500 is 'heavy'. To produce one kilogram of heavy water, 50 tons of ordinary water had to be treated for one year, consuming 320 000 kilowatt hours, and, then, the output had a purity no better than about ten per cent.

It was immediately clear to Tronstad that Norsk Hydro, possessing the largest electrolytic hydrogen plant in the world with a capacity of 20 000 cubic meters (7000 tons) per hour, was uniquely positioned for the industrial production of heavy water from the residues after electrolysis. Perhaps his interest in heavy water had originally been sparked by Eric Rideal, his supervisor during his Cambridge stay in 1931–1932 who himself took an active part in the Royal Society discussion on heavy hydrogen in 1933. (Another participant in that discussion was the German physical chemist Paul Harteck, whom we have briefly met and whose name will come to the fore when we get to the German heavy-water effort in World War II.) At any rate, Tronstad became a permanent consultant to Norsk Hydro from early in 1934, administering two 'Hydro-stipends' in deuterium chemistry at NTH, and generally espousing the case for heavy water. In fact, he was not the first to place a heavy-water proposal before Norsk Hydro. In that he was anticipated by the scientist Odd Hassel, in some ways as curious a personality as Kristian Birkeland before him.

For an eventual Nobel Prize laureate, Hassel is poorly known, even among his fellow countrymen. Born in Kristiania in 1897, the son of a physician, he was eight years old when his parents died [4-55]. Though he was not an outstanding student, he nevertheless passed his *Candidatus Realium* degree at Kristiania University at the young age of 23 with a thesis on chemical reaction kinetics. He then knocked about in France, Italy, and Germany for a few years, torn between physics and chemistry. Settling into chemistry, he only lasted seven months in Kasimir Fajans' laboratory in Munich, not getting along with Fajans. He did, however, obtain his doctorate in x-ray crystallography at the Kaiser Wilhelm Institute in Berlin in 1924, under Fritz Haber, who could cope better with Hassel. Ten years later he assumed the newly created chair in physical chemistry at Oslo University. (In 1925, the name of Norway's capital reverted to 'Oslo', as the city was originally called.) Hassel's backers for the chair had included Victor Goldschmidt, director of the Mineralogical Institute, and Fritz Haber, Hassel's mentor in Berlin and somebody who counted. Among other things, Haber's name is associated with the ammonia process that was to become so important for Norsk Hydro.

Hassel's odd personality stemmed partly from his being an albino, who spent most of his waking hours in shadows because light troubled his eyes. He has been described by a former student as 'rather difficult', though 'often kind and helpful', yet easily offended and sarcastic with people he didn't care for [4-56]. Even Sir Derek Barton, with whom he shared the Nobel Prize for chemistry in 1969 for work on the structure of organic molecules, found him hard to take, with a negative character 'as if he were not there' [4-57].

Be that as it may, no sooner was the discovery of heavy water announced than Hassel, in the Spring of 1933 and while still a Dozent at the university, approached chief engineer Antonius Foss at Norsk Hydro's Research Division, explaining that he was following up on the recent discoveries by Urey *et al.* He laid before Hydro a verbal proposal for deuterium production utilizing the natural enrichment of the alkaline solution in the electrolyzers at Vemork. In the event, he failed to convince Hydro management, or rather, Herr Foss, of the profitability of undertaking such a venture. Consequently, Hassel set about concentrating heavy water on his own in the unlikely setting of the Geologisk Museum's quarters at Tøyen in Oslo, using lye originally secured from the DENOFA electrolytic concern in Fredrikstad, and later from Hydro [4-58], employing largely homemade apparatus. By August, Hydro relented, providing him with an electrolysis cell drawn up for the purpose by the chemical engineer Jomar Brun [4-59]. In the event, Hassel's experiments took so much time that he was anticipated in several experiments with heavy water by an old acquaintance from his student days in Berlin, the German physical chemist Karl Friedrich Bonhoeffer, brother of the Protestant theologian Dietrich Bonhoeffer (who was executed for the attempt on Hitler's life in 1944). Bonhoeffer, too, had requested lye samples from Hydro as early as May of 1933 [4-60]. Nevertheless, his exertions during 1933–1934 would not be Hassel's only brush with Hydro *vis-à-vis* heavy water. Being interned by the German authorities in his homeland during World War II, he would figure in a minor way in the German heavy-water effort in the closing months of the war.

More persuasive than Hassel in steering Norsk Hydro into a new venture was Jomar Brun, then 29 and head of Hydro's hydrogen electrolysis plant at Vemork. Formerly fellow student with Tronstad at NTH, Brun was the son of customs inspector Lorentz H Brun. Upon completing his studies in electrochemistry in 1926, he stayed on at NTH for several more years, and joined Norsk Hydro in 1929. In August of 1933, as noted, Brun had been asked by the head office at Vemork to see to the construction of an electrolysis cell for use in Hassel's experiments [4-61]. Though the cell gave positive results, Hydro management decided nevertheless not to pursue the matter further. What spurred Hydro to take action was another German request for some water samples from the electrolyzers at Vemork, this time from Ernst Hochheim of I G Farbenindustrie. In his letter to Hydro Director Bjarne Nilssen on 11 July 1933, he specifically asked for lye from an electrolyzer that had operated for a long time [4-62].

This inquiry, coupled with Bonhoeffer's earlier request, puzzled Jomar Brun at Vemork. 'At that time, as far as I know', recalls Brun, 'none within Norsk Hydro realized that the electrolysis plant could serve as the basis for production of heavy water' [4-63]. However, he discussed the matter with Tronstad, who had been a consultant to Hydro off and on for some time. In a report to Hydro headquarters late that Summer, Brun warned that

> ... since we can well imagine that heavy water, or compounds containing heavy hydrogen, may prove to be of technical or therapeutic value, we recommend caution in divulging information concerning the amount of heavy water in our electrolyte, particularly since our electrolysis plant at Vemork ... is probably the largest existing source for the possible production of heavy water. [4-64]

Brun and Tronstad were familiar with the latest American experiments by Lewis and Macdonald. The decisive factor in convincing Tronstad and his Hydro colleague that Vemork was just the place for the production of heavy water was probably the results of Bonhoeffer's analysis of the potash-lye samples dispatched to him back in June, which indicated a heavy-to-normal hydrogen ratio of 1:2300, compared to 1:6000 for normal water, according to the best available data at the time [4-65]. They decided that Tronstad, with his scientific reputation, should approach Hydro with a specific proposal for the production of heavy water by the repeated concentration of water from the electrolyzers at Vemork. He presented their proposal to Hydro in the fall of 1933, and it was agreed by the administration that he and Brun should go ahead with some preliminary experiments, with the samples analyzed at NTH in Trondheim.

In the event, they got off to a poor start, using lead electrodes and sulfuric acid as the electrolyte, a combination which caused excessive pitting of the cathodes and the formation of fog. (Water by itself is a poor conductor of electricity; for an electrolytic reaction to take place, a conducting electrolyte must be added to the water.) It soon became apparent that soft iron, used routinely for the electrodes in the normal hydrogen process at Vemork, coupled with potash lye as the electrolyte, gave the best results. Brun and Tronstad settled for standard Pechkranz electrolyzers from the Swiss Fabrique d'Elektrolyseurs Hydroxygene SA. With potash lye, they gave a heavy water concentration six times that of normal water fed as input to the Vemork electrolyzers [4-66].

The reference to the giant German chemical concern I G Farben-industrie Aktiengesellschaft in this connection calls for an explanation. As early as 1906, Norsk Hydro had been in touch with Badische Anilin-und Sodafabrik in Ludwigshafen, where a promising nitrogen-fixation process similar to the Birkeland–Eyde process had been under study for some time, and German hydropower concessions in Norway were being explored. To avoid a prolonged and costly competition, Hydro and Badische agreed on a cooperative plan, according to which the two partners would own and develop certain hydroelectric sites with plants and lines of transportation of mutual interest. The joint venture was not appreciably affected by the outbreak of the first world war, Norway in principle remaining steadfastly neutral. (In practice, Norway proved a 'neutral ally'; its far-flung shipping aided the allied cause, while suffering grievously from German U-boats, and the nation maintained a tense relationship with Germany.) As part of the settlement after Germany's defeat in 1919, all the German shares in Norsk Hydro were confiscated by the Allies, and the partnership was effectively dissolved. However, during 1927–1928, Hydro entered into a new German collaboration, this time with the then newly created chemical cartel I G Farben, for the purpose of facilitating a switch to the less power-intensive Haber–Bosch method for producing ammonia from hydrogen [4-67].

Thus, it came about that by 1933 I G Farben, together with the Swiss-registered conglomerate I G Chemie, Basel, held 25% of Hydro's joint stock. And thus it was that heavy water came to the attention of Norsk Hydro again, soon after Hassel's rebuff, leading eventually to the first attempts to produce heavy water at Vemork. It was arranged that Tronstad would formally request samples of reconcentrated potash lye, channeled to him at Trondheim through Jomar Brun. Test results obtained in the winter of 1933, meticulously preserved in Tronstad's own hand in his personal journal for the period [4-68], all argued for the need to catch up with Urey and Washburn, and undertake serious deuterium production at Vemork. In a series of notes to Hydro management, it is recommended that Norsk Hydro do just that, using electrolysis in favor of fractional distillation (Urey's first method), that Tronstad be hired as permanent consultant to Hydro, and that further electrolyzer samples be shipped to Trondheim for analysis [4-69]. Reconcentration efforts using existing, albeit modified, apparatus were concluded in April 1934. The report of a four-day conference at Rjukan in May, devoted largely to heavy-water production, leads off with the following conclusion:

> On the basis of experience gained from tests conducted at Vemork..., as well as on the basis of recent information reported in the scientific literature, we deem it entirely justifiable to undertake, without delay, the production of heavy water on a technical scale. [4-70]

The heavy-water installation went on line in mid-1934. At the front end was a series of Pechkranz electrolyzers arranged in successive stages, each stage smaller than the one before it in step with the degree of concentration demanded of it. Stage 1 was fed water containing approximately 0.05% heavy water condensed from earlier plant electrolyzers. The condensate from the gas given off in Stage 1, somewhat richer in heavy water than the input condensate, was used as input to Stage 2, and so forth down the line. The cascade of Pechkranz electrolyzers terminated in the high-concentration stage, consisting of a series of specially constructed cells, each with cathodes of mild steel and the usual potash lye as the electrolyte. Water distilled from the electrolyte was fed from one cell to the next; hydrogen given off in the electrolysis was burned in oxygen, producing water richer in heavy water and again fed back into the system. (*Unburned* hydrogen from the earlier stages was fed to the synthetic-ammonia plant, as part of the first phase in the fertilizer production.) The end product was heavy water concentrated to better than 99% in purity.

Already in June of 1934, Tronstad could report, in a letter to *Nature*, production of large quantities of '1:300-water' and prospects for about 10 liters per day of 'pure' heavy water if there were sufficient demand. The efficiency of separation in both an alkaline and acid solution agreed with that found by workers abroad—e.g., by Harteck in Germany [4-71]. A full paper by Tronstad and Brun had already been submitted to *Zeitschrift für Elektrochemie* and appeared in print later in the year [4-72].

Hydro records show that the first quantities of relatively pure (in excess of 98%) heavy water were shipped abroad in January 1935, in 10 to 100 gram lots, to laboratories variously located in Scandinavia, Germany, France, and England. For instance, six shipments are recorded for 2 January alone, as follows:

Figure 4.6. *The original heavy-water, high-concentration cells at Vemork. Courtesy Norsk Industriarbeidermuseum, Vemork; negative No UF-113.*

2 g to Physikalisches Institut, Eidgenössische Hochschule, Zürich

10 g to Physikalisches Institut, Technische Hochschule, Darmstadt

5 g to Laboratorie de Physique Expérimentale de la Faculté des Sciences, Grenoble

5 g to Universitetets Institut for Teoretisk Fysik, Copenhagen

50 g to Institut de Biologie Physico-Chimique, Paris

200 g to Dr M P Appleby, Imperial Chemical Industries, Billingham.

Some forty-odd lots were shipped in January. Recipients included Leif Tronstad at NTH, Klaus Hansen and Odd Hassel at Oslo, and P M S Blackett at Birbeck College, London, as well as Harold Urey at Columbia. Urey received 500 gram of 99.2% material shipped free of charge on 1 February [4-73]. Though the initial stimulus for the Norwegian heavy-water initiative had come from Germany, the Germans showed only marginal subsequent interest in the Norwegian production, with England the main customer outside the Scandinavian countries during the first six months of production, and, despite the general interest evinced by the Royal Society discussion in 1933, it was generally felt that the most promising application of the new-found isotope lay in biochemical areas. This was the opinion of Harold Urey, and it was shared by Axel Aubert, General Director of Norsk Hydro since 1926 and himself a chemist. Aubert believed that heavy water might prove to be of special importance in cancer research, something much on his mind [4-74].

Possible therapeutic use of dilute heavy water was discussed in 1934 by, among others, E J Larsen and T Cunliffe Barnes at Yale University [4-75]. Indeed, Gilbert Lewis of Berkeley had, as mentioned earlier, fed his entire initial stock of heavy water, one cubic centimeter, to a mouse with, it might be added, neither helpful nor harmful consequences. Jomar Brun cites the equally preposterous case of Klaus Hansen, professor of medicine in Oslo. In 1935, before a group of journalists, he apparently drank a small quantity of heavy water, simply to show that it was harmless. More harmful, perhaps, was the stinging criticism of Hansen's demonstration by a large group of his academic colleagues at the university [4-76].

The Vemork plant was gradually expanded. By the end of 1935, it had grown to encompass five Pechkranz stages, each typically containing 114 cells, followed by a high-concentration stage of seven special cells, each designed for 300 A [4-77]. Even so, the quantity produced was modest. It rose from 40 kg of liquid concentrated to ~99% in 1936 to twice that amount in 1938, all the while remaining but a minor by-product of the regular hydrogen production line at Vemork. The price was high: initially 10 Norwegian kroner (or roughly 50 cents) per gram, though it dropped somewhat as the plant efficiency rose. Nevertheless, lackluster sales led Hydro to temporarily discontinue operation of the high-concentration plant in mid-1939. Sporadic production resumed at the end of January 1940. The final entry in Hydro's shipping records, before the German invasion of Norway, 100 gram of 99.6% heavy water to I G Farben, is dated 14 March 1940 [4-78].

As for Tronstad, he remained an active consultant to Hydro. During his regrettably few years of active research, before the outbreak of war claimed all his attention, he authored or coauthored some 14 publications on the physical, chemical, and biological properties of deuterium, deuterium oxide, and various other deuterium derivatives. The subsequent revitalization and development of the full-scale heavy-water concentration plant at Vemork, immortalized in the heavy-water saga of World War II, was largely the handiwork of Brun, assisted by an able team of Hydro engineers and technicians, with Tronstad consulting as long as he remained in Norway. We will come to that saga in due course. For now it suffices to echo some sentiments by N A Sørensen in his eulogy of Leif Tronstad following the latter's heroic demise in the waning days of World War II.

> If anybody had predicted, when Tronstad lectured on atomic chemistry while on a stipend to NTH in the fall of 1929, that industrial production of isotopes would be under way in Norway within 5 years, we, perhaps including Tronstad, would no doubt have displayed the greatest of skepticism. That industrial production of isotopes was in fact realized in Norway, was entirely due to the realization of the young lecturer that there are no limits

between basic and applied chemistry. The production of heavy water at Vemork was a technical and scientific triumph for our country.... [4-79]

Chapter 5

ARTIFICIAL RADIOACTIVITY

5.1. Another French miss, and triumph at last

The next development in our chronicle had its start at an elevation even higher than the Telemark peaks, in a setting even more forbidding than the Rjukan environs. Still smarting from having handed Chadwick the clue to the discovery of the neutron, Frédéric Joliot and Irène Curie lost no time in following up on Chadwick's new particle. At the same time, a new radiation phenomenon increasingly occupying colleagues could not be ignored, and might prove productive in ascertaining the properties of the neutron. Already before the world war, the mysterious form of radiation seemingly emanating 'from above' had been determined by the Austrian physicist Victor Hess, by studying the spontaneous discharge of balloon-borne electrometers at various altitudes, to originate in outer space. The term 'cosmic rays' for the highly penetrating radiation was coined by Robert A Millikan, himself a controversial head of an active school researching the rays in America. He referred to them as 'the birth cries of atoms', and 'music of the spheres', in keeping with his view that some of the radiation was a form of electromagnetic radiation accompanying nucleosynthesis in interstellar space. Another American, Arthur Holly Compton, found a different explanation, showing from its deflection in the geomagnetic field, that the radiation consisted mainly of charged particles. Nor did it fall uniformly on all parts of the earth's surface.

Thus it came about that, during late April and early May of 1932, the Joliot-Curies found themselves at the international scientific research station 3500 meters up on the Jungfraujoch in Switzerland. The station, reached via the railway that tunnels through the Eiger in the Bernese Oberland, clings precariously to the side of the mountain, overlooking the Jungfrau and Mönich peaks and the majestic Aletch ice fields crisscrossed with crevasses—the largest glacier in Europe. The French couple were the only visitors in residence just then, 'treated like royalty' and only sorry to be deprived of Hélène and Pierre who were left back in Paris with Frédéric's mother [5-1]. Searching for Chadwick's neutron in the cosmic radiation, they came up empty-handed; evidently it was not a major component of the ubiquitous radiation from space. All the same, the radiation had a surprise in store when later in the year, important news came from abroad—this time from America. Once again, an important discovery had eluded Frédéric and Irène by the narrowest of margins.

During their earlier study of the Bothe–Becker radiation, with the Wilson cloud chamber clamped between the poles of an electromagnet, the two had made an interesting observation. Among the tracks of high-speed electrons ejected by the energetic gamma-rays emanating from beryllium, some curved in the opposite direction from the bulk of the electrons. Evidently, the contrary tracks were the signature of electrons moving 'backwards' from the chamber volume toward the polonium source. Alas, the bombshell announcement from America brought a different explanation.

Among the protégés of America's plenipotentiary of physics, Robert Millikan, was Carl David Anderson. He was born in 1905 in New York to Swedish immigrants who later moved to Los Angeles. There the elder Anderson scraped out a living in the restaurant business. For want of funds, young Carl remained at home, and a good thing it was. It allowed him to complete his undergraduate work at Caltech in nearby Pasadena, after which he stayed on for graduate school. He obtained his PhD degree under Millikan at Caltech in 1930 on a study of photoelectrons with a Wilson cloud chamber. Marking time as a postdoctoral fellow in Pasadena, he then began, on Millikan's suggestion, assembly of a cloud chamber suitably modified for the study of cosmic-ray electrons, ostensibly to follow up on certain of his mentor's notions concerning the rays. Expecting the energy of the electrons to be very high, Anderson took advantage of the availability of a powerful dc generator for powering a wind tunnel in the nearby Guggenheim Aeronautical Laboratory, and incorporated in his chamber array one of the strongest research magnets to date—one capable of 25 000 gauss across the chamber volume. However, on first operating the chamber with the magnet energized, it revealed an unexpected and awkward result: nearly equal numbers of particles of positive and negative charges. Alternatively, and equally unplausible, some of the particles were apparently upward-moving negative electrons. Not very likely, thundered Millikan; 'everyone knows that cosmic-ray particles travel downward, not upward!' [5-2].

To sort out the confusion, a lead plate was mounted horizontally across the center of the chamber. This arrangement allowed not only determination of the sign of the charge of a particle traversing the chamber from the direction of curvature in the magnetic field of a given polarity, and its energy from the amount of curvature, but also made it possible to distinguish between upward and downward moving particles (that is, providing the plate was not so thick as to stop the particle entirely). In traversing the plate, a particle would expend some energy, and its track would be curved more strongly upon exiting from the plate than it was upon entry.

Behold! There was no longer any doubt. Among roughly a thousand 'Wilson photos' scanned, quite a few showed tracks that were produced not only by particles moving upward, but also by particles with a positive charge and a mass small compared to the mass of the proton. 'It seems necessary', Anderson cautiously worded his preliminary letter to *Science* after carefully weighing the alternatives, 'to call upon a positively charged particle having a mass comparable to that of the electron' [5-3]. The long-sought positive counterpart to the negative electron was at hand, and not in the guise of the will-o'-the-wisp positive electron chased on pure speculation by Jean Becquerel during 1905–1906 [5-4]. To Anderson's annoyance, the editor of *Science News Letters* in which a photograph of one of the renegade tracks was first published [5-5] suggested calling the new particle the positron, yet Anderson did not argue. More to the point, he was vaguely aware that the particle had been predicted in May of 1931 by the Cambridge theorist Paul Adrien Maurice Dirac as one of two energy states of the electron which pops out of his relativistic theory of the electron. Commenting years later on the possible influence of Dirac's work on his own discovery, Anderson allowed that 'Yes I knew about the Dirac theory. ... But I was not familiar in detail with Dirac's work. I was too busy operating this piece of equipment to have much time to read his papers' [5-6].

Hard on the heels of Anderson's announcement, other physicists besides the Joliot-Curies dusted off their old photographic plates; sure enough, there were positron tracks in abundance—tracks too faint to have been noticed, tracks misinterpreted or discarded for one reason or another. As if to make up for their averted attention, however, Frédéric and Irène came up with a valuable contribution of their own, soon after firing up their cloud chamber once more in April of 1933. Within a month they were able to produce the first photograph showing the creation of an electron–positron pair from a hard gamma-ray photon emitted from their Po–Be source. The parameters of their apparatus were well chosen. The creation of the two electrons requires a minimum photon energy given by the sum of the rest masses of the two electrons, or $2mc^2$ (assuming the two rest masses are identical). Any excess photon energy is shared as kinetic energy between the fragments thus created.

Every now and then, Irène's health demanded that she take a full rest, in the country or mountains; the sickness to which she would succumb was ever so slowly gaining on her. On 22 July 1932, Frédéric wrote her that he had obtained tickets on the night train to l'Arcourest in the nick of time; the train was fully booked.

I have wrapped the source of Po in a thin sheet of silver and placed them in a glass tube. But we have trouble with the electrometer because the thread keeps on sticking. That's due to the blasted vaseline that you've used. It really is a most impressive gaffe, my dear. Savel has the same inconvenience with his apparatus. [5-7]

Resuming their study of radiations emitted by elements bombarded with polonium alpha-particles, they found that medium-weight elements ejected mainly protons. However, they also found that certain light elements, notably aluminum and boron, sometimes ejected a neutron and then a positron. More remarkably, the positrons thus emitted appeared to have a continuous energy spectrum, analogous to that of negative electrons emitted in beta-radioactivity—the latter a long-standing puzzle high on the agenda of the upcoming meeting of the Solvay Council that October. The nearly simultaneous ejection of a neutron and a positron suggested to the Joliot-Curies evidence for a compound structure of the proton; that is, that the proton is a close combination of a neutron and a positron. If so, the mass of the neutron was greater than that of the proton, not less as hitherto supposed.

The Solvay Conferences ranked among the most prestigious of physics conferences. They were held every few years, with attendance by invitation only and restricted to a few dozen of the world's leading physicists. They were instituted by the Belgian industrial chemist Ernest Solvay, who had made a fortune on a process for manufacturing sodium carbonate; he used a substantial portion of this fortune in support of various philanthropic causes. Solvay happened also to be an amateur dabbler in theoretical physics, and in 1910 encountered the German physical chemist Walther Nernst, from whom he learned of the contradictions posed by the new quantum concepts of Max Planck and Albert Einstein. He saw in Nernst a means of bringing some of his own ideas on that subject to the attention of Planck, Einstein, and others. Nernst, for his part, saw Solvay's desire and enthusiastic philanthropy as the opportunity to organize an international conference to 'review current questions in connection with the kinetic theory of matter and the quantum theory of radiation'. The upshot was that some two dozen prominent physicists convened on the Hotel Metropole in Brussels during October–November of 1911; among them, besides Nernst, were Planck, Einstein, Rutherford, Jeans, Mme Curie, Kamerlingh Onnes, and Lorentz, the chairman. Thus was born an important tradition within the physics community, with the Conference meeting again in 1913, 1921, 1924, 1927, and 1930—always at the Metropole and each time with a different underlying theme.

The seventh *Solvay Conference* was scheduled to convene in Brussels from 22 to 29 October 1933, with the general theme the 'structure and properties of atomic nuclei'. It would be the last such conference to take place before World War II. (One repeatedly rescheduled but finally planned for 1939, on elementary particles and their mutual interactions, had to be canceled at the very last moment by ensuing military events.) Though a major discovery continued to elude them, the Joliot-Curies had by 1933 made a sufficient reputation to be honored with an invitation to the conference, along with Marie as well as their countrymen Maurice and Louis de Broglie, Francis Perrin (son of Jean), and Salomon Rosenblum—also of the Curie Laboratory but really a citizen of the world. Paul Langevin would be the man of the hour, serving as session chairman.

John Cockcroft presented the first report at the conference, comparing the fruits of his and Ernest Walton's nuclear disintegrations with proton bombardment at Cambridge with those more recently achieved at Berkeley with deuterons accelerated in Ernest Lawrence's cyclotron. (Lawrence was the only American participant at the conference, having been invited on Bohr's suggestion.) Cockcroft was followed by his colleague Chadwick, who reported not only on the Cambridge program of nuclear transmutations under α-bombardment, but also on his own experimental triumph, the discovery of the neutron. The Joliot-Curies came next on the agenda, reporting on the 'penetrating radiation from atoms under the action of α-rays'. Curiously, the discovery of the positron—another milestone in that *annus mirabilis* of particle physics— was related by P M S Blackett almost as an afterthought in the discussion period following their report. (Anderson himself was not among the invited participants. As it happened, Blackett and Giuseppe Occhialini had performed experiments suggesting electron–positron pair production even before Anderson's discovery, using a cloud chamber controlled by a Geiger counter.) Dirac came next, with his theory of the positron, followed by George Gamow on γ-ray spectra, and, finally, Werner Heisenberg on the structure and stability of atomic nuclei [5-8].

In the discussion following the Joliot-Curie presentation, virtually everybody spoke up, not only Blackett. In particular, Lise Meitner had much to say. She had performed similar experiments to theirs at the Kaiser Wilhelm Institute for Chemistry in Berlin-Dahlem, where she had built up a department fully on a par with the Curie Institute in Paris and, for that matter, the Cavendish Laboratory. Meitner ruled her institute with a firm hand, and had a reputation for exacting work. She now bombarded her French rivals with questions, and stated flatly that she had been unable to uncover a *single* neutron. Others in the audience seemed inclined to go along with what she said. Not surprisingly, Joliot was rather upset.

> In the end [Joliot recalls], the majority of the physicists present did not believe in the exactitude of our experiments. After the session we were quite downhearted, but at that moment professor Niels Bohr took us aside, my wife and I, to tell us that he found

our results very important. A little later Pauli gave us the same encouragement. [5-9]

For once, Meitner was wrong. (She later retracted her statement.) All the same, her critique, coupled with the encouragement of Bohr and Pauli, convinced the French couple that they were on the track of something important. Returning to the Radium Institute, they decided that the next step was to demonstrate that the emission of a neutron and a positron occurred simultaneously, irrespective of the energy of the bombarding α-particles. For a starter, Joliot tried the simple expedient of moving the polonium α-source a little back, increasing the atmospheric absorption and shortening the range of the α-particles. At a certain separation, the neutron emission fell off to zero, but the positrons kept coming. Even when he removed the polonium source altogether, the tell-tale signature of the positron emission continued in the cloud chamber, decreasing over a certain period of time in the same manner as radiation from a naturally radioactive element.

Joliot called in Irène, and switched to a Geiger counter for more quantitative measurements. When they bombarded aluminum with α-rays, and then removed the source as before, the counter kept crackling, only quieting down after several minutes. 'Everything became clear!' But wait. Had ordinary, stable aluminum truly become radioactive, or was it simply a matter of a defective counter? It was six in the evening, and Frédéric and his wife had an invitation to dinner that could not be put off. He summoned Wolfgang Gentner, a young German physicist spending a year with them, who was their resident counter expert.

In the morning, Joliot found a note from Gentner; there was nothing wrong with the Geiger counter [5-10]. He and Irène knew at once that they had chanced across an entirely new phenomenon: artificial radioactivity [5-11]. In swallowing the α-particle (2 protons and 2 neutrons), the aluminum nucleus, of 13 protons and 14 neutrons, transformed itself into an unstable nucleus of phosphorus (15 protons and 15 neutrons) plus a lone neutron, thus:

$$_{13}\text{Al}^{27} + {}_{2}\text{He}^{4} = {}_{15}\text{P}^{30} + {}_{0}\text{n}^{1}.$$

The phosphorus, in turn, spat out a positron and a neutrino, settling into a stable silicon nucleus of 14 protons and 16 neutrons:

$$_{15}\text{P}^{30} = {}_{14}\text{Si}^{30} + \text{e}^{+} + \nu.$$

Frédéric and Irène set about confirming that phosphorus was indeed created in the aluminum foil, with a half-life of only a few minutes. Quickly, they bombarded a piece of aluminum, and within three minutes confirmed, by standard radiochemical means [5-12], the existence of radio-phosphorus with a half-life slightly over three minutes. Joliot began to run and jump about in the cavernous basement laboratory. 'With the neutron we were too late', he exclaimed to Gentner. 'Now we are in time!' [5-13].

The same Friday afternoon, 15 January 1934, Joliot called up Pierre Biquard, his friend from his student days at EPCI. Biquard raced over to their cluttered sub-basement room from his nearby laboratory on the Rue Vauquelin. Joliot filled him in on the background, then performed the experiment again. No sooner had he finished, than in walked Marie Curie with Paul Langevin in tow. Joliot repeated the demonstration quietly once more; few words were exchanged. By this time, Marie was chronically ill. Though quite aware of the dangers of radioactivity, precautions could not be allowed to interfere with her work. She picked up the vial of newly created phosphorus with her radium-damaged fingers, and held it in front of a Geiger counter, which began to crackle. 'I will never forget the expression of intense joy that seized her', said Joliot much later. 'It was without doubt the last great satisfaction of her life' [5-14]. Within five months, Marie was dead from leukemia.

The very same day, Frédéric and Irène announced their discovery of a new type of radioactivity, involving positron emission, and the artificial formation of light elements, in a note to the Academy of Sciences in *Comptes Rendus* [5-15]. They also dispatched a succinct, one-page letter to *Nature*, dated 19 January [5-16]. Among the many letters of congratulations pouring in, none must have moved them more than one from Rutherford, dated 29 January.

> I congratulate you both on a fine piece of work which I am sure will ultimately prove of importance. I am personally very much interested in your results as I have long thought that some such an effect should be observed under the right conditions. In the past I have tried a number of experiments using a sensitive electroscope to detect such effects but without any success.... [5-17]

Their discovery brought the Joliot-Curies the Nobel Prize for chemistry in 1935, awarded jointly in the tradition of Pierre and Marie Curie, and the third Nobel Prize in the family. Characteristically, in their Nobel addresses, Irène, the chemist, chose to dwell on the physical aspects of their discovery, while Frédéric, considered to be the physicist, dealt with the radiochemistry aspects. The honor, coupled with Joliot's redoubled social concerns in the years to come, meant ultimately a decline in Joliot's active researches. The transition from a practicing physicist to an administrator and spokesman of science was hastened by his election, on Langevin's urging and by a narrow majority in the Academic community, to a vacated chair in chemistry, renamed 'nuclear chemistry', at the Collège de France—the pinnacle of French learning. (Irène was not overlooked: the same year she was elected professor at the nearby Sorbonne, but continued as research director at the Radium

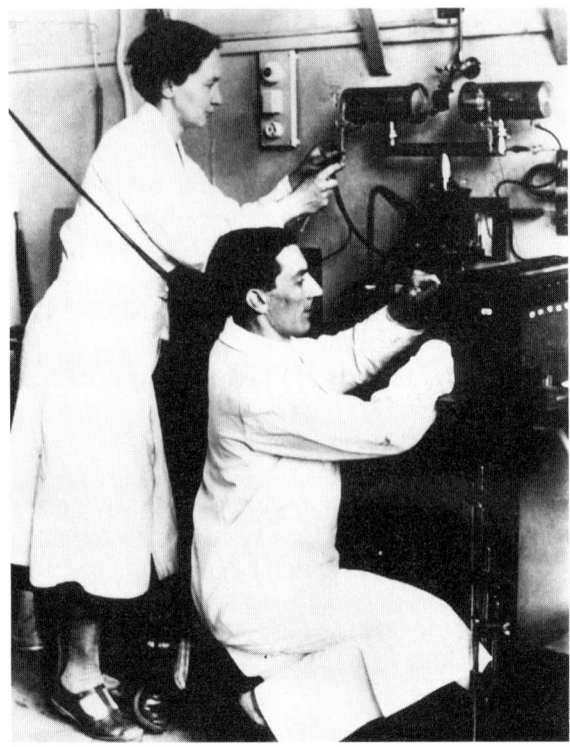

Figure 5.1. *Frédéric Joliot and Irène Curie bending to the apparatus. Courtesy AIP Emilio Segrè Visual Archives.*

Institute.) Even before the vote for Joliot's chair was final, prospects for funding for new laboratories and resources were bright, thanks to the newfound reputation of himself and his wife, and the authority of their close colleague Jean Perrin, who was becoming increasingly influential in the institutional development of science in France. Among the improvements in the wings was a brand new multi-story laboratory building crammed into a lot adjacent to the main college buildings off Boulevard Saint Germain; in its cellars ten meters below street level would be located the first cyclotron in Western Europe.

The discovery of artificial radioactivity had an interesting sequel. As noted earlier, Ernest Lawrence, too, had attended the Solvay Conference, if not as a featured speaker; he was the eighth American to do so, and the only one in 1933. Like the Joliot-Curies, he also had been stung by criticism from the floor after his comments (following Cockcroft's report) on deuteron experiments at Berkeley. The gist of his remarks was that the neutron and proton did not differ appreciably in mass—a conclusion hinging on the spurious notion that the deuteron is unstable. The culprit behind the notion was in due course traced, in another laboratory, to deuterium contamination in his cyclotron, but that is another story. It suffices to note that back in Berkeley, Lawrence received the latest copy of *Comptes Rendus* on 20 February 1934. It contained the announcement by the Joliot-Curies, that ended with the observation that 'long-lived radioactivity, analogous to what we have observed, no doubt can be created by several nuclear reactions. For example, the nucleus $_7N^{13}$ which is radioactive on our hypothesis, could be obtained by bombarding carbon with deuterons, following the emission of a neutron' [5-18]. Lawrence read this with alarm. Deuterium projectiles were their specialty at Berkeley, despite the dispute as of late, owing to the productivity of the electrolytic cell of Gilbert Lewis across the street in producing deuterium. Had they missed something big?

To Lawrence's chagrin, he and his team not only verified the production of the first microquantities of nitrogen-13 under deuteron bombardment of a carbon target, but found that virtually everything they bombarded turned radioactive. Said Robert Thornton, a Canadian member of the team, 'We looked pretty silly. We could have made the discovery at any time.' Added John Livingood, 'we felt like kicking each other's butts' [5-19]. The standard apology for Lawrence missing this epochal discovery has it that, in the interest of quicker cyclotron turnaround operating time, the same switch controlled the cyclotron and the Geiger counter. Heilbron and Seidel find it hard to take that the equipment was so peculiarly wired; their hunch is that 'it was not a question of labor-saving switches, but of labor-saving thinking' [5-20].

For our purpose, the most important impact of the *Compte Rendus* communication from the Joliot-Curies was on a young Italian physicist, just back from a skiing holiday in the Dolomites. Like Lawrence, he seized upon the concluding paragraph, hinting that sustained radioactivities, analogous to those observed in Paris, could be brought about by bombardment with projectiles other than α-particles [5-21].

5.2. Rome: another discovery, and a discovery missed

Enrico Fermi was the same age as Ernest Lawrence, and, like his American colleague, was the only representative of his native land, Italy, at the Solvay Conference in 1933. Fermi, too, led a team of young physicists who had not yet made their mark, but had systematically been building up their laboratory in the Institute of Physics on Via Panisperna in Rome. Unlike Lawrence, however, his team had not come up with a substantial program to tackle with the apparatus in hand. Nuclear physics, according to their patron, Orso Mario Corbino, held a particular promise in helping lend prestige to their university and nation. On the other hand, recalls Fermi's teammate Edoardo Amaldi, 'we had not yet found any [nuclear physics] problem to work on' [5-22].

The news from Paris of artificial radioactivity gave them their big opportunity. Here was a field hardly scratched, ripe for picking. Fermi decided at once to use neutrons instead of alpha particles for their projectiles [5-23]. Lacking an electric charge, they were not prone to be knocked about by negative electrons in traversing matter, nor was there a strong electrical repulsion to prevent them from slipping into an atomic nucleus and forming a heavier, unstable nucleus with interesting ramifications. Complicating matters, they had to be produced by α-bombardment of certain elements, and, since only one in 100 000 alphas found their mark, they were correspondingly scarce. Only experiment would show whether neutrons were actually good projectiles.

In point of fact, Fermi was a theorist, not an experimentalist. Only two months after returning from the Solvay Conference, he had completed a major theoretical undertaking, his theory of β-decay, absorbing much of what he had heard on the subject at Brussels. Though in time it would be hailed as 'his theoretical masterpiece... of fundamental importance for particle physics' [5-24], it had been rejected by *Nature* on the grounds that it was not quite suited for that journal. To be sure, it was consequently published in the little known *Ricerca Scientifica*, where Almaldi's wife worked, and later the *Zeitschrift für Physik*. Still, Fermi felt a diversion from theory to experiment was in order, and pounced upon artificial radioactivity. Apparently, Rutherford agreed. In April, he congratulated Fermi on his 'successful escape from the sphere of theoretical physics' [5-25].

Enrico Fermi was born in Rome in 1901, the son of an administrative employee of the Italian railroads [5-26]. His grandfather had served the Duke of Parma, one of the numerous small states dotting the Italian countryside in the wake of the Peace of Vienna, and was the first of the Fermis to make a transition from peasant agriculture in the Po Valley to eking out a modest living in the town of Caorso. Enrico's father entered the railroad business with little formal education, but by virtue of his willpower and determination rose steadily up the ladder, finally settling in Rome with the title of Capo Divisione, or division head, a civil rank usually reserved for persons with a university degree.

As was then not uncommon Italian practice, infant Enrico spent his first two and a half years with a wet nurse, because of delicate health. Joining the family in Rome, he was raised in the company of a sister Maria and a slightly older brother Giulio. The course of Enrico's career was set by a traumatic event when he was fourteen: Giulio unexpectedly died during a minor operation for a throat abscess. The tragedy was an enormous shock to Enrico, who had been very close to Giulio, but he kept the blow to himself. After a melancholy period, he threw himself into studying physical science out of second-hand books bought in Roman stalls. He also befriended Enrico Persico, one of his brother's schoolmates. The two Enricos would remain lifelong friends, becoming the first two professors in theoretical physics in Italy. Of equal importance for Fermi's development was a friendship struck up by his father with another railroad employee, who had considerable technical background and recognized the prodigy of the young man. He lent him his own technical books, and guided him through high school in Rome; he then suggested he try for admission to the Scuola Normale Superiore in Pisa. Located only a short distance from the Duomo and its leaning tower, the Scuola Normale was an offshoot of the University of Rome, which offered free university education. Fermi placed first, with an obligatory essay on 'Characteristics of sound'—an entry that astounded his examiner for its technical sophistication.

At the Scuola Normale, Fermi studied pretty much alone in the excellent library. One by one he mastered the classics of physics, from Poincaré's *Théorie des Tourbillions* to Sommerfeld's *Atombau und Spektrallinien*, telling Persico that he was soon considered the resident authority on relativity and quantum theory [5-27]. Having finished all preparatory course work, Fermi and two others, one of them Franco Rasetti, took up joint x-ray experimentation, building much of the apparatus with their own hands. Fermi was the natural team leader. In 1922, he prepared his doctor's dissertation on an original piece of x-ray diffraction—an experimental, not theoretical investigation. Though he was considered essentially a theorist, theoretical physics was not yet recognized as a university discipline in Italy of 1922 [5-28].

Upon receiving his doctorate *magna cum laude*, Fermi returned to his family in Rome. There he approached with some trepidation O M Corbino, Senator of the Kingdom, past Minister of Public Instruction, Director of the Physics Laboratory of the University of Rome, and someone who had great influence on the development of physics in Italy. He need not have worried. Corbino took him under his wing, and became his patron and confidant as long as he remained within the purview of Corbino's university influence. Fermi's immediate concern, however, was to broaden his first-hand physics experience beyond the provincial confines of the Italian academic community, and to establish personal contacts abroad. Receiving a traveling fellowship from the Italian Ministry of Education, he spent the winter of 1923–1924 at Max Born's Institute in Göttingen. Unfortunately, his stay there proved not as productive as it might have been, though he was on the best of terms with Born. Perhaps the Göttingen physicists, among them Werner Heisenberg and Pascual Jordan, were too preoccupied with their own problems and failed to appreciate the ability of Fermi—a situation undoubtedly exacerbated by his inward nature [5-29].

More successful was his interaction, during a subsequent short stint, with Paul Ehrenfest at Leiden. At any rate, at the end of his postdoctoral fellowship, Fermi took up an interim post as lecturer at Florence. In 1927 came his break. That year, mainly through the efforts of Corbino, the first chair in theoretical physics in Italy was established at the University of Rome. Fermi placed first in the competition for the position, Persico second, and the chair went to Fermi. (Persico went to Florence.)

Though Fermi was already nearing his peak as a leader in theoretical physics, experimental physics at Rome was not very secure. Nominally under Franco Rasetti, Fermi's fellow student at Pisa whom Corbino had taken on with the position of First Assistant, Rasetti was slower to develop. Equipment available was mainly spectroscopic. The library was excellent, but the shops were mediocre, and money was scarce. The only old-timer who counted beside Corbino was Antonino Lo Surdo, an experimentalist who had done good work in spectroscopy, but was an embittered old man by 1927. What students there were, were not up to Fermi's expectations. Before long, however, he and Rasetti recruited Emilio Segrè, who had been studying engineering, and Corbino added Edoardo Amaldi, also an engineering student. Others joined up, and a respectable group of physics students came into being. To broaden themselves, Rasetti, Amaldi, and Segrè spent time abroad, Rasetti in the laboratory of Millikan at Caltech and of Lise Meitner at the KWI in Berlin, Amaldi with Peter Debye in Leipzig, and Segrè with Otto Stern in Hamburg. In the midst of these hectic activities came news of artificial radioactivity from Paris, which, as noted, furnished just the sense of direction needed for the fledgling program.

Fermi suggested to Rasetti that, for a starter with neutrons, they look for effects similar to those seen by the Joliot-Curies. Rasetti already had in hand a neutron source prepared by himself by evaporating polonium on beryllium. Alas, it proved too weak to give positive results. At that critical juncture Rasetti left for an ill timed Easter vacation in Morocco, and Fermi took over. What to do? By happenstance, Professor Giulio Cesare Trabacchi, in charge of the physical laboratory, had in the basement of the physics building a plant for producing radon, the gas formed in the first step of radium decay. Owned by the Public Health Department, it consisted of a complex of vertical glass pipes, fed from a lead-lined safe containing a gram of radium, worth about 34 000 dollars at that time. Trabacchi went by the nickname 'Divine Providence' for his habit of lending Corbino's students bits of apparatus and supplies. Taking him up on his kind offer of the radon, Fermi prepared a much stronger radon-plus-beryllium neutron source by filling a small glass bulb with beryllium powder, evacuating the air, and replacing the air with radon. The fact that the source (unlike polonium) also emitted β- and γ-rays did not interfere with a *delayed* effect, and the half-life of radon, 3.82 days, was long enough. Some Geiger counters, albeit home-built and primitive, were already ready for use.

Assisted by Amaldi and Segrè, Fermi went to work, systematically bombarding elements in order of increasing atomic number, starting with hydrogen. All gave negative results until he reached fluorine. Success at last! Before long, the number of detectable activities began to accumulate. Fermi did most of the measurements and calculations, Amaldi took care of the electronics, as it is called nowadays, and Segrè secured targets, sources, and equipment. Fermi also cabled Rasetti to return forthwith from Morocco. Element after element was irradiated by neutrons, followed by emission of alphas, protons, and gammas. Some of the half-lives were very short. 'Then haste was essential', notes Laura Fermi, his wife and a student in general science at the University of Rome before they were married in 1928, 'and the time to cover the length of the corridor [between the irradiation and counter rooms] had to be reduced by swift running. Amaldi and Fermi prided themselves on being the fastest runners. ... They always raced, and Enrico claims that he could run faster than Edoardo. But he is not a good loser' [5-30]. The results were posted in letters to *Ricerca Scientifica* and to colleagues at home and abroad. Within a month, Rutherford acknowledged their work, as he had acknowledged the Joliot-Curies three months earlier.

Dear Fermi,

I have to thank you for your kindness in sending me an account of your recent experiments in causing temporary radioactivity in a number of elements by means of neutrons. Your results are of great interest, and no doubt later we shall be able to obtain more information as to the actual mechanism of such transformations. It is by no means clear that in all cases the process is as simple as appears to be the case in the observations of the Joliots. [5-31]

Finally, they arrived at uranium in the periodic table. It was tacitly assumed that its radiation (and that of heavy elements generally) should produce a heavier element by γ-emission, and then a transuranic element by subsequent β-emission. This was seemingly confirmed by the absence of any reaction products with atomic numbers between lead and uranium. In this they were seriously in error, as time would tell. Indeed, they failed to heed an article mailed to them by the German chemist Ida Noddack, co-discoverer with her husband Walter of the element rhenium, which offered a quite different suggestion [5-32]. At any rate, having reached uranium, the neutron irradiation program came to a halt, with the onset of summer vacation, 1934 [5-33]. Amaldi and Segrè brought a manuscript on their work to Rutherford at Cambridge, and he passed it on to the Royal Society [5-34]. On Segrè's polite inquiry whether the paper could be speedily published, Rutherford retorted, 'What do you think I was the president of the Royal Society for?' [5-35].

One morning in October of 1934, Amaldi and Bruno Pontecorvo, a newcomer to the group from Pisa, made a singularly important discovery. They noticed that when silver was irradiated on a wooden table, it became much more active than when it was irradiated on a marble shelf in the same room. Furthermore, with the neutron source and silver target placed inside a small lead box, the activity was much more intense when the silver was in a corner than it was when placed elsewhere. When removed from the box, the activity decreased considerably with increasing distance from the source—something that did not happen inside the box. The next step called for interposing a lead wedge between the source and detector. As it took a while to prepare the wedge, Fermi decided to try an attenuating filter of a lighter material first. They dug a cavity in a block of paraffin, put the source inside the cavity, irradiated the silver, and brought it to a Geiger counter. The counter clicked madly, nearly going off-scale. In a conversation with the astrophysicist Subrahmanyan Chandrasekhar many years later, Fermi told of the event that day.

> I will tell you how I came to make the discovery which I suppose is the most important one I have made. We were working very hard on the neutron-induced radioactivity and the results we were obtaining made no sense. One day, as I came to the laboratory, it occurred to me that I should examine the effect of placing a piece of lead before the incident neutrons. Instead of my usual custom, I took great pains to have the piece of lead precisely machined. I was clearly dissatisfied with something: I tried every excuse to postpone putting the piece of lead in its place. When finally, with some reluctance, I was going to put it in its place, I said to myself: 'No, I do not want this piece of lead here; what I want is a piece of paraffin'. It was just like that with no advance warning, no conscious prior reasoning. I immediately took some odd piece of paraffin and placed it where the piece of lead was to have been. [5-36]

In his own recollection of this historical occasion, Segrè cannot be sure but thinks paraffin was first tried on the morning of 22 October. It was in the middle of an examination period, so several of the group were in other parts of the building. At noontime, everybody was summoned to watch the black magic. Amazing!

> By the time we went home for lunch and our usual siesta, we were still extremely puzzled by our observations. When we came back at about three in the afternoon, Fermi had found the explanation of the strange behavior of filtered neutrons. He hypothesized that neutrons could be slowed down by elastic collisions and in this way became more effective—an idea that was contrary to our expectations. [5-37]

We have encountered experiments with paraffin earlier in the beryllium-radiation experiment of the Joliot-Curies, though there the substance played a different role. Paraffin, as does wood, contains a great deal of hydrogen. Since the nuclei of hydrogen, protons, are virtually the same size as neutrons, protons are particularly effective in slowing down neutrons by elastic collisions, in the same manner as a billiard ball is slowed down by colliding with a ball of its same size (as opposed to, say, a ping pong ball bouncing off a bowling ball with no loss in energy [5-38]). Slowing down a neutron by repeated collisions with protons in its path gives it more time to spend in the vicinity of, for example, a silver nucleus, with a greater chance of being captured. To make sure of his explanation, Fermi suggested, why not try water instead? With that, the party trooped down into Corbino's private garden behind the laboratory. There, in a lush spot shaded by an almond tree, was a small pond in which Rasetti had experimented with raising salamanders, and the physicists had whiled away time operating little candle-powered toy boats that had become the rage on the Italian market. Placing a neutron source and silver target in the pond, the brackish water proved even more effective than paraffin [5-39].

That very evening, at the Amaldi's, they prepared a short letter to *Ricerca Scientifica* [5-40], Fermi dictating, Segrè writing, and Rasetti, Amaldi, and Pontecorvo milling about, the three chatting in unison. The practical possibilities of slow neutrons were not lost on Corbino, who insisted that they take out a patent on radioisotope production by slow and fast neutron bombardment. Nobody could, of course, then suspect that slow neutrons held the key to nuclear energy—that haunting prospect first brooded about by Frederick Soddy and H G Wells, earlier in the century. Nor, for that matter, was the importance apparently foreseen of an intimately related development, namely the production of heavy water, initiated by Brun and Tronstad the previous spring. Heavy water, it would turn out, is even more effective than ordinary water in slowing down neutrons.

The study of slow neutrons would, in any event, dominate Fermi's program during the next several years. Alas, the onset of the Ethiopian War marked the beginning of the decline of the work at the Institute, hastened by the death of Corbino in 1937. The deteriorating political situation in Italy, and with it the Fascist racial laws of 1938 that threatened his wife, convinced Fermi to leave Italy. Upon receiving the Nobel Prize in December 1938, he proceeded directly from Stockholm to New York with his family. He was welcomed with open arms at Columbia University, where he had a standing invitation to join the faculty. By then, the rest of the team in Rome was spread before the four winds: Rasetti also in New York, Segrè in Sicily and thence to Berkeley, Pontecorvo with Joliot in Paris. Only Amaldi stayed on, inheriting the chair in Rome and doing his best in guiding Italian physics through the troubled years ahead.

5.3. Meanwhile, back in Paris

With all the excitement in Rome, Irène Curie had not been resting on her laurels back in Paris. After the Nobel Prize, her collaboration with

Frédéric came to an end. Frédéric became increasingly preoccupied with fund-raising and other administrative duties at the Collège de France. However, Irène's most important period of research lay just ahead, despite her temporary involvement in politics, serving for a while as Secretary of State in Léon Blum's Popular Front Government, and deteriorating health from worsening tuberculosis. When not recuperating in the Alps or the countryside at l'Arcouest, she often worked lying down at home in Paris [5-41].

In a tit-for-tat, Irène, whose work with Frédéric had stimulated Fermi's program in the first place, now brought her radiochemistry skills, inherited from her mother, to bear on sorting out the complex broth of elements and isotopes resulting from bombarding thorium and uranium with neutrons, starting with those of thorium in 1935. The substances making up the broth, all highly radioactive and transmuting into one another with varying half-lives, were too small in quantity to be discernible other than by their emitted radiations. Therefore, Irène and her students resorted to the standard radiochemical procedure of adding a chemically related, stable isotopic 'carrier' of the radioactive atoms. In this way, they extracted a sample mixture large enough to be manipulated, as the radioactive atoms followed the carrier through the chemical treatments.

Focusing on one of the elements with a pronounced half-life of 3.5 hours, Curie and her co-workers Hans von Halban of Austria and Peter Preiswerk (Swiss), added lanthanum (element 57, a rare earth) as a carrier to the solution from irradiated thorium, then precipitated the lanthanum back out. Behold! The 3.5 hour activity accompanied the lanthanum, instead of remaining in the solution. This particular activity, then, came from a substance chemically similar to lanthanum. To be sure, Curie and her team, and everybody else for that matter, did not for a minute believe that the substance actually *was* lanthanum, whose atoms are only a little over half the weight of thorium atoms. Actinium, element 89, very radioactive and nearly the weight of thorium, was a good candidate, and Irène left it at that [5-42].

Two years later, now joined by the young visiting Yugoslavian physicist Pavel Savitch, Curie turned to untangling the mixture of isotopes from irradiated uranium. This time, instead of relying on chemical separation, they concentrated on whichever isotope gave off the most penetrating β-rays. They filtered the radiation from the solution with sheets of copper, and studied the radiation of a single substance still coming through. Again it had a half-life of 3.5 hours, though they failed to identify it with the 'actinium' radiation from 1935, since actinium is substantially lighter than uranium. Instead, they ascribed the new radiation to an unsuspected isotope of thorium [5-43].

The next year, 1938, Curie and Savitch doggedly returned to the actinium problem. They now employed a laborious procedure devised long ago by Marie Curie for separating actinium from lanthanum. The 3.5 hour substance followed along with the lanthanum, not the actinium. Still they refused to believe it was indeed lanthanum. 'It seems', they concluded in their report to the Paris Academy that May, 'that this substance cannot be anything except a transuranic element, possessing very different properties from those of other known transuranics, a hypothesis which raises great difficulties for its interpretation' [5-44].

Amidst her radiochemical and other preoccupations, and despite her somewhat pensive attitude, Irène Curie maintained an informal atmosphere at the Radium Institute, mixing with students, technical staff, and employees alike, often in the entrance hall while she warmed her back against a radiator. She looked after all manner of details; thus she rented a gymnasium and everybody took judo lessons. The Institute was only a short walk from the Collège de France, where Joliot held weekly seminars. So did Jean Perrin every Monday in his Physical Chemistry Institute adjacent to the Radium Institute, where everybody of importance mingled with outside visitors, discussing the latest problems over tea in laboratory glassware, brewed over a Bunsen burner [5-45]. Personally, Curie continued to enjoy outdoor sports, even with her worsening health. Her favorite authors were Victor Hugo, Sidonie-Gabrielle Colette, Rudyard Kipling, and especially Margaret Mitchell and her 'Gone with the Wind'. Even with household help, she took pleasure in serving a simple dinner every Sunday at 1 p.m. She was not a great cook, but their table seated 20. To Joliot's regret, she usually served stew [5-46].

The same month that Curie and Savitch reported their latest results to the Academy, Joliot ran into Otto Hahn at the *Tenth International Congress of Chemistry* in Rome. It seems likely Joliot also came across Leif Tronstad on that occasion, who was representing NTH, Trondheim, and a featured speaker at the Congress. It was, in any case, the first time Joliot and Hahn met. Hahn and his partner, Lise Meitner, had also been hot on the trail of Fermi's heavy elements, as Joliot was only too aware. They, too, had published a paper in 1935 on neutron-induced processes in thorium and were 'a little angry with Mme Irène Curie for not having cited us properly . . . We regret this very much, for a scientific echo has never been as necessary to us as just now' [5-47].

Hahn had reasons for being on the edge. Known for his anti-Nazi views, he counted on scientific recognition to bolster his shaky political position [5-48]. As Director of the Kaiser-Wilhelm Institute for Physical Chemistry, moreover, he had his hands full with administration, and with industrial applications of radioactive tracers to chemistry. He had found it necessary to ask Fritz Wilhelm Strassmann, one of his assistants, to join him and Meitner in the uranium investigation. At the same time, he and Meitner had closely followed the experiments of their French competitors, though were not yet aware of the latest work in Paris on the actinium–lanthanum separation. They were of the opinion that Irène Curie's chemistry was not as good as their own. Still, Hahn now expressed his respect for Joliot and his wife, but professed to a certain skepticism over Irène Curie's results. (In Berlin, they dubbed the latest element out of Paris 'Curiosum'.) Hahn also told Joliot frankly that he was going to repeat her experiments.

He would do just that, with unprecedented consequences, as it would turn out. However, before he got around to it, several events intervened. The very same month, March of 1938, German troops poured into Austria, and Hitler proclaimed the *Anschluss* of his native land— reducing Austria to a province of the Third Reich. Before long, Hahn received news of the 3.5 hour substance from neutron bombardment in Paris. Hahn was dumbstruck. Unfortunately, his partner was no longer around to consult. Hitler's march on Austria had seen to that.

5.4. Escape, in the nick of time

On 13 July 1938, a train pulled out of a railway station in Berlin, bound for Holland. On board was Lise Meitner, huddled in a corner of her carriage in the company of Dirk Coster, the Dutch physicist. Her belongings filled two small suitcases; other than an old diamond ring forced upon her by Hahn, she carried only ten marks in her purse.

Only four months before, the *Anschluss* had imperiled the 60-year-old physicist, an Austrian of Jewish extraction. The fact that she was baptized and raised as a Protestant mattered not one bit. No sooner had the *Wehrmacht* rolled across the Austrian frontier, than Kurt Hess, an avid Nazi who headed a section for organic chemistry at the KWI, turned on her. 'The Jewess endangers the Institute!' he declared [5-49]. Hahn was not of much help. Himself an anti-Nazi, he was in a precarious position. He could easily lose the directorship—a position coveted by Hess. The problem of Meitner had already come to the attention of Rudolf Mentzel, a high official of the influential Reich Research Council (*Reichsforschungsrat*) of the Ministry of Education—an agency charged with funding atomic research under a plan administered by Hermann Goering himself. 'At that point I rather lost my nerve', confessed Hahn later, and he broached Heinrich Hörlein, treasurer of the Emil-Fischer-Gesellschaft, 'about Hess, about Lise' [5-50]. When informed of Hahn's meeting with Hörlein, Meitner was shocked. 'He was, in essence, throwing me out', she wrote in her diary [5-51].

Hahn and Meitner had collaborated for over 30 years. They first joined forces in 1907, when Meitner, 29, arrived in Berlin to study under Max Planck, and Hahn had just returned from a year with Rutherford at Montreal. What had been intended as a short stay for Meitner would stretch into decades.

Born in 1878, Meitner was the third of eight children of a Viennese lawyer [5-52]. She entered the University of Vienna in 1901, where she met with some rudeness from her male student companions, but was greatly inspired by Ludwig Boltzmann. In 1905, she became the second woman to receive a doctorate in science, under Franz Exner with a thesis on heat conduction in non-homogeneous materials. Perhaps she opted for an experimental thesis, since Boltzmann was lecturing in California at the time. After graduation, she remained in Vienna for a period, being introduced to the new subject of radioactivity by Stefan Meyer, then *Assistent* in Boltzmann's Institute. She still had her eye on theoretical physics, however. She had been impressed by Planck when he visited Vienna, exploring the possibility of succeeding to Boltzmann's chair after the latter's shocking suicide in 1906. With her father's approval and modest monetary support, she enrolled for Planck's lectures in Berlin, while also casting about for a place to do experimental work. That problem was solved when Hahn arrived, looking for a physicist to help him in his radiochemical work.

Hahn came from Rhenish peasant stock, though his father made a successful transition from farming to the glazing business in Frankfurt am Mein [5-53]. Indeed, Glasbau Heinrich Hahn is still a thriving enterprise in Frankfurt. Hahn was four months older than Meitner, his future companion. A sickly youth, he regained his health at 14, and was never seriously ill again. He became seriously interested in chemistry while attending some popular science lectures for adults during high school, and entered Marburg University in 1897. Though in later years he would express regret at not paying sufficient attention to physics and mathematics, he absorbed enough chemistry to earn his doctorate in organic chemistry in 1901. After two more years under his principal professor, Theodor Zincke, Zincke obtained a position for him in William Ramsay's laboratory at the University College, London. Ramsay, whose fame derived chiefly from his discovery of several inert gases, put Hahn to work as his radiochemical assistant, despite the latter's utter ignorance of radioactivity. That Hahn caught on in a hurry is an understatement; before long he found himself the discoverer of radiothorium, a new radioelement. Impressed, Ramsay urged him to forget about entering industry, as had been Hahn's inclination, and secured him instead a post with Emil Fischer's Chemical Institute in Berlin-Dahlem. Fascinated by now with radioactivity, Hahn readily agreed, but felt he first needed to broaden his background with a further year abroad—preferably under the master of radioactivity himself, Ernest Rutherford. His year at McGill University, Montreal, began on a somewhat strained note, as Rutherford had a low opinion of Ramsay's physics, and Rutherford's close associate at Yale, the radiochemist Bertram Borden Boltwood, regarded radiothorium as a 'compound of Th-X and stupidity' [5-54]. Hahn soon convinced them otherwise by adding radioactinium to his credit, and left for Fischer's Institute on the best of terms with Rutherford.

Within a year of arriving in Berlin, Hahn was appointed *Privatdozent* in Fischer's Institute. At the same time, he teamed with Meitner, though at first she was not allowed in Fischer's laboratory building proper; it seems that Fischer had had unfortunate dealings with women before. Instead, they equipped a basement carpenter's shop with an entrance from the street, and went to work. They began by entangling β-radiations from Hahn's plentiful stock of radioelements. When the newly formed Kaiser Wilhelm Gesellschaft opened its Institut für Chemie in Berlin-Dahlem in 1912, Hahn was made head of a small department of radioactivity, and Meitner stuck with him. Their work was interrupted by World War I, Hahn called to service in chemical warfare under Fritz Haber, and Meitner, much like Irène Curie, volunteering as a radiological nurse in the Austrian Army. Even then, they kept the collaboration going on rare occasions when their military leaves coincided in Berlin. They must have put those precious occasions to good use; even before the armistice they reported the discovery of the most stable isotope of element 91: the parent of actinium, which they named protactinium.

At the end of the war, Meitner was appointed head of the physics department of KWI, while also maintaining a tenuous academic connection with the University of Berlin; she became *Dozent* in 1918 and extra-ordinary professor in 1926. After 1918, she and Hahn largely went their own ways professionally. Hahn stuck with various aspects of radiochemistry, while Meitner returned to physics proper, using radioactive sources in the study of β- and γ-ray spectra and the attenuation of γ-rays in matter. Only in 1934 did they resume the collaboration, following up on Fermi's neutron-bombardment experiments in Rome.

The relationship between Hahn and Meitner bears some parallels with that of the Joliot-Curies. Hahn, like Irène Curie, was a superb chemist, unexcelled as an experimentalist and, as Rutherford once wrote of him, had a 'special smell for discovering new radioactive elements' [5-55]. However, unlike Irène, he relied on his physicist companion for generalizing and grasping the broader significance of his discoveries. Like Joliot, Hahn had the more outgoing personality. Meitner was reserved, even shy on occasion. In spite of these traits, which she shared with Irène, Meitner was indisputably the leader, inclined to be bossy with her partner and in managing the Institute. Compared with Irène, Meitner was a stickler for keeping the Institute free of radioactive contamination. That habit not only paid off in allowing them to measure extremely weak emissions, but also by and large saved their health, in sad contrast to Irène's misfortunes. (Joliot remained in good health all along.)

It fell on the Dutch director of the brand new KWI für Physik, Peter Debye, to take the first step in extricating Meitner from the political quagmire in which she found herself, following the annexation of Austria. He contacted Dirk Coster at the University of Groningen, and

Coster arranged with the Dutch authorities that Meitner be allowed to enter Holland, despite her invalid passport. Coster did more than that. He journeyed to Berlin to personally escort Meitner out of the country, arriving at her flat in the Institute Villa on the night of 12 July [5-56]. Stealth was of the essence; the notorious Kurt Hess, who had been the first to denounce her, occupied the flat next door. Betrayal was, in fact, nearer than they realized. Hess had notified the *Sicherheitsdienst* (SD) or intelligence arm of the secret police that Meitner was about to flee. His handwritten note was delayed, by friendly, unnamed individuals within the SD, by a few days—just long enough to keep it from doing any harm [5-57].

Awaiting Coster with Meitner in her flat was Hahn, and so was Paul Rosbaud, an old friend. On top of things by virtue of his unique position in Berlin, Rosbaud was scientific adviser to the publishing house Springer Verlag, and a tireless opponent of the Nazi regime [5-58]. Together, the men spent the night helping Meitner pack. Recalls Hahn, 'I gave her a beautiful diamond ring that I had inherited from my mother and which I had never worn myself but always treasured; I wanted her to be provided for in an emergency' [5-59].

Rosbaud picked Meitner up the next morning in his Opel, and drove her to the station. Overwhelmed by fear, she protested all the way, and it was all Rosbaud could do to get her on the train. Coster was waiting on board in a first-class compartment, and they were off [5-60]. As they halted at the Dutch border, Meitner had a momentary fright as a Nazi patrol went through the coaches, but all went well. Safe at last, Meitner remained in Holland for only a few days, then flew on to Copenhagen where she was warmly received by Niels and Margrethe Bohr at their Carlsberg House of Honor. She had a well deserved rest, while Bohr made final arrangements for her prearranged exile in Sweden. She would remain under a Nobel Foundation Grant with Manne Siegbahn, the 1924 Nobel Laureate, in his new Nobel Institute for Experimental Physics nearing completion on the outskirts of Stockholm. The Institute was jointly supported by the Swedish Academy and by the Wallenberg Foundation—the same Wallenbergs who had helped underwrite budding Norsk Hydro in 1904. Just as Siegbahn's Nobel Prize had not been without some controversy, so was his Institute, with certain observers within the Academy protesting excessive diversion of Nobel Foundation funds into the project.

In Copenhagen, Meitner also caught up with her 34-year-old nephew, Otto Robert Frisch. He, too, was a physicist, and had chosen not to remain in Germany when Hitler came to power in 1933. Like Tante Lise, he had obtained his doctorate at the University of Vienna, under the *enfant terrible* of nuclear physics, Karl Przibram [5-61]. He then gravitated to Berlin, spending three years mainly in the optics division at the Physikalisch-Technische Reichsanstalt, somewhat outside the maelstrom of the new physics Meitner was exposed to in Berlin's academic community. Following another three years as an *Assistent* in the laboratory of Otto Stern at the University of Hamburg, he did not pass up the offer of refuge extended by Britian's Academic Assistance Council in 1933. The AAC, headed by Rutherford, had been created as a result of a massive public rally in London's Royal Albert Hall on 3 October 1933, 'to provide a type of Fellowship or Professorship for ten or twenty of the most distinguished exiles whom we would like to keep in this country'. He (and several of his co-workers) had been summarily dismissed, even though he was then an Austrian citizen and presumably legally immune from Nazi persecution. Stern, too, for that matter, had been dismissed; however, the situation was not critical for him. He was a comparatively wealthy man, already famous through the classic Stern–Gerlach experiment proving the directional quantization of the angular momentum of atoms, and would have no difficulties finding a place for himself in laboratories abroad. However, he sounded out Marie Curie about taking on Frisch. She could not, but it developed that Patrick Blackett, then at Birbeck College in London, could. So off to London Frisch went, fortified with a one-year grant from the Academic Assistance Council. His London experience had its ups and downs; in any case, when the year was up, and still restless, he took Bohr up on his offer for a stay at his Institute for Theoretical Physics in Copenhagen, another haven for displaced scholars.

Among those who helped in the foundation of the Academic Assistance Council was a fellow refugee, Hungarian-born Leo Szilard— a man of uncanny intuition about physics and the world at large, who would play a singular role in events to come. He had approached Frisch's aunt about doing nuclear physics at the KWI a year before Frisch left Hamburg. Meitner had agreed, but German political upheavals of 1933 intervened. Within days of his meeting with Meitner, Szilard slipped across the Austrian border in the face of Nazi authorities, and soon found himself in London haranguing British officials and scholars of the importance of finding places for academic victims of Nazism—somewhat in the vein of his prodding President Roosevelt six years later to take serious the menace of German developments in nuclear physics.

Meitner and Frisch, meanwhile, had two weeks together in Copenhagen. He showed her the new cyclotron in the care of L Jackson Laslett, on loan from E O Lawrence in Berkeley, and they relaxed with the Bohrs at their Tisvilde summer villa by the sea. She then left for Stockholm, stopping *en route* for a few days with her friend, the physicist Eva von Bahr-Bergius; she had helped in getting Meitner out of Germany, and was then teaching at Uppsala. Settling in Stockholm proved not all that blissful—in fact, she was utterly miserable. The country was new and she was unable to speak the language; she was sequestered in a tiny hotel room with hardly space to unpack what few belongings she had brought along; she had almost no money; some of her family were still in Austria; she was cut off from everything and everybody at the KWI. Worse yet, Siegbahn's laboratory proved not up to expectations. His interest in nuclear physics was lukewarm at best. Though the Institute was brand new, established for the express purpose of introducing nuclear physics into Sweden, with even a cyclotron nearing completion under Sten von Friesen, Siegbahn seemed more interested in building equipment than putting it to good use. Back in Berlin, moreover, Hahn had trouble with the Reich Education Ministry in shipping not only her books, records, and papers from the Institute in Dahlem, but her personal belongings as well; medals, books, radio, linens, towels, clothes—even two toothbrush holders. 'Dear Otto!' she wrote Hahn

> Here everyone is very friendly, but I am always reminded of [the saying]: if one must rely upon the friendliness of people, one must either be very self-confident or have a great sense of humor; the first I never possessed, and the latter is difficult to call forth in my current situation. And then Max jun[ior] [von Laue] writes to me that I am to be envied. People just don't have the imagination to picture the situation that others really are in. ... Tell a bit what's happening in the institute. ... What is our good friend upstairs [Hess] doing? [5-62]

Meitner would have much preferred England, and attempts were indeed made in several quarters to secure a position for her there. C D Ellis, then at King's College at the University of London, petitioned strongly on her behalf, but was shot down by none other than Frederick Lindemann, alias Lord Cherwell, Churchill's adviser. No admirer of women, he was adverse to Jews in particular [5-63]. At Cambridge, Girton College for women was prepared to offer her a stipend, but for one year only. Though no longer in Cambridge, Ellis's old partner James Chadwick was a man of influence, and quite fond of Meitner. He, too, made overtures on her behalf, but John Cockcroft, whose opinion counted even more, dallied; Cockcroft was mainly suspicious of Otto Frisch who himself had feelers out about returning to England. Later in the upcoming war, Cockcroft, along with a certain Lieutenant Commander, Eric Welsh of a joint Anglo-Norwegian Intelligence Service, would have another reason to keep Meitner in Stockholm, as we shall see.

While the Nazi manifesto brought Meitner to Sweden, the anti-Semitic Manifesto della Razza promulgated the same summer in Italy performed a similar service in the case of Enrico Fermi. Early that fall, Fermi was confidentially tipped off by Bohr that he was a candidate for a Nobel Prize. Normally, such information is kept strictly secret, but the Swedish Academy thought it advisable to feel him out on the matter, in view of the political situation in Italy and the monetary restrictions imposed by the Italian authorities. Fermi had, in fact, already decided to emigrate, and had even accepted an offer, one among many, for a professorship at Columbia University. Then came 10 November, and along with it the providential call from Sweden. Within a month the entire family, including a maid, boarded a train in Rome's Stazione Termine, bound for Stockholm. Fermi was departing on a ruse of giving a lecture series at Columbia, thereby sidestepping certain American immigration restrictions and Italian fiscal restrictions levied on citizens leaving for good. As with Meitner, the only nervous moment in the journey was a perfunctory passport inspection at the German border. Fermi pointed out the visa from the German Konsulat in the passport. 'The German guard saluted and smiled. Germans and Italians were good friends, were they not!' [5-64].

In Stockholm, Fermi gave his Nobel Lecture one month to the day after taking the call from the Swedish Academy. Though it contained reference to the supposed transuranic elements 93 ('ausonium') and 94 ('hesperium'), the prize, in the words of Professor H Pleijel in his Presentation Speech, was awarded for 'Your discovery of new radioactive substances belonging to the entire range of elements and for the discovery you made in the course of this work of the selective power of slow neutrons' [5-65]. Little could Fermi suspect that even as he spoke in Stockholm's Concert Hall, ongoing efforts in Hahn's laboratory would soon dictate dramatic reappraisal of the experimental work begun in Rome five years earlier.

Chapter 6

NUCLEAR FISSION

6.1. Berlin: December 1938

The crowning irony of pre-World War II efforts to appease Hitler was Prime Minister Neville Chamberlain's 'triumphal' return on 30 September 1938 from his negotiating mission with Hitler in Munich, waving the Anglo-German peace declaration before his countrymen. 'My friends', he declared, '. . . there has come back from Germany to Downing Street, peace with honour. I believe it is peace in our time.' In fact, the document was useless. Czechoslovakia was effectively doomed. Hitler had already secretly agreed with Mussolini that they would have to join in the eventual invasion of the British Isles.

At the time of the Munich crisis, another crisis was brewing in Berlin-Dahlem—an intellectual crisis destined to reach a dramatic climax in the waning hours of 1938. Curie and Savitch's latest paper on the mysterious '3.5-hour substance', first thought to be thorium, then actinium, and finally a transuranic with lanthanum-like properties, reached Hahn on about 20 October. Not knowing what to make of it, his first reaction was that the Paris team was 'muddled up' [6-1]. On second thought, he couldn't simply ignore the paper, and so he passed it on to Fritz Strassmann for his opinion. Strassmann had already acquired a reputation for his skill as an analytical and physical chemist, and was by 1938 acknowledged as an equal partner in his collaboration with Hahn and Meitner [6-2].

Strassmann went over the paper with care, inclined to take seriously the purported substance, since, in contrast to earlier papers from Paris, the latest communication contained detailed decay curves, which spoke for themselves. He suspected that since Curie had used potassium lanthanum sulfate as a carrier, the precipitate might well contain a mixture of radium and its decay products. It did not take much to convince Hahn of his hunch, and of the need for following up on the experiment with all due haste. Wrote Hahn to Meitner:

> Dear Lise! Despite dreadful weariness I want to answer your
> letter of the 23d, at least briefly. ... Toward the end of last week
> a new paper by Curie and Savitch about the 3.5-hour substance
> appeared. ... Curves are also shown. We are now working
> on reproducing them and we do now believe in their existence.
> According to Curie's results we have found the substance, perhaps
> even better than Curie and Savitch. ... Curie makes remarkable
> claims about its properties. The properties do in fact appear to
> be remarkable. The substance is certainly not identical to the 2.5-
> hour substance [ekaPt observed in Berlin]. (Perhaps a Ra isotope
> has something to do with it. I tell you this only with great caution
> and in confidence!). ... A great pity that you are not here with us
> to clear up the exciting Curie activity. [6-3]

Hahn and Strassmann proceeded on several parallel paths: (a) using a
lanthanum carrier, as had Curie and Savitch, to precipitate rare-earth
elements, such as actinium, (b) a barium carrier to precipitate alkaline-
earth elements, such as radium, and (c) throwing in a zirconium carrier
for good measure. On 8 November they reported their preliminary
findings to *Naturwissenschaften* [6-4]. The most startling result was the
appearance of three radium isotopes and their actinium decay elements.
Strassmann supposed that a two-step process was at work. First, the
uranium, absorbing a slow neutron, decayed by alpha emission to a
very short-lived thorium isotope. The thorium isotope, in turn, quickly
decayed by α-emission to an isotope of radium. That is,

$$_{92}U^{238} + _{0}n^{1} \rightarrow _{2}\alpha^{4} + _{90}Th^{235}$$

$$_{90}Th^{235} \rightarrow _{2}\alpha^{4} + _{88}Ra^{231}.$$

The radium isotope, finally, decayed by beta emission into an actinium
isotope which, also being a β-emitter, led back to thorium, completing the
isomer loop. If true, this was a highly unusual process; (n, α) reactions
had been observed in Berlin before, but only with *fast* neutrons. In the
experience of Hahn and company, slow neutrons were simply captured.

While following the developments in Berlin-Dahlem with a mixture
of incredulity and skepticism in Stockholm, Meitner had other matters
on her mind. Laura Fermi recalls her, from their brief encounter during
the Nobel festivities, as worried and tired 'with the tense expression
that all refugees had in common' [6-5]. A visit to Copenhagen, shortly
before, had done some good, but also added to her worries. Early in
November, on the same day that his and Strassmann's short paper went
to *Naturwissenschaften*, Hahn had been in Vienna to give a lecture and
look into the possibility of an Austrian successor to Meitner at Berlin-
Dahlem. While there, he had dined with Meitner's remaining relatives,
among them an older sister Gusti and her husband Jutz Frisch. The next

day, he heard that Jutz had been arrested, but could learn no more than that. Greatly worried, Meitner prepared for a quick visit to Copenhagen to consult with Frisch. By chance, Hahn was also scheduled to be in Copenhagen just then, ostensibly on Bohr's invitation to give a talk, and she looked forward to seeing him as well as the Bohrs. In the event, the trip went as planned. Among other things, it afforded her a chance to discuss with Hahn the KWI experiments in progress; in particular, to go over the proffered explanation for double α-decay caused by slow neutrons which convinced neither her, nor Bohr, nor Frisch. Mainly, though, she implored Hahn to reexamine their results with the greatest of care [6-6].

Returning to Berlin, Hahn did more than that. He and Strassmann began their final, grueling round of tests for the alleged radium, in an effort to separate it from its barium carrier by the elaborate fractional crystallization procedure ('fractionation') devised by Marie Curie. Strassmann was not privy to Hahn's clandestine meeting with Meitner in Copenhagen, but surmised that the directive for the final push had come by letter from Meitner. He still regarded their absent partner as the intellectual leader of their threesome team [6-7].

The attempt to separate the supposed 'radium' from barium proved a daunting challenge. Most problematic was the feebleness of the radiation—at best the activity of a few thousand atoms. That few emitters could easily be carried away by the overwhelming number of inactive barium atoms without detection. Hahn therefore diluted some natural uranium off the shelf down to the intensity of the preparations separated by Strassmann, using barium chloride as a carrier on the suggestion of Strassmann, since it formed 'beautiful little crystals' of exceptional purity [6-8], and passed it through the whole radiochemical procedure. It separated without a hint of trouble. Ergo, the fractionation procedure was not at fault. Gradually it dawned on Strassmann that there were no radium isotopes in the reaction broth; the alleged radium not only behaved like barium, but *was* barium—element 56, slightly more than half the atomic weight of uranium.

One more test deemed essential, completed on 19 December, clinched the matter. If the 'radium' actually was radium (element 88), it should, as noted earlier, decay by β-emission to a spot one step higher in the periodic table, namely to actinium (89). If, on the other hand, it was truly barium, it should decay by beta emission to lanthanum (57), which fortuitously could be separated from actinium by fractionation. The same day, Monday evening, 19 December 1938, Hahn wrote Meitner as follows.

> Of course, there is something about the 'radium isotopes' that is so remarkable that I will tell it only to you. The half-lives of the three isotopes are rather exactly ascertained. They are separable from *all* the elements except barium. All the reactions are correct, except

one . . . : The fractionation does not work. Our radium isotopes behave like barium. We do not get a clear enrichment with barium bromide or chromate etc. . . . On Saturday Strassmann and I fractionated one of our [alleged] 'radium' isotopes with Msth 1 [a stock radium isotope] as indicator. The mesothorium was enriched as prescribed, our radium was not. . . . We more and more come to the terrible conclusion: Our radium isotopes do not behave like radium, but like barium . . . [6-9]

The text of the short paper of Hahn and Strassmann, written by Hahn, had been largely finished by the time the final actinium–lanthanum experiment was over. Time was running out. Christmas was upon them, including the annual party at the KWI the next day, 20 December, and the Institute would close immediately after that. That evening, not waiting for a reply from Meitner, Hahn called Paul Rosbaud at Springer Verlag. Hahn, of course, did not yet realize that the atom had actually been split, but Rosbaud sensed at once that an important breakthrough in nuclear physics was in the making. He warned Hahn that he had to have the manuscript in hand no later than Friday, 23 December for it to make the next issue of *Naturwissenschaften*, a Springer journal. Rosbaud also called Fritz Süffert, editor of the journal, and had him pull a less urgent article already being typeset.

With the latest results, a revision of the paper, originally titled 'On the radium isotopes produced by the neutron bombardment of uranium and their behavior', was clearly in order. However, for lack of time for more substantial rewriting, they simply substituted 'alkaline earth metals' for 'radium isotopes' in the title, since barium, like radium, is an alkaline-earth metal.

Most of the report had been written, and it would have had to be rewritten completely, especially since in view of [the barium] result its major portion was not especially interesting any more. [6-10]

The paper was delivered to Rosbaud at the Springer offices on Linkstrasse on Thursday, 22 December. Full digestion of the final results would have to wait until after the New Year. For now, the paper simply ended as follows.

We had to conclude that our 'radium isotopes' have the chemical characteristics of barium. Speaking as chemists, we even have to say that these new substances are barium, not radium. [6-11]

Meitner received Hahn's letter of 19 December in Stockholm on the 21st. She replied right away.

Your radium results are very startling. A reaction with slow neutrons that supposedly leads to barium! By the way, are you

quite sure that the radium isotopes come before actinium? ...
And what about the resulting thorium isotopes? From lanthanum
one must get cerium. At the moment the assumption of such a
thoroughgoing breakup seems very difficult to me, but in nuclear
physics we have experienced so many surprises that one cannot
unconditionally say: it is impossible. [6-12]

Meitner had, meanwhile, made arrangements to spend the Christmas
holidays with Eva von Bahr-Bergius and her family in Kungälv (King's
River) outside Göteborg halfway up Sweden's west coast, and left
Stockholm on Friday the 23rd. Frisch, who had been in the habit of
spending Christmas with his aunt in Berlin, had been invited along as
well. He came up from Copenhagen by way of the ferry across the
Sound to Malmö, joining Meitner for breakfast in an inn on Västra Gata in
Kungälv the next morning. She was poring over Hahn's letter, obviously
puzzled by it. Frisch was eager to discuss a new experiment he was
planning on the magnetic properties of neutrons, but she would not
listen. 'I had to read that letter. Its content was indeed so startling that
I was at first inclined to be skeptical' [6-13]. They took a long walk in
the fresh snow, across the Kungälv and into the woods beyond—he on
skis, she keeping up on foot. Finally they sat on a log, and pondered the
results.

Gradually they pieced together the puzzle. It was not a matter of
chipping protons or alphas off a nucleus. Rather, the explanation seemed
to lie in Bohr's liquid-drop model of the nucleus, dating from ~1935
[6-14]. This model is the antithesis of the 'independent particle' model
of the nucleus often invoked, in that it treats nuclear matter as consisting
of protons and neutrons interacting via strong, short-range forces in a
collective manner. As a drop of liquid, the model is characterized by a
certain temperature, surface tension, and modes of vibration. The total
energy, or mass, of the droplet is determined by the mean binding energy
per nucleon, modified by a surface energy and an electrical (Coulomb)
energy between the protons alone [6-15]. The surface tension holding
the drop together is opposed by the electrical repulsion of the protons;
the heavier the element, the more intense is the repulsion. The uranium
nucleus is so heavy as to be on the verge of instability, easily upset by
the addition of a single neutron. Upon capture of the incident neutron,
the imparted energy is shared among the constituents in the nucleus.
The nucleus goes into an unstable, oscillatory mode, one of which is an
elongated ellipsoid. When the dumbbell-shaped ('Kernwurst-shaped',
the Germans would say) deformation reaches a critical point, the system
breaks into two fragments of more-or-less equal mass, which fly apart
with great energy. The energy driving the fragments apart is given
by Einstein's $E = mc^2$, alias the 'packing fraction' for the system:
the two smaller nuclei weigh less combined than the parent nucleus
plus neutron. To be sure, the 'mass defect' is small—only about 0.1%;

however, multiplying by the factor c^2 gives a sizable energy output. It should amount, Frisch estimated, to about one-fifth of a proton mass, or about 200 MeV.

Letters went back and forth between Berlin-Dahlem, Stockholm, and Kungälv, Hahn not realizing at first that Meitner was away on vacation. A copy of the Hahn–Strassmann paper also reached Kungälv via Stockholm. Though Meitner purposely remained evasive about her and Frisch's dawning explanation in her correspondence, Hahn too, perhaps reading between the lines, had reached essentially the same conclusion, that the uranium nucleus had split in two. At the eleventh hour, he managed to alter the page proofs from Rosbaud, though the final version is still couched in language specific to the false 'transuranics', and contains some unfortunate confusion between atomic charge and weight [6-16]. Strassmann seems, at the time, to have been more receptive to the notion of fission than his boss [6-17]. In any case, Hahn also revised a statement pitting barium results 'against all previous laws of nuclear physics', to read 'against all previous experience'.

On 1 January 1939, Meitner returned to Stockholm and Frisch to Copenhagen, Meitner to go over the literature and Frisch to sound out Bohr on their conclusion just as he was preparing to sail for the United States. The same day Meitner received Hahn's revised proofs, and acknowledged them on an optimistic, yet melancholy note.

> Dear Otto! I am now almost *certain* that the two of you really do have a splitting to Ba and I find that to be a truly beautiful result, for which I most heartily congratulate you and Strassmann. … Both of you now have a beautiful, wide field of work ahead of you. And believe me, even though I stand here with very empty hands, I am nevertheless happy for these wondrous findings. [6-18]

6.2. More neutrons?

January 1939 was an uneasy month in European affairs. In Germany, Hjalmar Schacht was dismissed as President of the Reichsbank for his blunt warning that the huge armament program must be curtailed if catastrophic inflation were to be averted. In Warsaw, Joachim von Ribbentrop, Germany's Foreign Minister, had arrived to open the diplomatic offensive against Poland. To the south, Count Ciano, Italy's Foreign Minister, paid a visit to Belgrade across the Adriatic Sea to seek closer political, economic, and cultural ties between Italy and Yugoslavia. In southwestern Europe, the conquest of Catalonia by the Spanish fascists, supported by the Italians, brought France face to face with a new dictatorship on its frontiers. Across the Atlantic, President Roosevelt asked congress for $552 000 000 for defense, while openly declaring the sympathies of the United States for the European democracies.

Even as Roosevelt spoke, scholarly representatives of the European democracies were arriving on American shores. Among the physicists of note stepping off the liner in New York that month was Niels Bohr. He had left Copenhagen via Göteborg aboard the *Drottingholm* on 7 January. Otto Frisch had caught his attention for all of five minutes on the eve of his departure, telling him of the conclusion of himself and Meitner. As Frisch recalls it, Bohr slapped his forehead with his hand.

> Oh what idiots we all have been! [Bohr exclaimed]. Oh but this is wonderful! This is just as it must be! Have you and Lise Meitner written a paper about it? [6-19]

Not yet said Frisch, but he assured Bohr they would at once, and Bohr promised not to talk about it before the paper was safely in the press. Arriving in New York on 16 January, Bohr, traveling with his son Erik and his Belgian collaborator Léon Rosenfeld, was welcomed at the Swedish American Line pier by the Fermis. Enrico had joined Columbia University exactly two weeks earlier. Keeping company with him in New York was Leo Szilard, residing at the King's Crown Hotel, accustomed as he was to living out of hotel rooms and suitcases. Unconventional and unwilling to accommodate himself to the University routine, he came and went at nearby Columbia, but held no regular position there.

Bohr was scheduled to spend the second half of the 1938–1939 academic year as a Visiting Professor at Princeton's Institute for Advanced Study, in the company of Einstein, John von Neumann, Eugene Wigner, and other eminent newcomers to American society. Bohr was also slated to attend the fifth in a series of *Conferences on Theoretical Physics* in Washington, DC. They were the brainchild of George Gamow, who had modeled them after the annual meetings at Bohr's Institute and made his employment at George Washington University in 1934 conditional on their taking place. They were co-sponsored by GWU and the Carnegie Institution of Washington, and organized by George Gamow, Merle Tuve, and Edward Teller, a compatriot of Szilard who joined Gamow at GWU in 1935.

In England that month, James Chadwick was struggling to get a cyclotron into operation at Liverpool with the help of another machine physicist on loan from E O Lawrence, Bernard Kinsey. The effort had consumed well in excess of the budget allotted by the University and the Royal Society; Chadwick bore the balance out of his Nobel Prize money. With the clouds of war gathering over the Continent, Cockcroft had his hands full with apparatus of a different kind—radar developments. Indeed, by 1939, scientists at both the Cavendish and the Clarendon were deep into radar, and other more immediate problems than nuclear physics. Late January found Cockcroft preparing for a meeting at Cambridge with Robert Watson-Watt, the 'father of radar' in charge of developments at the Air Ministry. The meeting had been arranged by Sir Henry Tizard who worked tirelessly to ensure scientific input in laying out the chain of radar stations along England's east and south coasts [6-20].

Back in Copenhagen, Frisch was on the phone composing, with his aunt, the draft of a letter to *Nature*, as Bohr had admonished them to do [6-21]. Frisch also asked an American biologist working with Hevesy on a Rockefeller grant at the Institute, William A Arnold, what biologists call the process whereby cells divide. 'Binary fission' was the answer, and with that the lexicon of nuclear physics acquired an important addition, shortened to fission [6-22]. On 7 January, the eagerly awaited article by Hahn and Strassmann appearing in print the day before, arrived in the Institute library, and Frisch shared it with George Placzek, a Bohemian theorist whose opinion counted. The stories of Placzek's great wit and disorganization—mainly when it came to catching trains, which he did a lot—are legion. He was, in any case, like Frisch, then enjoying the Institute's hospitality, but not for long. He would catch up with Bohr at Princeton within a month *en route* to Cornell University where he had landed a position with Hans Bethe—another New World refugee from Nazi persecution.

Skeptical as always, Placzek urged Frisch to look for himself for the tell-tale, fast-moving fragments signaling uranium fission. 'Oddly enough that thought hadn't occurred to me', muses Frisch, 'But now I quickly set to work' [6-23]. All it took was a piece of uranium-coated foil, an ionization chamber and proportional amplifier biased to exclude the natural alphas from uranium in favor of the ionization bursts from the massive fission fragments, and a cathode-ray oscilloscope. On 13 January, burning midnight oil, 'pulses at about the predicted amplitude and frequency' were observed on the scope without difficulty [6-24]. Frisch scribbled a short note on the measurements [6-25], and on 16 January mailed it to *Nature*, along with the short joint note composed over the phone with Meitner. Ideally, the theoretical paper should have been submitted first; however, Frisch had shared a draft of it with Bohr as he boarded the boat-train for Göteborg, and, Bohr had, as noted, promised to respect its confidentiality. In the event, the theoretical note, by happenstance, proceeded the experimental note into print by a week.

Placzek, the theorist, was not above sharing a laboratory workbench as well, and had earlier collaborated with Frisch on the absorption of slow neutrons in gold, cadmium, and boron—a hot topic ever since Fermi's discovery that slow neutrons had much more effect on certain nuclei than fast ones. Placzek wanted thick layers of gold, and it was his inspired innovation to use several Nobel Prize medals, belonging to German colleagues and left for safe keeping at the Institute when Hitler came to power. Hevesy did Placzek one better when Denmark was invaded in 1940; he dissolved the medals in acid and stored them on a dusty chemical shelf. After the war, they were converted back into metal and the Swedish Academy struck them into medals once more [6-26].

The paper by Hahn and Strassmann reached Frédéric Joliot's office at the Collège de France on 16 January, the very day Bohr landed in New York with the same, if sketchier, news from Berlin-Dahlem. Sequestering himself for several days with the paper, Joliot then dropped the bombshell announcement, that uranium atoms might have been split in two, before an emotional meeting with Irène Curie and his two research assistants, Hans von Halban and Lew Kowarski. 'What fools we have been!' Irène exclaimed, according to Kowarski [6-27]. Once again the Joliot-Curies had missed out on a major discovery, despite Irène Curie and Paul Savitch having skirted the notion that uranium might break up.

All the same, here was welcome relief from the administrative chores that had occupied Joliot, to the exclusion of experiments, for several years, and he pounced on the new phenomenon. First order of business was obviously confirming the tentative finding of Hahn and Strassmann. The experiment, performed on 26 January, did not take long. Joliot assembled a concentric arrangement of a (Rn + Be) source of neutrons, a small amount of uranium oxide (U_3O_8), and a clean bakelite surface on which to catch the fission fragments. Tests showed that the surface would not be activated by neutrons in the absence of uranium, or in the absence of neutrons but in the presence of uranium. However, when both source and sample were in place, the bakelite collected a complex mixture of radioactive fragments with a distinctive half-life similar to that in the solutions concocted by the Curie–Savitch and Hahn–Strassmann teams. The energy was about right: 200 MeV [6-28].

Kowarski, looking over Joliot's shoulder, congratulated him on a simple experiment that took all but an hour. Professing modesty, Joliot replied that he had no doubt the experiment had already been done elsewhere. He was right. The very day before, John Dunning and his student, Herbert Anderson, had seen fission fragments at Columbia University, just as Fermi was leaving for the Washington Conference [6-29]. The Columbia team had been tipped off by Léon Rosenfeld at Princeton on 16 January, the day he and Bohr arrived in New York. On board the *Drottingholm*, Bohr had led Rosenfeld to believe that the note by Frisch and Meitner had already been submitted to *Nature*, and so, going right to Princeton on landing, Rosenfeld broke the news at an evening meeting of the Physics Journal Club. Sitting in on the meeting was at least one from Columbia, Willis Lamb, who hurried back to New York. The next day, the day of Joliot's experiment in Paris, Bohr took the floor during the opening session of the *Washington Conference on Theoretical Physics*, ostensibly devoted this time to low-temperature physics. Knowing by now that the secret was out of the bag, he announced the Hahn–Strassmann experiment and the interpretation of Frisch and Meitner, still ignorant, however, of the confirmatory Copenhagen experiment [6-30].

In the audience, Richard B Roberts and Lawrence R Hafstad of the Department of Terrestrial Magnetism (DTM) under Tuve at the Carnegie Institution listened intently as Bohr and then Fermi spoke. Quietly they slipped out of the lecture room of George Washington's Building C on G Street, NW, and headed back to the DTM grounds in nearby Chevy Chase. There they prepared to confirm the discovery of fission, inaugurating the brand new 5 MV pressure-insulated Van de Graaff accelerator as a source of neutrons by deuteron bombardment of deuterium (in contrast to the ubiquitous Be sources in Rome, Paris, and Columbia). In so doing, one might add, they capitalized on an earlier snafu plaguing Lawrence's Laboratory in Berkeley [6-31]. The venerable 1 MV open-air machine next door, exploited so successfully by Tuve, Hafstad, and Odd Dahl in their nuclear structure studies, would have been equal to the task, but had a burned-out filament. On 28 January, Roberts, Hafstad, and R C Meyer completed the test successfully [6-32], hard on the heels, it turned out, of a similar confirmatory experiment the same day by R D Fowler and R W Dodson at nearby Johns Hopkins University [6-33]. Owing to the proximity of the DTM to the *Washington Conference* site, that evening just about everybody of note from the meeting crowded the circular target room beneath the DTM generator and was treated to a historic demonstration of fission by Tuve and his colleagues [6-34].

Luis Alvarez in far-off Berkeley was not far behind. Though the *Washington Conference* was void of west-coast participants, Alvarez learned of the discovery of fission from the *San Francisco Chronicle* while having a haircut in the student union. He sprinted back to the 'Rad Lab' up the road, wired Gamow in Washington for more details, and on 31 January he and George K Green observed the same ionization pulses [6-35]. The DTM, Johns Hopkins, and Berkeley experiments were reported in back-to-back letters in the February issue of the *Physical Review* [6-36], and furnish an interesting contrast between research styles on the east and west coasts [6-37].

Joliot reported his results to the Paris Academy at its regular Monday morning meeting on 30 January [6-38]. Shortly thereafter, he repeated the experiment, substituting his beloved cloud chamber for the counter. Once again, Joliot partly made up for missing out in a fundamental discovery by resorting to his trusty low-pressure Wilson chamber: he was the first to obtain photographs of the paths left by the fission fragments diverging from a common point in the chamber—much as he and Irène had photographed electron–positron pairs in the wake of Anderson's discovery of the positron.

Joliot was by no means the first to ask the next question: is there more than one neutron liberated in the act of fission, and, if so, are there enough to repeat the process in a self-sustaining chain reaction? The general idea of release of energy through a chain reaction went back at least to Frederick Soddy's broodings and, as we recall, was picked up by H G Wells for public consumption in his *World Set Free*. Jean Perrin had agonized over it, too. The concept had returned to haunt Leo Szilard—a fan of Wells—soon after he slipped across the Austrian border in 1933 and landed in London. That September, he chanced upon a review in *The Times* of a new book by Wells, *The Shape of Things to Come*, containing the author's latest forecasts for the future. Within a fortnight, *The Times* carried news of the British Association Meeting in Leicester, agog with prophecies of atomic energy and Rutherford's 'moonshine' denial. Szilard was not so sure; 'might not Rutherford prove to be wrong?' [6-39].

Joliot himself, in his Nobel Prize address of 1935, had predicted that 'scientists, disintegrating or constructing atoms at will, will succeed in obtaining explosive nuclear chain reactions. If such transmutations could propagate in matter, one can conceive of the enormous useful energy that will be liberated' [6-40]. More recently, the question of uncontrolled neutron multiplication had become the subject of banter at Bohr's Institute in Copenhagen. By such a chain reaction, Christian Møller had remarked to Frisch, 'the neutrons would multiply in uranium like rabbits in a meadow!' [6-41]. Frisch's immediate reaction was to dismiss the notion. 'In that case no uranium ore deposits could exist: they would have blown up long ago by the explosive multiplication of neutrons in them' [6-42]. He soon saw the naiveté of his argument, however; the neutrons in such an ore would either be swallowed up by contaminants, or escape through thin seams. In any case, on closer deliberations, the spectre of a chain reaction appeared unlikely to Møller and his colleagues, and the subject faded from lunch conversations. The discovery of fission reopened the question in altogether a new light, not only in Copenhagen.

The problem, as Joliot saw it, was how to detect an occasional neutron amidst the torrent of bombarding neutrons needed to induce fission in the first place. As it happened, Joliot had just the man to assign the problem to, Hans von Halban, their resident expert in neutron measurements. Appropriately, Halban, too, had spent some time at Bohr's Institute, and had returned to Paris mindful of the talk of a chain reaction.

Hans von Halban had been attracted to the Radium Institute in Paris by the reputations of Joliot and the Curies. Born in 1906 of Austrian parents residing in Leipzig, he took after his father, who was a physical chemist as well as being fond of cultural pursuits of every sort. His grandfather had been financial advisor to Emperor Franz Joseph, and had changed his name from Blumenstock to von Halban upon being ennobled by the Emperor. (Joliot made him drop the 'von'.) Young

Halban obtained his doctorate in atomic physics at the University of Zürich, where Halban senior became professor of chemistry in 1930. From there, Halban went straight to the Radium Institute, where he learned radiochemistry under Irène Curie, and even lent a hand in her early studies of the elusive '3.5-hour substance'. He honed his skills in the techniques of nuclear physics during a year with Frisch and others at the Institute for Theoretical Physics in Copenhagen, then followed Joliot to the Collège de France. By 1939, he had some two dozen publications to his credit, and was an acknowledged expert in experimental neutron physics.

Joining von Halban was Lew Kowarski, an altogether different type: huge of stature, 'with a deceptively innocent face and an astonishing appetite' [6-43]. He was born in Saint Petersburg in 1907, the illegitimate son of a Jewish businessman and a Russian Orthodox opera singer [6-44]. In time, his parents drifted apart. Young Kowarski was raised as a Christian, growing of age with a feeling of 'not quite belonging to the core of things'. During the Russian revolution, his father smuggled him and his brother through the remnants of the German army in Russia to Vilnius, an insecure Polish city that often changed hands. From there he migrated to Lyons, obtaining a degree in chemical engineering, then ended up in Paris, where he found a part-time job with a firm manufacturing gas mains. Determined to do better, he joined Jean Perrin's laboratory, preparing for a doctor's degree in molecular physics. Earning his doctorate in 1935 opened the doors to the Radium Institute to him, and Joliot took him on as a part-time assistant. On moving to the Collège de France, Joliot brought him along as his personal secretary. There Halban took pity on Joliot's 'little typist', as he was jokingly referred to because of his bulk [6-45], and undertook to introduce him to the intricacies of experimental nuclear physics. Joliot concurred, and well that he did so. When Halban took off on a skiing holiday, Kowarski and Joliot went over the scientific problem at hand, neutrons from fission. Kowarski right away put a good idea on the table, as we shall see. Thus came into being the team of Joliot, Halban, and Kowarski, with Kowarski the electronic expert among them.

At the outset, the team agreed on a work-ethic, or, as Halban put it, 'a kind of conspiracy agreement' [6-46]. All ideas, theories, and discoveries originating in their collaboration should be considered products of the team, not of any one individual. This ethic was rigorously adhered to throughout the experiments, as is apparent in the regularly shifting order of authorship of the surviving records and papers. It is nevertheless also clear that much of the credit for the novel concepts stemming from the collaboration goes to Kowarski, who long felt an intruder for having gotten into physics at the comparatively late age of 31.

For a start, however, they borrowed Fermi's old innovation in the Institute garden off the Via Panisperna in Rome, immersing a source

Figure 6.1. *Lew Kowarski, Hans Halban, and Frédéric Joliot manning a cloud chamber, 1939. Courtesy Radium Institute and AIP Emilio Segrè Visual Archives.*

of neutrons in a tank of water, if not as romantic as Fermi's pond. The source, as in their experiments all along, consisted of radium and beryllium; alphas from the radium struck the beryllium and ejected neutrons. (The alternative, a radon–beryllium source like Fermi's in Rome, produced neutrons too energetic for the intended experiment.) The source was placed in the center of a tank a half meter in diameter, filled with water but at first without any uranium added. Emerging from the source, the neutrons bounced back and forth off the hydrogen and oxygen nuclei, gradually slowing down as they fanned out from the source. To detect them at any given point, a small strip of dysprosium, a silvery metal, was placed there. The radioactivity induced by the neutrons in the strip was subsequently measured with a Geiger counter. Moving the strip from spot to spot, a neutron density-distribution curve, similar to that shown in figure 6.2, was gradually mapped out [6-47]. An actual figure from Halban's notebooks is reproduced in figure 6.3 [6-48].

In both figures, the neutron density Ir^2 is plotted against radial distance r from the center of the tank at $r = 0$. Here I represents the intensity of radioactivity induced in the strip detector; hence, Ir^2

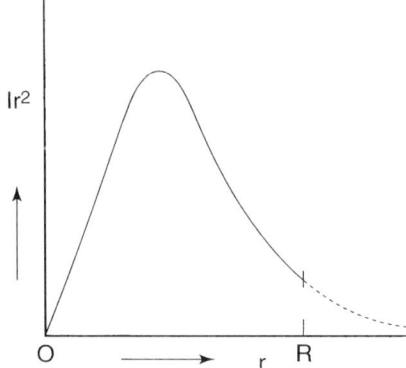

Figure 6.2. *Calculated neutron density-distribution curve, a convenient indication of the feasibility of a chain reaction.*

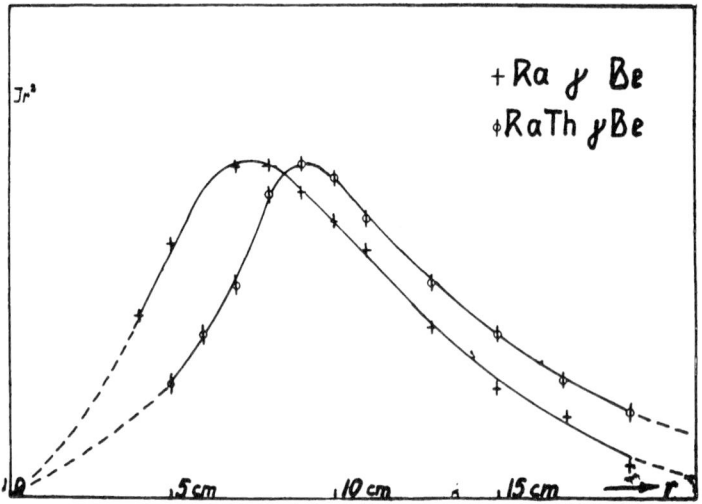

Figure 6.3. *Experimental neutron density-distribution curves, for two different neutron sources. Courtesy Halban Papers, 1939–1940, Microfilm Collection (MI30), Niels Bohr Library at the American Institute of Physics.*

is a measure of the number of neutrons captured in a spherical shell surrounding the source. As can be seen, the density distribution reaches a maximum some distance from the source, then drops off and would approach zero exponentially at some point sufficiently far away, were it not for the edge of the tank R, as indicated in figure 6.2. The maximum in the distribution agrees with Fermi's original finding. Neutrons were more effectively absorbed by the nuclei in the detector strip as they were slowed down by elastic collisions; still farther out, their number dropped off with distance. The area under the curve, obtained by counting squares on graph paper, is given by the integral $\int I r^2 \, dr$ and, thus, is proportional to the total number of neutrons at any given time.

Adding uranium to the water should have two possible effects. The capture of neutrons by uranium nuclei would decrease the area under the curve, while the release of neutrons from fission of uranium would have the opposite effect of increasing it.

The measurements began on 19 February, and lasted two weeks. In some runs, uranium nitrate was dissolved in the water; in others, the tank was filled with ammonium nitrate as a uranium-free calibration check. Alas, the area under the curve remained the same, with or without uranium. Clearly the uranium absorbed too many neutrons. But wait! Kowarski called their attention to a difference in *shape* of the two curves. The peak of the curve with uranium was slightly lower, but the curve fell off more slowly at large radii—something only possible if neutrons were emitted by the uranium. To be doubly sure, the team obtained a stronger source of radium from the Radium Institute, in the care of Irène Curie but over the vigorous protestations of a functionary at the Institute. Everything checked; the splitting of a uranium nucleus under the impact of a single neutron is accompanied by the emission of 'more than one' new neutrons.

Kowarski personally carried their paper with a cover letter to Le Bourget airport on 8 March, so that it could leave by the first plane to London, and appear as soon as possible in *Nature* [6-49]. They decided against *Comptes Rendus*, since inclusion of a curve illustrating the neutron density distributions would have delayed publication in the French journal. Speed in publication seemed of the utmost importance; no doubt their colleagues abroad were in hot pursuit.

They were right, if only partially so. In Berkeley, realizing full well the importance of possible neutrons emitted in fission, Alvarez had placed a bottle of uranium oxide in a beam of slow neutrons from their cyclotron, but a perfunctory test for fast neutrons came out negative [6-50]. At any rate, the priority at Berkeley was new and better cyclotrons; they had little time to spare for fission [6-51]. What about Berlin or Copenhagen? In Berlin-Dahlem, a half-hearted search for neutrons from fission yielded nothing; but then, their priority was the radiochemistry of fission fragments, not neutron production. In Copenhagen, Frisch too thought of looking for fission neutrons, but the threat of war unnerved him; what good would a new, involved experiment do if it might soon be stopped by external events? His mind was more on getting out of Denmark while it was still possible.

Only at Columbia University were serious experiments under way— experiments remarkably similar to those in Paris. Perhaps one reason was that Placzek was in and out of Columbia on a regular basis, and kept Halban informed by letter about Fermi's program. In any case,

Fermi's group, too, set about mounting a neutron source in a tank of water in the basement of Pupin Hall, home of the Columbia physics department. They, too, used a neutron-strip detector, in their case rhodium foil. Their setup differed from the French mainly in that the uranium oxide was packed in a bulb surrounding the neutron source, not dissolved in the water. Unfortunately, the only source available at Columbia was of the radon–beryllium type. Radon ejects much more energetic neutrons than radium, and could have caused neutrons from reactions other than fission. Generously, Szilard came to the rescue, renting a gram of radium from a commercial source out of his own pockets. He had actually been working upstairs at the Pupin on another neutron experiment with Walter Zinn, a Canadian postdoctoral student at nearby City College. Procuring the radium, and arranging for insurance and other sundry details took time. As a result, Joliot and his team, with a copious supply of radium on hand, beat the Columbia group, but by no more than a week. The paper by Anderson, Fermi, and H B Hanstein carries 16 March as the date of submission [6-52].

Both groups pursued an alternative experiment as well, namely, the one proposed by Kowarski in his private talk with Joliot, and the one started by Szilard and Zinn upstairs at Columbia [6-53]. The idea here was simply to bombard uranium with slow neutrons and see whether *fast* neutrons came out; the latter neutrons could only come from uranium fission. The next paper from Paris went to *Nature* on 7 April [6-54]. With their latest measurements, Joliot and colleagues had confidence that neutrons are produced in fission: 3.5 on the average, they concluded—a number about 20% too high, due to 'excessively slap-dash arithmetic', as Kowarski puts it [6-55]. The number 2.5 is, in fact, closer to the accepted value. Moreover, some neutrons are emitted with high velocity. 'With this positive information', recalled Halban, 'we were thoroughly convinced that the conditions for establishing a divergent chain reaction with neutrons could be realized' [6-56].

Meanwhile, Frisch was right to worry about the developing political situation on the Continent, growing more ominous month by month. The tensioning bubble burst in March, as the experiments in Paris and New York came to fruition. On 15 March, spearheads of the German army entered Prague, two hours after the capitulation of the Czechoslovakian government, and the next day Hitler decreed Bohemia and Moravia German protectorates. In France, the mounting tension over Germany's expansion to the east came to a head when Édouard Daladier, Premier, asked for and received from Parliament power to govern by decree without express limitations—a situation unprecedented under the Third Republic. France assumed an attitude of quiet determination; at the same time, on the last day of March, she joined with Britain in a pledge to aid Poland in the event of attack, marking the end of the sad policy of appeasement.

On the scientific side, there were grounds for anxiety about the developments in Germany as well. Cockcroft, more concerned than most, was in intermittent correspondence with Meitner in Stockholm and Hahn in Berlin-Dahlem, who kept him somewhat informed about what was happening in Germany. On 10 March 1939, he met for lunch at the Athenæum in London with Rosbaud, who filled him in, with his masterly overview at Springer Verlag, on progress in nuclear physics in the Reich [6-57]. At Hahn's own KWI, a number of physicists besides Hahn and Strassmann were at work on the uranium problem; thus Gottfried von Droste and Siegfried Flügge, theorists, were independently arriving at similar conclusions to those of Meitner and Frisch. In Göttingen, Georg Joos and Wilhelm Hanle had set wheels in motion that would lead in late April to an organizational meeting for an *Uranverein* ('Uranium Club'). In Vienna, fission experiments of Jentschke and Friederich Prankl were beginning to show results. As they parted, Cockcroft, somewhat shaken, implored Rosbaud to keep him informed, and Rosbaud could only agree. A nuclear weapon, he knew full well, was the one weapon that had to be kept out of Nazi hands.

6.3. Prospects for a chain reaction on the eve of war

In late January 1939, a secretary at Columbia University's Pupin Laboratories opened a telegram by mistake. The sender was Halban in Paris, and it was addressed to George Placzek, who came and went at Columbia. Since it landed on what was regarded as Szilard's desk, he learned of its cryptic message, which read: 'Joliot's experiments secret'. Apparently Halban was admonishing Placzek, who had just returned from one of his regular jaunts to Paris, to keep mum about what he had seen there. Yet it could not have been about fission experiments, as Szilard assumed, since, as Weart points out, they got going only after Placzek left Paris [6-58]. Whatever its meaning, Szilard was right in believing fission experiments were under way in Paris, and he wrote accordingly to Joliot on 2 February.

> When Hahn's paper reached this country about a fortnight ago, a few of us got at once interested in the question whether neutrons are liberated in the disintegration of uranium. Obviously, if more than one neutron were liberated, a sort of chain reaction would be possible. In certain circumstances this might then lead to the construction of bombs which would be extremely dangerous in general and particularly in the hands of certain governments.
>
> It is of course not possible to prevent physicists from discussing these things among themselves, and, as a matter of fact, the subject is fairly widely discussed here. However, so far, every individual exercised sufficient discretion to prevent a leakage of those ideas into the newspapers.

In the last few days there was some discussion here among physicists whether or not we should take action to prevent anything along this line from being published in scientific periodicals in this country, and also ask colleagues in England and France to consider taking similar action. No definite conclusions have so far been reached in these discussions, but if and when definite steps are being taken I shall send you a cable to tell you what is being done.

We all hope that there will be no, or at least not sufficient, neutron emission and therefore nothing to worry about. Still, in order to be on the safe side, efforts are made to clear up this point as quickly as possible. Experiments at Columbia University are in charge of Fermi and will perhaps be the first to give reliable results.

Perhaps you have also thought of the same things and have contemplated or started such experiments. Maybe you are able to get definite results at an earlier date, which, of course, would be very valuable help towards ending the present disquieting uncertainty. Whatever information on the subject you might care to transmit by letter or cable at some later date will, I am sure, be greatly appreciated. Also, should you come to the conclusion that publication of certain matters should be prevented, your opinion will certainly be given very serious consideration in this country. [6-59]

Two days later Fermi, too, wrote Joliot, explaining merely that he was 'presently engaged, like, I think, in every nuclear physics laboratory, in trying to understand what goes on in the catastrophical disintegration of Uranium' [6-60]. He added nothing about keeping things secret.

Szilard's letter caused some stir and puzzlement at the Collège de France, particularly in view of Fermi's follow-on letter, which gave the impression that the Columbia team was hot on their track, and might beat them into print, if they did not move fast. Then again, another cable was apparently coming. Biding his time, Joliot chose not to reply. He, too, was right about activities at Columbia, however. By mid-March the preliminary experiments under Fermi on neutron production in fission were completed, and a report was dispatched to *The Physical Review* on 16 March [6-61], just as German troops crossed into the remnants of Czechoslovakia. Growing more uneasy by the day about the implications of nuclear fission amidst the approach of war, Szilard sat down with Fermi and Edward Teller, and he and Teller urged Fermi not to publish. Fermi countered that unfettered exchange of knowledge is basic to the scientific spirit. However, if the majority was against publication, he would go along with that. Accordingly, he contacted John Tate, editor of *The Physical Review*, and told him to hold their paper until further notice.

Just then, the 18 March issue of *Nature* arrived at Columbia—that being the issue with the letter of Halban and company to the editor dated 8 March [6-62]. Fermi took this to mean that the secret was out of the bag, and that they were free to publish. Szilard quite disagreed; the essential datum—the precise number of secondary neutrons liberated in uranium fission—was yet to be determined and made public. Fermi, consulting with George Pegram, senior spokesman in the Columbia physics department, agreed to stall somewhat longer. Meanwhile, Szilard talked the matter over with other like-minded colleagues, among them Victor Weisskopf, an Austrian fellow émigré physicist at the University of Rochester, then visiting Princeton. Weisskopf needed no convincing of the continued need for secrecy. But how best to get Joliot's attention? As it happened, Weisskopf knew Halban rather well from the time when Halban was a student in Zürich, so he and Szilard drafted a cable to him, not Joliot, and sent it over Weisskopf's signature.

'It was the morning of April 1st', recalled Halban, 'and I was sitting in my bath, when round the door came a hand holding a telegram' [6-63].

KINDLY INFORM JOLIOT THAT PAPERS RELATING TO SUBJECT OF YOUR JOINT NOTE TO NATURE HAVE BEEN SENT BY VARIOUS PHYSICISTS TO PHYSICAL REVIEW BEFORE PUBLICATION OF YOUR NOTE STOP AUTHORS AGREED HOWEVER TO DELAY PUBLICATION FOR REASONS INDICATED IN SZILARDS LETTER TO JOLIOT FEBRUARY SECOND AND THESE PAPERS ARE STILL HELD UP STOP NEWS FROM JOLIOT WHETHER HE IS WILLING SIMILARLY TO DELAY PUBLICATION OF RESULTS UNTIL FURTHER NOTICE WOULD BE WELCOME STOP IT IS SUGGESTED THAT PAPERS BE SENT TO PERIODICALS AS USUAL BUT PRINTING BE DELAYED UNTIL IT IS CERTAIN THAT NO HARMFUL CONSEQUENCES TO BE FEARED. [6-64]

The idea, then, was that, instead of simply withholding manuscripts with sensitive material, authors submit them to a journal to ensure priority in discovery. Weisskopf also cabled P M S Blackett the same day, explaining their action 'in view of remote but not negligible chance of grave misuse in Europe'.

IS IT POSSIBLE FOR YOU TO OBTAIN COOPERATION OF NATURE AND [ROYAL SOCIETY] PROCEEDINGS? [6-65]

Weisskopf followed up his telegram with a longer letter to Blackett, going over the situation in greater detail. Tuve was willing to cooperate in delaying manuscripts, and they thought Lawrence would go along, but knew not yet of Joliot's attitude on the matter [6-66]. A reply from Paris was not long in coming, however. On 5 April, a cable signed Joliot, Halban, and Kowarski reached Szilard.

SZILARD LETTER RECEIVED BUT NOT PROMISED CABLE
STOP PROPOSITION OF MARCH 31 VERY REASONABLE
BUT COMES TOO LATE STOP LEARNED LAST WEEK THAT
SCIENCE SERVICE HAD INFORMED AMERICAN PRESS
FEBRUARY 24 ABOUT ROBERT'S WORK STOP LETTER
FOLLOWS. [6-67]

The Science Service release of 24 March alluded to reported that Richard
Roberts and colleagues at the Carnegie Institution had seen neutrons
produced in uranium fission [6-68]. These were, in fact, *delayed* neutrons
which, it was suspected even then, are inadequate to sustain a chain
reaction. Science Service was, moreover, a sore point with Joliot and
his group; on 16 March, the day the French report appeared in *Nature*,
a news release by the agency had erroneously implied that their results
had been anticipated by the DTM team. Szilard promptly cabled Joliot on
6 April, assuring him that Roberts' delayed neutrons were 'harmless' and
of Tuve's promised cooperation [6-69]. To underscore his point, Szilard
wrote Joliot again on the next day, enclosing for his information a copy
of a note written by himself and Zinn [6-70], explaining that

> its publication is being delayed for reasons of which you know.
> A definite policy in this respect was formulated on March 20th
> and I was on the point of cabling you on that day when I was
> told about your note to *Nature*. I have cabled to you yesterday in
> reply to your cable to Weisskopf . . . and when we shall have your
> cable reply the matter of withholding publications may come up
> again for discussion. [6-71]

Alas, the French answer, dated 6 April, was blunt.

QUESTION STUDIED MY OPINION IS TO PUBLISH NOW
REGARDS JOLIOT. [6-72]

Joliot explained their own position in a more tactful letter on 19 April.
He fully understood the American concerns, but felt, on principle, that
open publication had to continue, or else Hitler could claim success in
destroying another precious freedom [6-73].

> We have continued research on [the liberation of neutrons] and
> you will find herewith the manuscript text of a letter we sent
> to *Nature*. Unfortunately it is too late for us to add your
> communication as reference; however, we shall certainly do this
> in a general article to be published in the near future.
>
> I was very troubled about the question of postponing publication
> on this subject, for I am certainly among the first to understand
> your reasons. But you can understand that we, as well as others
> you may have warned, are not the only ones to be concerned with
> the question and we soon saw articles in scientific publications

and the newspapers, in France and abroad, in which the energy consequences of the phenomenon was clearly explained. These are the only reasons for the wording of my last cable. I certainly agree with the principle of an entente, but if it is to be effective it must be extended to *all* the laboratories which could concern themselves with this question. [6-74]

With Joliot's reply, the effort of the American refugees to restrict publication came to an end, and the Columbia scientists gave *The Physical Review* their go-ahead to print their papers, both the one by Anderson, Fermi, and Hanstein, and the paper by Szilard and Zinn. Indeed, the effort at secrecy was doomed with the appearance in print three days later of the latest one-page note to *Nature* by the French team— the one accompanying Joliot's letter—in which they estimated (actually overestimated) the number of neutrons per fission [6-75].

It appears that Joliot did not share Szilard's concern for a nuclear bomb in the foreseeable future, nor, obviously, his preoccupation with the need for secrecy. However, in spite of Joliot's abhorrence of censorship, he came to accept one particular limitation on open publication—namely with regard to prospects for a controlled nuclear chain reaction for power production. Perhaps Halban, the resident expert on slow-neutron physics, deserves credit for pushing reactors over bombs, though many other scientists thought it at least an even chance that a sufficiently large mass of uranium could yield heat and power in unprecedented amounts. Nor could the possibility of a uranium bomb of extraordinary potency be dismissed out-of-hand as well. To help them sort out fact from fiction, Joliot asked Francis Perrin for his analytical assistance in their chain reaction studies. Professor at the Sorbonne since 1935, Perrin was on good terms with Joliot and his colleagues, and well qualified to lend a hand. He had known Kowarski since the burly Russian worked on his doctorate in his father's laboratory, and knew Joliot even better through his early companionship with Irène in Paris and l'Arcouest. He was a theorist, with a background, like his father, in Brownian motion, and in nuclear physics, including one particular detail of paramount importance: resonance capture of neutrons.

During March–April, Perrin applied himself to the idealized problem of a sphere of uranium oxide or U_3O_8 (as distinct from the uranyl nitrate in water deployed in the actual experiments) surrounding a neutron source. From a simple equation for the diffusion of neutrons in the solid uranium, with a term added for secondary neutrons from fission, he deduced a critical mass for the onset of an explosively diverging chain reaction [6-76]. His estimate was a sphere of uranium some nine feet in diameter, weighing fully 40 tons. A tamper of neutron-reflecting iron or lead around the uranium core, preventing neutrons from slipping out through the periphery of the core, might reduce the mass to 12 tons—still too unwieldy for a practical weapon, it would seem.

To be sure, Perrin's calculation was a first pass through a complex problem, and he had resorted to gross simplifications. Thus he assumed only fast neutrons were involved, that they did not slow down, and that every neutron impact caused a fission. (Introducing slow neutrons, with the moderator and its energy-absorbing characteristics taken into account, would have complicated the calculation enormously.) Moreover, he assumed a mass of *natural* (unseparated) uranium which, it would soon be shown, does not work [6-77]. Still, the calculation seemed to hint at the possibility of an uranium bomb, albeit one highly inefficient. Perhaps it was no more promising than the equivalent weight of high explosive [6-78].

At the end of April, the team shelved bomb calculations and turned to controlled nuclear chain reactions for power production—a seemingly more tractable problem and one surely more appealing conceptually than dubious bomb contemplations. Since slow neutrons were a must, a practical device would have to consist of a mass of uranium mixed with a moderator of some kind of light element for retarding the neutron speed—perhaps hydrogen. Again, the critical mass of uranium might be reduced by surrounding the core with a shield of heavy metal for bouncing back escaping neutrons. A constant worry was keeping the reaction under control. One apparent solution involved periodic insertions and withdrawals of neutron-absorbing cadmium sheets as the reaction sped up or slowed down. Regulating the temperature of the uranium–moderator array was another possibility, capitalizing on the fact that only slow neutrons are effective in causing fission. Heating the core would speed up the neutrons, stabilizing a divergent chain reaction.

In effect, Joliot and his team anticipated virtually all the major features of a modern reactor, including control rods, cooling fluids, and shielding for radiation protection. As an added bonus, the conceptual plant would not only produce heat and power, but also radioactive isotopes for medical applications.

The question of taking out patents on nuclear devices first came up at a party celebrating a grant from the Caisse Nationale de la Recherche Scientifique (CNRS, soon to be renamed Centre National de la Recherche Scientifique), the agency spearheaded by Jean Perrin for administering research budgets in university and Government laboratories alike. Joliot had been strongly against patents, in line with Marie Curie's views from the time of the discovery of radium; absence of strings of any kind had been crucial in the formative stages of the burgeoning radium industry. The Joliot-Curies had adopted the same attitude with regard to artificial radioactivity: no patents, unfettered access to the field by everybody. However, someone pointed out that Marie Curie had been forced to beg radium from the world, and that nearly half of the radium acquired at the Radium Institute had been a gift from the women of America [6-79]. More to the point: if France did not take out patents, foreign competitors would undoubtedly do so when they caught up with the French in this field of uranium research, and exploit their hard-won achievements.

Then again, taking out patents was fast becoming accepted practice in the scientific community. Pierre Curie had patented scientific instruments. For that matter, nuclear physics—even nuclear energy—was already becoming embroiled in patent registration. In 1934, Szilard had lodged patents on his ideas for a chain reacting system, following the example of E O Lawrence, who assigned his patent for the cyclotron to the Research Corporation, a not-for-profit business endowed to serve men in academic positions to exploit useful and patentable inventions. The same year, Fermi and his collaborators had, on Corbino's advice, taken out patents on their method for producing artificial radioactivity with slow neutrons.

At any rate, Joliot's team and the CNRS eventually came to a mutually satisfactory agreement, and three secret patents on aspects of chain reacting systems (including one for explosive charges) were registered in the name of the CNRS. A major portion of the profits was assigned to the CNRS for the advancement of scientific research, according to certain guidelines by a commission established for the purpose.

One who figured prominently in the patent discussions was Henri Laugier, the first director of the CNRS and, in effect, the executive officer for French science. He saw to it that Joliot's team received a fair share of the CNRS budget, over and above the disposable funds in Joliot's fixed laboratory budget. However, money was one thing; if the team was to proceed from merely patenting ideas to serious prototype reactor development, an adequate supply of uranium was equally critical. Here the French had the definite advantage of a long-standing, healthy relationship with the Union Minière du Haut-Katanga, the Belgian firm mining uranium ore in the pitchblende deposits in what was then the Belgian Congo. In 1939 the firm was the world's leading producer of uranium, and of radium which forms in uranium by transmutation. It had, on numerous occasions, supplied the Radium Institute and Jean Perrin's adjacent Institute of Physical Chemistry with uranium and radium at nominal prices.

The advantageous interplay between Union Minière and CNRS in supplying uranium to the French would, before very long, have its fortuitous counterpart in the role of Norsk Hydro *vis-à-vis* Banque de Paris et des Pays-Bas with respect to an equally vital material (deuterium oxide, a scarcer material yet) in supporting a chain reaction, as we shall come to shortly.

On 4 May, the day the third of the patents was registered with the CNRS, Joliot wrote for an appointment with Union Minière management, and on 8 May he met at the firm's headquarters in Brussels with Edgar Sengier, the chief administrator of the company, and Gustav Lechien, head of the radium division. Joliot explained their ongoing work on nuclear power production at Collège de France, and inquired discreetly about the firm's stock of fairly pure uranium oxide in Belgium. It appeared that, in view of the limited industrial demand for uranium (mainly used in the ceramics industry) that the stockpile presently on hand was more than adequate for his team's potential demands—even enough for Perrin's pessimistic critical mass estimates. Equally reassuring, the physical configuration of the stockpile posed no danger! Joliot's presentation, and his proposal for a Franco-Belgian cooperative venture, must have impressed Sengier, an engineer by training; he and Lechien agreed to come to Paris for further discussions at the end of the week. Before then, however, Sengier had a high-level appointment in London.

On 10 May, Sengier sat down in the City of London with Lord Stonehaven, a British vice-president of Union Minière, Baron Carter, the Belgian Ambassador in Britain, and Sir Henry Tizard, chairman of the Air Ministry's Aeronautical Research Committee. The main outcome of the meeting was Sengier's agreement to notify the British authorities of any abnormal purchase of uranium, and the acquisition by the British of a single ton of uranium oxide. (The material was to be used in fission experiments under G P Thomson, then in the planning stage.) As Tizard rose to leave, he turned and said quietly to Sengier: 'Be careful and never forget that you have in your hands something that may mean a catastrophe to your country and mine if this material were to fall into the hands of a possible enemy!' [6-80].

Three days later, a Saturday, deeply impressed by Tizard's grim assessment of the strategic value of the Katanga ores, Sengier and Leichen met in Paris with Laugier and Joliot, along with Halban, Kowarski, and Perrin. They drew up arrangements for Union Minière to supply a total of 5 tons of uranium oxide for preliminary reactor experiments; if successful, 50 additional tons would follow for a full-scale test. Any material not destroyed would be returned to Union Minière, with its associated products. Union Minière also put its Technical Division at the disposal of the CNRS. Meanwhile, simply on the basis of a 'gentlemen's agreement' signed by Joliot and Lechien, the French group was promised the first installment of uranium oxide for immediate use.

The discussion in Paris even touched on plans for the deployment of an uranium fission bomb in the Sahara. In this connection, we note that a French nuclear device was indeed tested in the Sahara—some 24 years later!

In the event, signing of the formal convention was repeatedly put off, because of the complexities of the necessary arrangements, and European events in September, 1939 would put a final end to the discussions. Nevertheless, the Union Minière honored the gentlemen's agreement, forwarding at the end of May one hundred barrels, each containing 50 kilograms of uranium oxide. The firm also loaned Joliot one gram of radium in the form of a radium–beryllium mixture.

Amidst the high-level discussions, experiments with uranium oxide mixed in water continued in Paris, and with varying degrees of enthusiasm elsewhere. The singular exception was Germany. One might have expected a strong push there, particularly in light of a major review article in June's *Naturwissenschaften* on prospects for harnessing atomic energy, by Siegfried Flügge of Hahn's own institute [6-81]. Not so, in fact. The German research effort was gaining momentum in the spring of 1939, but only in the talking stage and curiously slowly. The sluggish pace was partly due to the pecking order of science policy in Nazi Germany between the armed forces, industry, and other parties [6-82], as we shall see.

In Britain, as in Germany, physicists appreciated the potentialities of nuclear power but, as noted earlier, had their hands full with more pressing matters—mainly radar. One who was persuaded to take action, having kept up with the steady trickle of French reports in *Nature*, was G P Thomson, son of J J Thomson and professor of physics at London's Imperial College of Science and Technology. He was also a member of the Aeronautical Research Committee, and thus it is not surprising that a memorandum he wrote on the potentials of nuclear fission came to the attention of Sir Henry Tizard, chairman of the ARC; by an interesting happenstance, Tizard was also Rector of Imperial College.

Thomson dropped Tizard a hint that he could put to good use about a ton of uranium oxide, an amount 'not much more than a pottery firm could use in a year' [6-83], and before he knew it one ton of the black material arrived at the RAF depot in Hammersmith, dispatched from Union Minière. That summer, Philip B Moon, on loan from Birmingham and under Thomson's direction, undertook a series of experiments with hollow cast iron spheres of various sizes containing uranium oxide, a neutron source at the center, and a moderator of either water or paraffin. For scarce paraffin they scoured shops for ordinary children's night-lights. Alas, by the outbreak of hostilities in September, all they could conclude from the effort was that 'an endless chain was *not* possible using uranium oxide and ordinary water or paraffin' [6-84].

In the United States, too, enthusiasm for fission work was flagging, with the notable exception of Fermi's team at Columbia, which kept pace with Joliot and company, if, it seemed, always a week behind them. (The uranium oxide for the escalating Columbia experiments came from Eldorado Gold Mines, Ltd, a Canadian company and the only serious competitor to Union Minière.) As always, the objective was the tell-tale sign for a chain reaction, whether steady or divergent. A difference between the two experimental configurations in Paris and New York, at first seemingly of no great moment, would soon become the key to a successful chain reaction in a uranium–moderator mixture. It was tied to a serious technical difficulty first brought to Fermi's attention by Placzek, sharp as always, who also alerted his friend Halban to the problem. The problem was the so-called resonance capture of neutrons by uranium—the selective non-fission capture by U^{238} of slow neutrons in certain discrete portions of the neutron energy spectrum. That was not the only problem. Hydrogen, the key ingredient in the water moderator, turned out to have a small but finite neutron absorption cross section [6-85] as well. Thus, the uranium and hydrogen might conspire to consume neutrons faster than they were produced by fission.

There was a ray of hope in Fermi's experimental arrangement. Rather than mixing the uranium oxide with the water, they distributed it in a number of sealed cans. This heterogeneous arrangement, Fermi now saw, might have an advantage over the French homogeneous mixture of powder and water, with respect to the neutron capture/absorption problem. Another way out would be to use a moderator less prone to absorb neutrons. Among other light elements besides hydrogen, there were few choices. Helium was impractical, since it is a gas and forms no compounds. Lithium, and especially boron, have a strong tendency to absorb neutrons. Graphite was a possibility, however, as it has a small neutron capture probability. Quietly the Columbia team members went about their business, keeping their thoughts to themselves and, by the same token, largely ignorant of developments in Paris and London.

In Paris, Halban and Kowarski discussed the bad news from Placzek. Kowarski hit upon the solution while sitting in his bath, as seemed to be the wont of the French scientists: keeping the neutrons away from the uranium as they were slowing down [6-86]. A heterogeneous geometry of segments of moderator dispersed throughout the uranium oxide in a kind of lattice might work. Unfortunately, by this time the experiments with their usual homogeneous mixture were well along, and it was too late to change. In fact, they had reached the point, in preparing for the 5 ton model reactor, where they simply ran out of space at the Collège de France in Paris. Not to worry; Joliot had foreseen just such an eventuality. At about the time he assumed his chair in Paris, he had secured space for a larger laboratory facility from the Compagnie Générale d'Électrocéramique at Ivry to the southwest of Paris. There he had established the Laboratory for Nuclear Synthesis in an enormous hangar, large enough for the erection of particle accelerators and other massive hardware, and equipped with overhead cranes and all the necessary services. An added attraction was an annex of the Radium Institute at nearby Arcueil, built for treating large lots of radioactive minerals.

The group relocated its operations to Ivry in June, and there began the erection of a relatively modest hollow copper sphere, 30 cm in diameter, filled with uranium oxide and water—not unlike the experiment at Imperial College. Anything on a larger scale, Halban had optimistically cautioned, ran the risk of exploding upon assembly. The sphere was sunk in a larger water-filled tank 3 m in diameter, which served the dual purpose of a neutron reflector and of allowing neutron density measurements far out in the tail of the density distribution, beyond $r = R$ (figure 6.2). Patiently they plotted density distribution curves under a variety of combinations of uranium, moderator, and cadmium neutron arrestors. Clear-cut signs of a self-sustaining chain reaction, much less a divergent one, proved difficult to pin down with the unreliable Geiger counters coaxed along by Kowarski. To be sure, the area under the neutron density curve grew with the gradual addition of water, but then the uranium oxide saturated. At that point, Halban surmised that the number of neutrons produced was about equal to the number escaping. They were at least making progress! The 30 cm copper sphere soon gave way to a 50 cm sphere. By August, while writing up a short report on results to date [6-87], they were preparing for a 90 cm sphere.

While thus preoccupied, all four—Halban, Kowarski, Joliot, and Perrin—took turns alternating between Paris and Ivry, with time off for vacation at l'Arcouest. Increasingly, however, the clouds of war dampened their spirits; time was running out. Mobilization was a real possibility, and plans were being hatched within the complex French establishment, should it come to that, to turn laboratories into semi-military entities with the team leaders on special assignment from the military [6-88]. This was all very well for Joliot and Francis Perrin, but it put Halban and Kowarski, neither being French citizens, in a more precarious position. Pulling strings, Joliot, Jean Perrin, and Laugier were able to push through special legislation nationalizing both of them— Halban in January 1939 and Kowarski in August. Their effort was in the nick of time. Within a month, France was at war.

6.4. A moderator of choice

Hitler launched World War II, at 10 a.m., in a speech before the Reichstag, on 1 September 1939. Beaming with self-assurance, he announced he was throwing all the might of Germany against Poland, to avenge a 'cowardly attack' on the Reich cleverly staged by his *Schutzenstaffen* (protection squad). In fact, the first phase of Hitler's military campaign against the Poles had already begun some hours earlier in the darkness before dawn.

1 September developed into a balmy Friday in western Europe, and nobody paid much attention to Hitler's rantings. Berliners sipped coffee and schnapps on sunny café terraces; Frenchmen and Britons, too, savored the bitter-sweet pleasures of a late summer holiday. Two days later, the BBC broadcast Prime Minister Neville Chamberlain's brief announcement that Britain was at war with Germany. Within minutes of his remarks, Londoners experienced their first wartime air-raid alert—a false one, to be sure. The same thing happened in Paris: declaration of war at the Élysée Palace, followed by the wail of sirens. In Berlin, too, sirens sounded that day. That was all, but no one could doubt that the dog days of summer were over, and war had begun.

In Paris, life went on with a modicum of normalcy. Weekend boulevardiers took their customary stroll and apéritif, though some prudently carried gas masks. Wrote Irène to Hélène, sent to l'Arcouest for safety with her brother Pierre,

> there are a few Metro trains, almost all the lines are shut down. There are no buses and very few in the suburbs. ... Everywhere windows are painted in blue to prevent lights being seen at night, and sticky paper is put on the large windows so that there is less likelihood of them breaking in case of an explosion. I have seen groups of children being evacuated; they were carrying on their clothes a large ticket with their destination on it, looking like little parcels walking by themselves. [6-89]

Joliot, along with Halban and Kowarski, were called up for military service, while remaining in their laboratories on Special Assignment. (Francis Perrin, a reserve officer, was commissioned Lieutenant in the Army and was at first assigned to an antiaircraft searchlight battery.) Joliot held the rank of Captain, and headed the Army's Group 1 of Scientific Research. The CNRS requisitioned his laboratories; however, except for added red tape, the experiments continued pretty much as before.

The figure of merit in the chain reaction process was no longer taken to be simply v, the number of neutrons per fission. Rather, they seized upon the concept embodying the number of neutrons per *primary neutron*, or what came to be called the *multiplication factor k*: the ratio of the number of neutrons produced by nuclear fission to the number of neutrons originally present [6-90]. In other words, k signified the number of neutrons produced for every neutron absorbed or otherwise lost; if k drops below 1, the nuclear fission ceases to be a self-sustaining chain reaction and will eventually die away [6-91]. The multiplication factor, it turned out, was amenable to calculation, which the team proceeded to do. Conditions being what they were, they wrote up their latest thinking *vis-à-vis* a chain reaction theory in a sealed document deposited at the Académie des Sciences on 30 October. The document remained sealed for ten years [6-92].

By August prospects for achieving a successful chain reaction with a uranium–water reactor were fading, in light of the more refined neutron multiplication calculations. Nevertheless, the team went ahead with the planned 90 cm sphere at Ivry. This time, blocks of uranium oxide were assembled with scaffolding into a nearly spherical form, and the whole drenched with water, acting as always as the moderator. As expected, the multiplication factor remained below unity. A self-sustaining chain reaction with hydrogen and natural uranium was evidently all but out of the question.

One way out of the apparent impasse, assuming Bohr and John Wheeler were right about slow-neutron fission being caused by U^{235} alone, was enrichment of that isotope; to do so, however, would demand costly resources and, more importantly, would take time. Another possibility was some sort of heterogeneous arrangement of moderator and uranium, say lumps of moderator embedded in uranium, as the group already knew. With Perrin no longer around to perform analytical calculations on the optimum ratio of moderator to uranium, they had to approach the problem largely empirically, except for a few rudimentary calculations by Halban, assisted by occasional letters from the ever helpful Placzek. Their counterpart team at Columbia had the advantage of strong theoretical support—by Szilard, Wheeler, and even Fermi himself. Interestingly enough, the American approach ran counter to the French, being predicated on a lattice of uranium lumps embedded in a moderator core.

By the end of 1939, Joliot and his colleagues settled on a dual approach, of comparing a heterogeneous distribution of paraffin cubes in dry uranium (assembled in the old 50 cm copper sphere) with the latest homogeneous system of wet uranium oxide. Comparing neutron density distribution curves for the two systems, as well as measuring the relative proportions of U^{239} resulting from neutron capture in U^{238}, they concluded that a heterogeneous approach favors decreased neutron capture. Equally important, a new moderator was clearly called for; hydrogen absorbed too many neutrons. But what should it be? Besides having less appetite for neutrons, it had to be light, i.e., comparable in mass to the neutron, and available in very pure form. Among the lighter elements, only deuterium, beryllium, and carbon appeared to fit that description. Deuterium, however, was extremely scarce and expensive; beryllium, too, was expensive to produce in bulk. That left carbon, which, in the form of graphite, could be produced industrially in a purer state than almost any other substance.

With that and allied matters in mind, Joliot called on Raoul Dautry, the French Minister for Armaments. Dautry was charged with the supply of raw materials and military fabrication; Joliot came under his jurisdiction since the CNRS had newly been placed under Dautry's control at Dautry's insistence; formerly it belonged to the Ministry of Education. Dautry was a very busy man and therefore, as the saying goes, just the man to get something done.

Raoul Dautry was an interesting person by all accounts [6-93]. He was a visionary in some respects, not so in others; thus he had little faith in the aeroplane as a weapon of war, but an abiding faith in science and in technology as forces for the advancement of mankind. In his opinion, a modern war would be effectively won in the laboratory, not on the battlefield. A graduate of the École Polytechnique, he made his reputation building railroads. However, he did not share the socialist leanings of the Popular Front that came into power under Léon Blum in 1936, and when the new government nationalized the railways in 1937 he resigned his position. Though at odds with socialism, he was widely respected as an indefatigable worker and responsible administrator. In September of 1939, as France went to war, he was approved unanimously to head the vitally important Ministry of Armaments.

Dautry was well aware of Joliot's scientific reputation, having crossed paths with him on several occasions; first during a tour of the Ivry facility, and later when they had served together on some research coordination committee. He listened attentively to Joliot when the latter came calling that autumn day in Paris. In fact, Dautry needed no convincing of the importance of nuclear power, whether for military or industrial purposes. He well remembers having once read in a popular article that the energy locked up in a mass the size of a table was enough to blow up the world. Joliot did not disappoint him. He outlined in considerable detail their ongoing uranium research program, explaining that while the potential for a nuclear weapon could not be excluded, nuclear power could equally well be used for submarine propulsion, say, or as a bountiful source of electrical energy.

Dautry was impressed by the lucidity of Joliot's presentation, as was Sengier the previous May, and he promised all possible help—never mind the expenses. High on Joliot's priorities was carbon. However, before settling on carbon as a moderator, its neutron absorption cross section would have to be determined more accurately. An impressive stack of carbon dioxide (dry ice) had already proved inadequate for a reliable measurement. Besides, it entailed the complication of having two oxygen atoms per carbon atom, which introduced an unknown effect into the analysis [6-94]. True to his word, Dautry soon located a large stock of industrial graphite in a factory near Grenoble, and by Christmas Joliot's team was at work there, in charge of Halban. With the help of assistants, they erected a pile of relatively pure graphite five feet square and ten feet high. In the event, only an upper limit for the absorption cross section was obtained. The value was encouraging, but not low enough to extrapolate with confidence to graphite of the highest degree of purity.

A reliable calculation was urgently needed to settle the matter. Unfortunately, for all his weight, Dautry was unable to extricate Francis Perrin, the theoretical physicist, from his searchlight battery. Instead, Kowarski, with Halban's help, came up with an approximate formula which indicated a strong preference for deuterium over carbon as a moderator. Halban was not surprised. During his stay in Copenhagen in 1937 he had, with Frisch and Jørgen Koch, measured the neutron absorption cross section of deuterium [6-95]. It was, indeed, very low. However, the substance was correspondingly scarce; for their measurements they had borrowed sixty liters of water containing but 10% of deuterium oxide from Norsk Hydro.

Just how scarce, Joliot emphasized in a memo to Dautry calling his attention to heavy water, and again in a memo to the High Commissioner for the National Economy. Writing to the High Commissioner on 8 December, he repeated the need to establish 'in the present and future interests of French scientific research, a reserve of a rare chemical product manufactured abroad, whose future supply is uncertain in the light of existing circumstances. It concerns a product known as "heavy water" (deuterium oxide), manufactured in Norway by electrolysis of ordinary water at a cost price impossible to secure in France' [6-96].

As with uranium, Joliot knew just whom to see about deuterium, which, again by fortunate circumstances, his team was in a favorable position to lay its hands on. He needed to have a talk with Jacques Allier at the Banque de Paris et des Pays-Bas, chief shareholder in Norsk Hydro. In fact, they could ill afford to be complacent on the matter, as Joliot would learn from Allier. On 14 January, a Norwegian officer of Norsk Hydro had stopped over in Paris at Banque de Paris' venerable headquarters on 3 Rue d'Antin, and had some disquieting news to report [6-97].

Chapter 7

HEAVY WATER REVISITED

7.1. The Allier mission

On 4 March 1940, Lieutenant Jacques Allier stepped off the train from Stockholm in Oslo's Eastern Central Station. A Reserve Officer drafted by the French Ministry of Armaments, Allier was a senior official of the Banque de Paris et des Pays-Bas, the bank that coordinated the French holdings in Norsk Hydro. He was also an agent in the Deuxième Bureau, France's secret service, presently entrusted in a mission of the highest importance. He carried a passport identifying him by his mother's maiden name, Freiss [7-1]. In Oslo, he made contact with the French Minister, arranging for accommodations on the Legation premises, without revealing the nature of his mission. From a nearby public phone, he then contacted Chr W Garben, General Secretary of Norsk Hydro, agreeing to meet on a certain street corner. Before long, Allier sat down for serious discussions with Axel Aubert, Norsk Hydro's General Director, in his office at Hydro's headquarters.

Allier's mission had its origins in a meeting he had with Joliot on 14 February in the Hotel Majestic, not far from the Arc de Triomphe. In charge of the meeting was Minister Dautry, accompanied by his Chief of Staff, Jean Bichelonne. Dautry explained for Allier's benefit the significance of heavy water in the developing race for harnessing nuclear energy among the western powers, Joliot filling in technical details. Allier for his part, being a member of the Deuxième Bureau, had news to report as well. One month earlier, Bjarne Eriksen, then Chief of Norsk Hydro's legal department and second in command at the giant concern, had passed through Paris on a routine business trip. He had informed Allier of a sudden, heightened German interest in buying up heavy-water stocks in Norway—something viewed with considerable alarm in Hydro circles.

As early as December of 1939, Norsk Hydro had received indications from several quarters of renewed German interest in heavy water. Hydro management looked upon the matter with distrust, unable to ascertain the exact reason for such interest, but suspecting some kind of 'deviltry' [7-2]. Routine heavy-water production, shut down since the previous summer, had, in fact, resumed on 29 January 1940, and shipment of heavy water to Germany continued well into March 1940. The final shipment listed in the Hydro archives is dated 19 March [7-3].

Earlier in January, meanwhile, prodded by General Director Axel Aubert, Director Nicolai Stephansen, head of Hydro's P-Division at Oslo headquarters, had asked Jomar Brun at Vemork in a confidential note for his opinion concerning the heightened interest in heavy water among the warring powers [7-4]. Brun could only guess, suggesting some application for bacteriological warfare [7-5]. Brun was also requested by Stephansen to prepare a plan for expanding the production capacity to ~1.5 kilogram per 24-hour period—a fivefold increase over the production rate one year earlier. Brun was admonished to treat the plan as strictly confidential; not even Hydro's local director at Rjukan was to know [7-6]. Among the interested parties was I G Farbenindustrie Aktiengesellschaft, part owner of Norsk Hydro. In January, Farben placed an order for up to 100 kg/month of heavy water. General Director Aubert replied personally that the request could not be met, being beyond Hydro's production capacity. (Hydro's production rate was then ~10 kg/month.) Farben persisted in its request, hinting they could use as much as two tons in the longer run, without offering a reason for their demand [7-7]. Hydro's management insisted on an explanation; I G Farben remained tight-lipped, and on 7 February their request was formally denied by Director Aubert [7-8].

Appraising the situation, Dautry asked Allier whether he would be willing to lead a team of French agents in a secret mission to Norway, for the purpose of negotiating a loan from Norsk Hydro of as much heavy water as possible—the entire stock, if it could be arranged [7-9]. Dautry had already cleared his request with the Prime Minister, Edouard Daladier and Paul Reynaud, the Finance Minister. Allier readily agreed, and Joliot then briefed him on certain technical precautions to be taken in transporting heavy water to Paris. Metal canisters of a particular size would have to be arranged for in Norway to fit suitcases specially prepared in planning for the operation. The canisters could be of most any metal, say stainless steel or copper, though preferably of aluminum to minimize weight. However, it was absolutely essential that neither cadmium nor boron be used in welding the canisters; the slightest trace of either element would render the heavy water useless as a chain-reaction moderator free of neutron capture. By the same token, should Allier fail in his primary mission, he should do his utmost to ensure that existing stocks of heavy water, whether at Rjukan or *en route* to Paris, were contaminated with either cadmium or boron, or certain other metals, rendering them useless for the Germans.

Allier left by train for Amsterdam on 28 February. He carried with him a letter of credit for one and a half million Norwegian kroner, as well as two operational orders. One was signed by Dautry, charging him with a secret mission on behalf of the French government, and authorizing him to call on any French secret agent in Norway or Sweden for assistance, as needed. The second order was signed by Daladier, empowering him to negotiate with unnamed parties for securing the largest possible quantity of 'the material that is the object of this mission'. Despite the secrecy shrouding the mission, French Intelligence intercepted a German message to agents in Norway, as Allier learned in Oslo. It warned them to be on the lookout for a suspicious Frenchman traveling under the alias of Freiss. However, he proceeded without incident by plane to Malmø, Sweden, and thence by train to Stockholm, arriving there on 2 March.

Meanwhile, while the mission was still in the planning stage, Dautry had second thoughts about Halban and Kowarski being party to the unfolding top secret operation. To be sure, both were naturalized French citizens. However, they came from countries now formally hostile to France, particularly with the Soviet invasion of Finland then in full swing. To placate Dautry, Joliot asked his two colleagues to lie low on enforced 'holidays' on islands off the coast of France until Allier's mission was concluded. They agreed. Joliot chose the island of Porquerolles on the Mediterranean coast for Halban, with his taste for good living; Kowarski ended up on Belle-Ile off Brittany, a more bracing setting for the hardier physicist. Joliot himself bore the expenses, at least initially.

In Stockholm, Allier contacted three countrymen, all agents of French Intelligence: Captain Jean Muller, an Army Officer; Lt Fernand Mossé, in peacetime a Professor in Germanics at the Sorbonne; and Jehan Knall-Demars, a reserve officer with the Compagnies des Wagons-Lits. They remained in the dark about the nature of the mission and would remain in Sweden until Allier sent for them. Two days later, he joined Aubert in his office at Solligaten 7 in Oslo.

Accounts of the historic meeting between Allier and Aubert note apprehension on the part of the French about perceived German sympathies of Aubert—a feeling fostered by the French Minister [7-10]. Axel Aubert, who became general manager of Norsk Hydro in 1926, had, after all, been instrumental in Hydro's switch to the Haber–Bosch ammonia process in collaboration with I G Farben in the late 1920s. In view of its close ties with I G Farben, Hydro, if not Aubert himself, might prove less neutral in its dealings with the Germans, in the case of a new war, than it had been during World War I and the 'Badische affair'. While on a visit to France during the summer of 1939, Aubert had been asked pointedly about Hydro's attitude in the eventuality of a war between France and Germany. He had stated that, if war came to pass, he would do his utmost to ensure that stocks on hand went to France [7-11]. Soon afterward, Bjarne Eriksen, on a visit to London, had been similarly asked about Hydro's policy and Aubert's leanings in particular.

Eriksen expressed surprise about possible doubts concerning Aubert's loyalty, and immediately put the inquirer at ease on the matter. As to Norsk Hydro's policy, there was after all, he noted, only one German member on Hydro's Board of Directors [7-12]. Moreover, specifically German shares of Hydro's joint stocks amounted to only 12.5% [7-13].

In the event, the opening round of discussions at Solligate 7 began well—cordially and to the point. Aubert repeated German demands on heavy water, and expressed his concern over I G Farben's refusal to explain their sudden and urgent need for such large stocks. He impressed Allier as a man totally devoted to the Allied cause.

> I well knew the sense of responsibility, the great integrity of this man [wrote Allier to Dautry in a later report], as well as his quite legitimate wish to ensure that his dignity and freedom in the matter was respected. Thus the only attitude I was able to adopt, and the one best likely to succeed, was to appeal to his better feelings. By dealing with the matter on this level, and by appealing to his friendship for France, I gained his assent to more things than we had hoped. [7-14]

On 5 March, Allier and Garben journeyed to Rjukan without giving any advance warning of their coming. The next day, Aubert joined them there—he, too, traveling incognito. Factory management was understandably upset about being unable to extend proper hospitality to such important visitors, and when the party stopped at Notodden on their way back to Oslo, they found, to their dismay, the French tricolor merrily flapping in the breeze [7-15].

Back in Oslo, the negotiations resumed, and were completed on 9 March. Aubert agreed not only to lend France the entire stock of heavy water, 185 kg or about 410 pounds, for the duration of the war, but also assured Allier that France would have priority claims on production in the near future. Only at the end of the negotiations, with everything so amicably settled, did Allier reveal the military implications of the transaction. Aubert ended the discussion on the following note.

> Please tell Prime Minister Daladier on my behalf that my company does not wish to receive one centième for the product you will deliver to him until the victory of France. As far as I am concerned, I know that if the experiments of which you have told me succeed and if later France has the misfortune of losing the war, I will be shot for what I am doing today. But it is a matter of pride for me to run that risk. [7-16]

At midnight the same day, 9 March, the heavy water was brought into the yard of the Vemork plant on the instructions of Jomar Brun, who had been alerted in advance by Director Bjarne Nilssen, the Hydro director responsible for Rjukan. There it was poured into 26 five liter cans. They had been manufactured to Joliot's specifications by a local craftsman in his private welding yard on the outskirts of Rjukan, under the guidance of chief engineer Arne Enger [7-17]. Had they been made by one of the large Oslo welding firms, the unusual welding instructions might well have revealed their purpose to German agents. Somewhat over half of the filled cans, containing 100 kg of the heavy water, were driven east over the icy Vestfjord Valley road by one of Hydro's cars, accompanied by Brun; from Notodden the transport continued via Kongsberg and Drammen, arriving in Oslo on the following day. A second of Hydro's cars, bringing the remaining cans, had left Vemork later the same evening. In Oslo the cans were received by Garben. There they were off-loaded and stored in a house belonging to the French Embassy, amid considerable nervousness, for good reasons: a pavilion next door belonged to the *Abwehr*, German Military Intelligence, and remained brightly lit all night. The night passed without incident, however, and the next day, 10 March, Allier's colleagues slipped in from Stockholm to lend a hand, still in the dark about the real purpose of the operation. Only in 1945 were they appraised of the contents of the grey metal canisters claiming their attention.

The question now was: how to spirit the cargo out of the country? A submarine from Brest or Cherbourg making landfall somewhere along the Oslofjord coast, or a destroyer out of Tönsberg at the mouth of the fjord, were quickly ruled out, in part, due to possible repercussions from military operations in neutral Norway. In the end, they settled on an aerial evacuation scheme, planned to throw *Abwehr* agents off by a clever, if old-fashioned ruse. Tickets were booked for three consecutive days on two scheduled flights out of Fornebu Airport, Oslo's new airport inaugurated only ten months earlier. One was bound for Perth in Scotland, and one for Amsterdam; both were scheduled to depart at nearly the same time, early in the morning. Late into the night before, Allier and his companions went carefully over their movements on the morrow, keeping a wary eye on the pavilion next door.

The next morning, 12 March, dawned bright and clear in Oslo, with two Junkers Ju-52 aircraft parked side by side on the tarmac at Fornebu. In the departure hall, Allier and Mossé openly made arrangements for boarding the plane bound for Amsterdam. Earlier in the morning, they had secured seats as well, under other names, on the flight for Scotland. The gates swung open, and the two headed for the Amsterdam plane. Just then, with all propellers turning, a taxi drove up to the departure gate, with Knall-Demars and loaded down with suitcases. Knall-Demars made a nervous fuss at the gate and the cab was allowed onto the tarmac. It stopped between the two aircraft, out of sight of the terminal building. The suitcases were manhandled onto the Scottish plane, and Allier and Mossé jumped on board just before the door slammed shut. Knall-Demars headed back to the gate. The Amsterdam plane taxied onto the runway, took off, and headed out over Oslofjord. A few minutes later, the plane for Scotland was airborne as well, winging westward with the two Frenchmen and 13 of the steel-grey canisters.

It appears, in retrospect, that the remarkably expeditious get-away of Allier and his cargo at Fornebu, in the face of alerted German agents, owed its success in part to Frank Foley, an agent with the British Secret Intelligence Service (known as SIS or MI6). Since August of 1939, he had been posted to the British Legation in Oslo, assigned to Section V as their expert on the *Abwehr* and related Norwegian matters. He is believed to have been instrumental in arranging for clearance at the airport, and assistance in handling the bulky heavy-water canisters [7-18].

Cloud cover was mixed over the south-Norwegian coast that morning. It was clear over the Skagerrak, where the Amsterdam-bound plane was intercepted by *Luftwaffe* fighters and forcibly diverted to Hamburg, as related earlier. In the second plane, too, an unidentified aircraft was observed on their tail off the Norwegian coast. Allier explained to the pilot that he and his companion were French officers in civilian dress, on a mission of great importance for the Allied cause. The pilot agreed to stay as much as possible within the cover of clouds. In fact, the fog over the North Sea proved so dense that the flight was diverted to Montrose in Scotland. Curiously, there was no customs inspection upon arrival. More curiously yet, with wartime Britain on the alert for suspicious visitors, the two French agents were not even questioned by British authorities.

They spent the night in an Edinburgh hotel, and the next day were joined by Knall-Demars and Muller bringing the remaining 13 canisters. All four caught the train for London with their precious cargo stashed overhead in their compartment. At all times, one of the party maintained an armed guard beneath the unobtrusive suitcases. In London, personnel from the French Military Mission assisted in transferring the cargo to the train for Paris, and the party continued on its way. On 16 March, Joliot signed a receipt for the cargo, safely lodged in the cellars of the Collège de France. He had a double reason for rejoicing. No sooner was the heavy water under lock and key, than Halban and Kowarski turned up, too, none the worse for their three-week enforced island holidays.

The mission was over, except for one last formality. On 27 March, Mossé returned to Oslo, carrying two letters to Axel Aubert. One conveyed the gratitude of the French Government for Norsk Hydro's splendid cooperation. The other, from Allier, asked that, in the event of a successful German invasion of Norway, any remaining heavy-water stocks be destroyed. On 8 April, Muller too arrived in Oslo late that evening, bringing certain confidential documents for the French Minister and Norsk Hydro [7-19]. Because of the late hour, he retired early. Turning up at the French Legation the next morning, he found the building empty. By the time he called Aubert at Norsk Hydro without response, he knew what was afoot. The German invasion of Norway had been under way since midnight.

7.2. Attack on Norway

8 April 1940 found Niels Bohr in Norway. He was there to lecture on nuclear physics before the Norwegian Engineering Society, as well as to discuss the worsening political situation in Europe [7-20]. That evening he dined with King Haakon VII and an intimate group at the Royal Palace in Oslo. A gloomy mood pervaded what should have been a stimulating evening, amid speculations about German intentions on the morrow. Bohr then boarded the night train for Copenhagen. At daybreak the next morning, as the train ferry across Øresund stopped in Helsingør (Elsinore), Danish police knocked on the compartment doors. Denmark was at war. *Weser* Day, start of *Weserübung*, the invasion of Denmark and Norway, was in full swing.

Hours earlier, the German Ambassador Cecil von Renthe-Fink had roused the Danish Foreign Minister, Peter Munch. Despite a nonaggression pact with Germany in force since the year before, he presented Munch with an ultimatum: accept German occupation, or Copenhagen would be bombed within the hour. At 6:20 a.m. King Christian X, recognizing the hopelessness of the situation, broadcast an appeal to the Danish people not to resist. To be sure, the directive was purely *pro forma*. Even before the King and his Cabinet had concluded their talks, the German troopship *Hansestadt Danzig* was discharging men and materiel in Copenhagen's ice-choked harbor. By the time Bohr and his fellow passengers were awakened, 40 000 German troops were rolling across the Danish border with Schleswig-Holstein to the south. *Luftwaffe* aircraft were already marshaling on Danish airfields in support of the attack on Norway.

In order to make simultaneous landfall in the farflung Norwegian coastal cities of Oslo, Kristiansand, Egersund, Bergen, Trondheim, and Narvik, elements of Admiral Raeder's Naval armada comprising the main force of *Weserübung Nord* had to sail from German ports at different times, some weighing anchor as early as 3 April. The units of the warship echelon were organized into six groups, one for each of the targeted cities, supplemented by several echelons of merchant ships carrying additional troops, equipment, and fuel. Simultaneously with the sea assault, air assaults were to be made on Sola airfield near Stavanger and at Fornebu outside Oslo [7-21].

The attack on Norway's west coast was an audacious undertaking, since all ports were exposed to the superior British fleet, with Bergen only eight hours steaming from Scapa Flow. Oslo, and to some extent Kristiansand, was less risky, these ports being more readily defended and requiring shorter voyages. On the other hand, whatever might happen elsewhere, Oslo was the key element in the German invasion of Norway, politically and strategically: politically, with a *coup de main* offering the possibility of seizing the King and his Ministers; strategically, due to the short sea passage and Oslo being the hub of the rail and road network leading west and north [7-22].

Naval Group 5 sailed from Kiel at 3:00 a.m. on the 8th, bound for Oslo with Germany's newest heavy cruiser *Blücher* in the van. The flagship of Rear-Admiral Kummetz, the *Blücher* carried the special units with administrative and technical skills needed in coordinating the occupation of Oslo, plus elements of the Gestapo. Following the *Blücher*, in column astern, came the pocket battleship *Lützow*, the light cruiser *Emden*, three torpedo boats, and an assortment of smaller craft. Near Færder light at the entrance to Oslofjord there was a brief skirmish with a Norwegian patrol boat, leaving it dead in the water and on fire. A little later the darkened ships were spotted by two small look-out forts, Rauøy and Bolærne; they fired warning shots and passed the alarm up the fjord. The Germans ignored the challenge, and continued steaming north at slow speed through the moonless night. Two of the torpedo boats and several minesweepers were detached to land troops to take the small Naval base at Horten.

The main defense of Oslofjord lay some two-thirds of the way to Oslo, at the Drøbak narrows, only 600 yards wide between Kaholmen Island and the mainland to the east. Oscarsborg fort on Kaholmen dated from the Crimean War, but had been modernized at the turn of the century; even so, it was listed as a museum on German charts. It commanded three 11 inch guns and a torpedo battery, though its searchlights were inoperative just then due to generator failure. On the Drøbak mainland was a battery of smaller guns and two Bofors antiaircraft guns. Colonel Birger Eriksen and his crew had been alerted by the entrance forts to the south. If nothing else, they hoped to get off a few symbolic shots.

At 5:15 a.m., when the silhouettes of a column of darkened ships hove into view at the entrance to the narrows, a searchlight on the mainland opened on them, illuminated the *Blücher* for a few minutes and then went out. That was all Colonel Eriksen needed to give the order to commence firing. The guns of the fortress opened up at point-blank range. Shells crashed into the *Blücher*, and then into the *Lützow* just behind her. The *Blücher's* steering gear and aircraft hangar were wrecked, and she blundered forward erratically while flames fed by the aviation fuel raged aft [7-23]. They burned away the fog, revealing a torpedo battery not even shown on the German charts. Again Eriksen barked orders to fire. At 7:23 a.m., the main magazine of the *Blücher* exploded. She rolled over and sank, taking with her more than 1000— half of them soldiers. The *Lützow*, also damaged, withdrew with the *Emden* to land troops on the eastern side below Drøbak.

While the engagement in Drøbak Sound was barely under way, the German Minister in Oslo, Kurt Brauer, presented the Norwegian Foreign

Minister, Halvdan Koht, with the German ultimatum: accept instantly and without resistance the protection of the German Reich. Half an hour later, he had Koht's reply, 'We will not submit voluntarily: the struggle is already under way'. And while Brauer paced impatiently on the Oslo quayside for ships that did not come, the airborne attack on Oslo proved almost a fiasco as well. Two Special Duty Wings of *Luftwaffe* Ju-52 troop transports and their escorts ran into thick clouds on the way to Oslo, causing some inexperienced pilots to abort, and then stubborn resistance by Norwegian Gladiators out of Fornebu. Only by noon could the irritated Air Attaché, Captain Eberhard Spiller, report Fornebu in German paratroopers' hands (the first such troops ever deployed in battle), and by mid-afternoon the capital fell to the Germans—the last of the targeted Norwegian cities to do so [7-24].

The debacle in Oslofjord had, however, provided King Haakon, the government, and members of the Storting the opportunity to escape. The Cabinet had initiated a partial mobilization at 3:35 a.m. At 7:23 a.m. a special train carrying the King, Crown Prince and Princess, their children, and members of the Cabinet and Parliament pulled out of the Eastern Central Station, bound for Hamar seventy miles to the north. (Half an hour earlier, the first of 30 trucks had lumbered out of the garden court of the National Bank of Norway, headed for a bombproof vault in Lillehammer, 30 miles further north of Hamar on the main road to Trondheim. It carried the first load of the 50 tons of gold bullion, packed in 1500 crates and barrels, comprising the government gold reserves [7-25].) Before the day was over, the ancient city of Hamar was transformed into an international center overrun with Ministers, Charge d'Affaires, and other members of Oslo's Corps Diplomatique.

Once in Hamar, the Storting went into session, but when word was received that an advance German party was only ten miles away, it adjourned to Elverum, twenty miles to the northeast and closer to the Swedish border. There, the Storting dispersed, having handed the King and government plenary powers to conduct the war [7-26], quite ignoring Major Vidkung Quisling's attempt back in Oslo to form a new government and countermand mobilization. King Haakon and his retinue continued north, hotly pursued by a motorcade of German paratroopers under Captain Spiller. (Spiller was killed in the ensuing battle for Lillehammer.) The King would spend the next 60 days moving from place to place without capture, in so doing acting as the catalyst for the military and civil resistance that disrupted Hitler's plans for a swift subjugation of the country. He left northern Norway in June aboard a British cruiser, in order to set up a government in exile and continue the fight from abroad.

Among those on hand at the railway station in Hamar to greet the royal party on 9 April was Frank Foley, the agent to the British Legation in Oslo, and Margaret Reid, his secretary and cipher clerk

[7-27]. Equipped with a wireless transmitter, the two of them would accompany Major General Otto Ruge, the commander in charge of the Norwegian forces retreating up the Gudbrandsdal and Østerdal valleys toward Åndalsnes, attempting to link up with British forces disembarking there in mid-April. Ruge, promoted to General over the heads of his peers, became, with King Haakon, the pillar of Norwegian resistance. The poor subsequent showing of the British expedition would lead to the fall of the British government, with Winston Churchill replacing Neville Chamberlain as Prime Minister. Foley and Reid proved invaluable for General Ruge, however, and subsequently for the British Military Attaché Lieutenant-Colonel E J C King-Salter, in ciphering and dispatching wireless messages to the War Office in London. The messages went via Lillehammer to Vigra wireless station and thence to Wick at the tip of Scotland [7-28]. Foley ultimately embarked for Scapa Flow with the British when they evacuated southern Norway at the end of April.

Before sailing with the evacuation fleet, however, Foley would as well have something to say about the fate of Norsk Hydro's heavy-water plant as it was about to fall into German hands.

7.3. The battle for Rjukan; Tronstad goes into action

Norsk Hydro's Rjukan plants figured prominently in the German invasion of Norway, in April 1940, for two reasons. One was the industrial importance of the hydrogen-electrolysis plant itself at Vemork, and with it the heavy-water, high-concentration cells. The second reason was the strategic importance of the Vestfjord valley, cutting through the central mountain massif between east and west Norway.

In the spring of 1938 Norsk Hydro had, at its own expense, begun establishing an antiaircraft defense zone in the Vestfjord valley, manned by Hydro personnel [7-29]. It eventually included Bofors antiaircraft cannon and Gatling guns, a bomb-proof command post, and splinter-proof air-raid shelters. There were even antiaircraft 'curtains' suspended across the valley at several places, in the form of heavy cables with thinner vertical strands dangling from them to within 100 m of the floor of the valley.

Military concerns were one thing. Another concern in case of war was Norsk Hydro's position *vis-à-vis* its foreign partners. On the one hand, I G Farben, together with I G Chemie of Basel, held 25% of the joint stock, besides being party to various other agreements dating from the accord of 1927. On the other hand, during the intervening decade, Norsk Hydro had played a central role in a three-way international cartel between German and British nitrogen producers and Hydro—the so-called DEN-Group—so there were grounds for optimism about Norway maintaining a neutral stance should it come to war between the contractual partners [7-30].

And so it did. In the event, Norsk Hydro gave notice of its agreements with Germany, and it fell on Hydro as well to wind up the international cartel. The consequences were relatively minor, as it turned out. Nitrogen exports by Hydro did not amount to much in the ensuing period of 'phony war' in the winter 1939–1940; what exports took place were mainly to the US, Spain, and Egypt. As noted earlier, heavy water was even less in demand; about 40 kg was sold during 1934–1938. Only in 1940 was there a sudden, nearly simultaneous interest shown in the isotope by German and French parties, initially evidenced through furtive German feelers to Hydro, and dramatically concluded in Allier's coup that March. Then came 9 April. Of all places in southern Norway, land-locked remote Rjukan anticipated war in Norway more than most.

Only about half of those conscripted for active military duty at Rjukan, or about 110 men, were actually on hand on 9 April. However, on 13 April the 3rd Infantry, or Telemark Regiment, surrendered at Heistadmoen drill-ground near Kongsberg, half-way between Oslo and Rjukan. As a result, many defecting soldiers joined up with Major Lowzow in his retreat to Rjukan, where by the 16th he disposed of a full infantry battalion. No sooner had he dug in, than orders came from Central Army Headquarters for all remaining active forces to withdraw to the north. Just then came counter-orders from the 4th Command District Headquarters (Bergen), instigated by Jacques Alliers in Paris, to the following effect:

> Rjukan must be defended by all means. The intervention of Allied forces demands that Norsk Hydro's plants remain in Norwegian hands. The commander in chief will ensure, with the aid of Telemark troops and resources of the plant, that the entrance to Rjukan is barred. [7-31]

Shortly afterwards, on 21 April, the French Military Attaché to Norway, a 'fierce-looking type with a patch over one eye' [7-32] by the name of Bertrand-Vigne, took Frank Foley to task at General Ruge's Headquarters in the Gudbrandsdal. He argued strenuously that the British had an obsession about preserving the iron-ore supply route through Narvik, quite overlooking resources far more important and closer at hand: the heavy-water production facilities at Rjukan. Bertrand-Vigne underscored his point by citing the recent French heavy-water coup, something Foley was more than aware of though heavy water was hardly his most pressing concern. On the contrary, the Frenchman insisted, the British should make the defense of Rjukan a top priority. Foley countered that he saw no way for the beleaguered British expeditionary troops to come to the aid of the Rjukan installation situated, as it was, in a most forbidding region of Telemark [7-33].

Amidst the conflicting orders, the 220 or so defenders of Rjukan held off initial German attacks, though by the end of April German forces numbered well over 1000, backed by aircraft and artillery. On the night of 2 May, in reply to a request for direction from Major Lowzow, came the following answer from HQ, 6th Command District in Troms county, far to the north: 'Can give no direction. Act as you best see fit!' The following day, having appraised local officials of his intentions, Major Lowzow dissolved his meager forces in the dark of night. The enlisted men doffed their uniforms and slipped quietly away to their home locations, while the officers disappeared northward over the inhospitable Hardangervidda plateau ('Vidda' in Norwegian for short). With the Allied expeditionary forces already evacuated from Åndalsnes, Rjukan thus became the last town to surrender in southern Norway, at the cost of one killed, six wounded, and one missing—presumed drowned in icy Lake Tinnsjø. German troops marched in on 4 May, and negotiations with Norsk Hydro management began at once, though production at Rjukan was all but stopped for the time being [7-34].

Clearly the Germans could have pushed harder in the Rjukan engagement. That they did not was no doubt due to their interest in preserving the installations with a minimum of damage—something the Norwegian commanders fully appreciated as well [7-35]. More surprising is the tardiness displayed by the German authorities, in view of their newfound interest in heavy water, in calling on Hydro's offices in Oslo. Only in June did they make an official appearance, and even then to arrange for an inspection of Hydro's facilities at Herøya, not at Rjukan or Notodden—the only plants relevant to heavy water. (Herøya would play a role quite as important as Rjukan and Vemork for the German industrial armaments initiatives in Norway, namely for the production of aluminum and magnesium. That chapter of wartime Hydro, a dramatic one at that which we touch upon later, is, however, another story.) Not until the end of 1940 did Germany signal continued, serious interest in SH-200, their code name for heavy water. The occasion was a visit to Rjukan by Karl Wirtz, a German authority on deuterium and an influential members of the Uranium Club, who then met Jomar Brun for the first time [7-36].

What about Leif Tronstad in all of this? His personal war began in Trondheim on that grey Tuesday morning, 9 April, with the arrival of the heavy cruiser *Hipper* in Trondheim harbor. As may be recalled, Tronstad was a Second Lieutenant in the Norwegian Army Reserves. The Army comprised a small regular cadre and a national army called out on partial or full mobilization. Men were called up for recruit training, and subsequently for periodic service. Amidst all the confusion on 9 April, full mobilization at Trondheim was sketchy at best, as most everywhere else. Admonishing his students at NTH to report for mobilization wherever practical, Tronstad himself took off for Oslo by car [7-37]. *En route* he learned that Oslo had fallen, turned around, and headed back north. Being intimately familiar with the mountainous

Dovre terrain, he set about improvising as best he could a defensive position at Hjerkinn—a sensitive and undefended spot at the summit of Dovrefjell in the northern confluence of the Gudbrandsdal and Østerdal valleys—assisted by the local voluntary rifle-corps. Among other things, his little band plowed up furrows in the snow blanketing the ice-covered lakes dotting the wintry landscape, spoiling landing opportunities for ski-equipped Junkers 52 transports. Though the Dovre-Dombås region proved relatively quiet, these efforts were nevertheless a factor in ensuring the successful evacuation of Allied troops in the closing days of the battle for southern Norway.

As can be imagined, Tronstad was still pretty worked up when he rejoined colleagues and students at NTH. He was still spoiling for a fight. His next chance came in September with the return to Trondheim from England of former student Erik Welle-Strand, who had fled Norway after partaking in heavy fighting at Voss in the mountain approaches to Bergen. Welle-Strand was put ashore by submarine, returning ostensibly to continue his studies in rock-engineering, but also to set up a clandestine radio station. The operation, conducted in the Bymarka woods outside town, was soon in full swing, with the assistance of fellow student and ex-Sergeant Bjørn Rørholt and code-named Skylark [7-38]. Before long Tronstad was in regular contact with the British War Office.

At first, keeping track of German warships coming and going in Trondheimsfjord took priority. Before long Tronstad became, in addition, the principal source of Allied information about German industrial initiatives in occupied Norway in general, and heavy water in particular. Naturally, he was in constant touch with Jomar Brun at Vemork. As time went by, however, Tronstad became increasingly involved in organizing an underground resistance network in the Trondheim area, as well as what became known as Milorg—the national military resistance organization. In the fall of 1941 the Skylark code was broken by the Germans, and he had to leave Norway on short notice [7-39]. His dramatic flight to England is chronicled in volume 1 of his personal wartime journal, preserved in the State Archives (*Riksarkivet*), Oslo [7-40]. Leaving word at NTH that he was off on a routine visit to Rjukan, he had scarcely time to say good-bye to 'Bassa' (his wife Edla) and their two children before heading for Oslo and thence slipping across to Sweden. He arrived in London about a month later.

Awaiting Tronstad at King's Cross Station in London was none other than Sergeant Rørholt, who himself had escaped across the North Sea shortly before. Rørholt escorted his colleague to the Admiralty Annex near SIS Headquarters, where Tronstad was introduced to spymaster Eric Welsh. Tronstad's many wireless reports—the 'Trondheim cables'—had routinely landed on Welsh's desk. On one occasion he was rumored to have exclaimed, 'What in heaven's name is this stuff called heavy water?'

[7-41]. In fact, Eric Welsh was more than aware of the importance of heavy water in the secret war unfolding in Europe.

Lieutenant Commander Eric Welsh had reputedly a background in chemical engineering. In fact, his background is as elusive as the clandestine operations he oversaw during World War II. It suffices to note that before the war he held a leading position with the Bergen firm Internasjonal Farvefabrikk A S and became an expert in chemical-resistant tiles and paints. In that connection, he visited Jomar Brun at Vemork on several occasions, advising on certain corrosion problems associated with the heavy-water production process. He became a close friend of Brun, and well informed about heavy water.

Chapter 8

THE BRITISH INITIATIVE

8.1. MAUD

Among the physicists awaiting Bohr's return from Oslo at his Institute for Theoretical Physics on that fateful morning of 9 April was none other than Lise Meitner. She had arrived from Sweden the very afternoon before to use the cyclotron at the Institute, since Siegbahn's cyclotron was not yet operational.

The routine at Blegdamsvej 17 was at first not seriously disturbed by the German authorities, who may or may not have been aware of the prominent Jewish scientist's temporary presence at the Institute. Experiments continued by and large as before, except for rationing of heat and electricity, and the end of American journals and most visitors from abroad. If anything, despite the strained circumstances, the war proved 'a relatively good period from the point of view of work' at Blegdamsvejen, at least during the first year or so [8-1]. Characteristically, Bohr's first concern was protecting those on the staff whom the Germans might have reason for persecuting. As for Meitner, it took him until the end of April before he was able to slip her back across Øresund. Before she left, he asked her to send a telegram from Stockholm to Owen Richardson at King's College, University of London, assuring him that all was well with the Bohrs. The somewhat cryptic telegram caused a stir among members of a committee of British physicists hastily organized just then:

> MET NIELS AND MARGRETHE RECENTLY BOTH WELL BUT UNHAPPY ABOUT EVENTS PLEASE INFORM COCKCROFT AND MAUD RAY KENT MEITNER. [8-2]

The committee in question first met at the Royal Society quarters in London's Burlington House off Piccadilly, on 10 April, the day after the invasion of Norway and Denmark. Formed as a sub-committee of the Committee for the Scientific Survey of Air Warfare, it was initially known simply as the Thomson Committee after G P Thomson. In view of its sensitive mandate, in-depth review of atomic energy from uranium fission for military purposes, a less revealing name was urgently called for, and was conveniently provided by Meitner's curious telegram, which Richardson had passed along to Cockcroft on 16 May. The reference to Cockcroft made sense in itself, but in conjunction with the inexplicable 'Maud Ray Kent' the whole message took on incongruous overtones. Just who or what was Maud Ray Kent? The key seemed to be the word 'Ray'. Obviously it was some kind of coded message, possibly garbled, relating to radiation or uranium research. One interpretation had it signifying 'radium taken', another as 'U and D may react'. (No one seems to have thought of contacting Meitner herself.) Only in 1943 was it learned that the telegram was perfectly innocuous, since Miss Maud Ray of Kent had once been a governess of the Bohr children. At any rate, Thomson's committee appropriated the name Maud, and was henceforth known as the MAUD Committee, which acronym has itself been the subject of debate, signifying perhaps 'Military Applications of Uranium Detonation' or something to that effect [8-3].

The charter members of the MAUD Committee were G P Thomson, its Chairman, Cockcroft, Chadwick, Philip Moon, and Marcus ('Mark') Oliphant, an old disciple of Rutherford and then chairman of the physics department at Birmingham. The German invasion of Norway was much on their minds as they first met in the afternoon of 10 April. By chance, Jacques Allier showed up in Cockcroft's office in the Ministry of Supply that morning. Cockcroft had a talk with Allier and brought him along to the meeting later in the day. Raoul Dautry had dispatched Allier on another mission, this time to London, armed with a letter of introduction from Paul Montel, director of the Franco-British mission in the French Ministère de l'Armement to Dr H J Gough, then Director of Scientific Research at the War Office, to further collaboration between Britain and France in uranium research. In this capacity Allier was authorized to share with the British French progress in nuclear physics, as well as a list drawn up by Joliot and colleagues of leading German scientists to watch. He was also there to report on the heavy-water situation in Norway—all the more ominous in view of the latest, shocking news from Norway that greeted him upon his arrival in London.

From Allier's perspective, the discussion with the uranium subcommittee was only moderately successful. Despite the French heavy-water coup in Oslo the month before, Joliot and his team were anxious to locate more sources of deuterium; alas, his British hosts had already searched at home and abroad with scarcely one kilogram to show for their effort. Cockcroft did agree that an exchange of information would be useful, particularly with the participation of Joliot himself, who until then had been less than forthcoming about chain-reaction results at the Collège de France. However, the British scientists also professed some skepticism about chances of success for a controlled chain reaction, with or without heavy water. Sir Henry Tizard, for one, was definitely skeptical, declaring that 'uranium disintegration is not in the least likely to be of military importance in this war' [8-4].

About all that British chain-reaction experiments by Thomson and Moon had shown was that uranium oxide, with water or paraffin as a moderator, was not very promising, as the French had learned as well. What triggered the MAUD Committee was a theoretical flicker of hope in the form of a memorandum by Otto Frisch and Rudolf Peierls, reporting calculations on the critical mass of U^{235}. The memorandum signalled a new, long-shot approach to a bomb based on fast neutrons in separated uranium, in contradistinction to the French approach predicated on thermal neutrons in natural uranium and heavy water—in effect a power-producing engine or 'boiler'.

Frisch, for all his worries, had left Denmark without any difficulty in the summer of 1939. Mark Oliphant, chairman of physics at the University of Birmingham, had visited Denmark that spring. Frisch had shared with him his concerns for the immediate future, and Oliphant brought him over to Birmingham for the summer term, with a view to seeing what could be done on a more permanent basis. At first there was not much for Frisch to do but laze in the sun, occasionally assist Oliphant with his lectures, and get to know Peierls, who had been on the Birmingham faculty since 1937. Peierls and his Russian-born wife Genia invited him to stay with them and made him feel at home. While marking time, he cast about for some research problem to tackle. Nuclear physics, his expertise, was rife with exciting problems. Unfortunately, Birmingham had no cyclotron, and Oliphant was concentrating departmental resources on radiolocation, or radar as it became known—a subject too sensitive for foreigners to touch [8-5].

Not to be put off, Frisch came up with a problem that seemed feasible even with the strained resources available at Birmingham, yet was of fundamental importance to the uranium fission problem: confirming Bohr's conclusion that uranium fission by slow neutrons is solely due to the rare isotope U^{235}. To be sure, doing so required preparing a relatively pure sample of U^{235} by isotope separation—not something undertaken lightly. However, on going through the literature, Frisch had come across a method surprisingly simple, devised in 1938 by the Munich physical chemist Klaus Clusius and his younger colleague Gerhard Dickel [8-6]. Their 'Trennrohr' (separation tube) was, in its simplest form, just that: a vertical tube with two concentric surfaces which are kept at different temperatures, filled with the gas mixture to be separated. As a result of thermal diffusion, the lighter molecules tend to get more concentrated near the hot surface, where they are carried upward by the convection current. Exchange with the current moving downwards along the cold surface produces a fractionating effect, and a state of equilibrium is reached when the gas near the upper end contains more of the light molecules than the gas near the lower end.

Though simplicity itself, and Oliphant saw to it that Frisch was able to avail himself of a laboratory bench in a small unused lecture room, the glass blower had his hands full blowing tubes for the radar boys—prototypes for the cavity magnetron that became the most closely guarded secret of British military technology. And so Frisch turned to the theory underlying Clusius's principle instead. If anything, the mathematics of thermo-diffusion proved more challenging, but he followed it well enough to learn that for the only gaseous component of uranium, uranium hexafluoride, the effect happens to be practically nil.

About the only other person at Birmingham, or in Britain for that matter, seriously interested in uranium-235 and chain reactions was Rudolf Peierls, a Berliner who had arrived in Cambridge on a Rockefeller Fellowship as Hitler came to power With the onset of the Nazi purge of German universities, he elected to remain in England, and ended up at Birmingham [8-7]. Earlier in the summer of 1939, Peierls had gone over and sharpened Francis Perrin's formula for calculating the critical mass of uranium; he concluded, like Perrin, that some tons of natural uranium were needed to sustain a chain reaction—far too much for a practical bomb. In February or March of 1940, Frisch, whom Peierls had consulted about his calculations, resurrected Peierls's formula on a hunch. Since the prospects for separating uranium-235 anytime soon appeared slim, why not calculate its critical mass instead with the formula conveniently at hand? Unfortunately, a critical piece of datum was missing: the cross section for fission in pure U^{235} with medium-energy neutrons. With nothing better to go on, Frisch made the plausible assumption that the cross section was about equal to the geometric cross section of the U^{235} nucleus (roughly 10^{-23} cm^2), or that every neutron striking a U^{235} nucleus causes fission. The order-of-magnitude resultant critical mass, though obviously an over-estimate, was shocking in its paucity: about one or two pounds! Conceivably 100 000 Clusius–Dickel tubes could do the job in weeks, even with the feebleness of the effect for uranium hexafluoride. Not only that; a 'back-of-the-envelope' calculation by Peierls, with whom Frisch shared his result at once, produced another shock. The fraction of uranium-235 fissioning before the developing pressure blew the uranium lump apart corresponded to 80 generations of neutrons born. Each generation, multiplied by 200 MeV per fission, corresponded to a total energy release equivalent to kilotons of TNT. 'At that point', recalls Frisch, 'we stared at each other and realized that an atomic bomb might after all be possible' [8-8].

Frisch and Peierls wrote a two-part memorandum on their conclusions, one entitled 'On the construction of a "super-bomb"; based on a nuclear chain reaction in uranium', and one less technical, 'Memorandum on the properties of a radioactive "super-bomb"' [8-9].

Not entrusting the document to a secretary, they typed it themselves, making but one carbon copy. Only a few pages long, its typing could not have taken very long, though they had a momentary fright while mulling over the exact wording. A head appeared in the window, 'seemingly out of nowhere' [8-10]. The interloper turned out to be a laboratory technician tending tomatoes underneath the window, which faced south. The first part, the longest, dwelt on the bomb mechanism: on how to separate the uranium, and on radioactive effects of the explosion. Two small hemispheres had to be brought together rapidly by mechanical springs to preclude predetonation, which would have destroyed the bomb but not much else. The second, even briefer part, explored the strategic issues associated with the weapon. Since the weapon would probably kill large numbers of civilians, it was impractical as a British offensive weapon, but worth developing simply as a deterrent [8-11].

What to do with the document? In the end Frisch and Peierls delivered it to Oliphant, who passed it up to Tizard, along with a covering letter. Tizard, in turn, passed it to Thomson, the chairman of the committee which, by that time, was on the point of disbanding after reviewing some desultory experiments with uranium and graphite. (Owing to neutron capture, they gave no immediate prospect of success.) The memorandum brought the committee a new lease of life, but also a problem. Peierls was only recently naturalized and still classified as an enemy alien. Frisch had collaborated with his aunt, and through her had strong ties to Hahn in Nazi Germany. Cockcroft for one had, as may be recalled, reservations about both Frisch and Meitner. The upshot was that at first the authors were kept in the dark about the fate of their manuscript, and about the committee itself.

> This ruling seemed ridiculous to us [notes Peierls]. I wrote a letter to the chairman, whose identity I did not know as yet, pointing out that Frisch and I had thought a great deal about the problem already and might well know the answers to important questions. Others, who had missed the point made in our memorandum, could perhaps miss other important points as well. The problem seemed urgent to us, at least until it had been shown that building a weapon was not possible. [8-12]

The letter brought results. Though Frisch and Peierls could not partake in the committee deliberations *per se*, they could be fully consulted.

The same problem confronted Franz (Francis) Eugen Simon of the Clarendon Laboratory at Oxford. With separation of U^{235} shaping up as a major challenge, yet the Clusius–Dickel process seemingly ruled out, gaseous diffusion through porous membranes was the only viable alternative. Simon, a physical chemist and thermodynamicist of the first rank, would clearly have to be brought into the picture. He too, however, was a refugee from Germany only naturalized in 1939 [8-13].

On the other hand, he had the ear of Lord Cherwell, née Lindemann, Churchill's intimate adviser. In due course, Simon not only became a member of the MAUD Policy Committee, but would generate a report on uranium separation nearly as crucial to the uranium effort as the Frisch–Peierls memorandum. Trusting the mails even less than secretaries, he duplicated no fewer than 40 copies of his report, and hand-delivered the original to Thomson during the height of the London blitz [8-14].

By the summer of 1940, experimental work under the aegis of MAUD was picking up speed in several British quarters. One was Liverpool under Chadwick, where Frisch had recently been posted from Birmingham. Back in Birmingham, besides Peierls, chemical work on the production of gaseous uranium compounds and metallic uranium was pursued in the chemistry group under William Haworth. At Oxford, the brunt of the investigation of isotope separation fell on Simon and his cryogenic team largely transplanted from Breslau. At Cambridge, the leaders were Egon Bretscher and Norman Feather, later joined by Eric Rideal, Tronstad's old mentor. Finally, the Bristol group under Alan Nunn May and Cecil Powell had been enlisted to help in the physics work, augmented by scientists from King's College, London.

Two experienced physicists suddenly appeared at Cambridge's Cavendish Laboratory late in June. In contrast to Leif Tronstad's friendly, albeit hushed-up greeting at London's King's Cross Station somewhat later, Joliot's colleagues Halban and Kowarski stepped off the train at Paddington Station one morning in June of 1940, to a decidedly lukewarm reception. Remarked a senior War Office official on hand to receive the two, to the Earl of Suffolk who was accompanying them, 'Oh, is that the lot?' The incident at Paddington marked, in fact, an auspicious occasion in the affairs of the MAUD Committee and a milestone of sorts in the British program generally. The two travelers brought with them 185 kg of heavy water. The world's sole supply of the precious liquid was seemingly at last in safe hands, having been spirited away for a second time past hoodwinked *Abwehr* agents.

8.2. Broompark

Late in the evening of 9 May 1940, two men met in the blacked-out Bendlerstrasse in Berlin, huddled on the doorsteps to the headquarters of *Oberkommando der Wehrmacht*. One was Colonel Jacob Sas, the Dutch military attaché to Berlin. The other was Hans Oster of the *Abwehr*, of all things a staunch anti-Nazi. 'The invasion is to begin', said Oster quietly. *Fall Gelb*, the German offensive against the Low Countries and France was set to begin at 5 a.m. the next morning [8-15].

Precisely on schedule, the vanguard of an army of two million armored, airborne, and motorized infantry troops poured across the borders of Holland, Belgium, and Luxembourg at daybreak.

Simultaneously, hundreds of *Luftwaffe* Stukas and low-level bombers swept in over Allied air bases and communication centers in northern France. The military armada deployed for the strike on the West extended from the Hook of Holland to the Swiss frontier. Half of its might, Army Group B, some 30 *Wehrmacht* divisions marshalled along the front from the North Sea to Aix-la-Chapelle, was poised to overrun Holland and Belgium, then swing south into France. Army Group A, 44 divisions strong, would constitute the main thrust into France through the Ardennes, with supporting units stretching from Aix-la-Chapelle to the Moselle. With the Channel in sight, Army Groups A and B planned to pivot north and south, respectively, entrapping nearly one million Allied troops on the Franco-Belgian border.

Long before dawn on the 10th, 150 miles of front facing Holland and Belgium were aflame. In a single day, all the outer lines of the Dutch defenses were mastered by the Armies under General von Bock. Fort Eben Emael, the key Belgian defense position, fell to gliders swooping directly onto the roofs of the fortress. Four days later, on the evening of 14 May, the Commander-in-Chief of the Dutch forces, still holding the Hague, ordered a general cease-fire.

Though Allied units in northern France had begun advancing with King Léopold's army along Belgium's Dyle River, they were, in effect, walking into a trap. Unknown to the Allies, the German attack through the Low Countries was only a feint. General Karl Rudolf Gerd von Rundstedt's Army Group A was set to deliver a deathblow through the Forest of Ardennes in southern Belgium. The following day, the first German assault boats crossed the river Meuse into France between Namul and Mezières. At two-thirty next morning Premier Paul Reynaud was awakened with the news that the Germans were advancing rapidly and that the government must prepare to evacuate Paris.

The same day, 16 May, Minister Dautry spoke to Joliot on the phone. The heavy water lodged in the cellars of the Collège de France must be moved immediately to another place of safety. In the event, the German army granted Paris a reprieve, driving west toward the Channel coast instead, reaching the seaside village of Noyelles on 20 May.

Paul Reynaud had been named Premier of France on 21 March, after the ignominious fall of the Daladier government for its failure to act on the eve of war, and a victim of the party warfare surrounding the Nazi–Soviet Pact and the winter war in Finland. On 3 April Allier paid Reynaud a call, describing Dautry's and Joliot's nuclear energy initiatives, and the new Premier promised state support as needed for the enterprise [8-16]. Indeed, there was, in the aftermath of the audacious heavy-water affair, considerable enthusiasm for the French chain-reaction effort among high-level government officials from Reynaud to Dautry and Laugier and their subordinates, though confusion reigned as to its primary goal. Dautry, in his recollections, singled out a nuclear bomb as the ultimate mission, whereas the scientists themselves felt that a power-producing plant took precedence [8-17].

At any rate, Joliot's team now had heavy water in hand. To be sure, not enough for a full-blown reactor, but enough for preliminary experiments on homogeneous and heterogeneous mixtures of deuterium and uranium oxide. In the event, both experiments ran into problems. For a heterogeneous experiment, they proposed pouring the heavy water into aluminum cans dispersed throughout a uranium-oxide lattice. Alas, aluminum sheet for producing the cans was less easily obtained in France than in Norway at war, despite Joliot's cajoling Minister Dautry about the urgency of the matter. As for a homogeneous experiment with a spherical configuration of uranium oxide suspended in deuterium, preliminary tests showed that the uranium oxide powder would slowly settle to the bottom of the sphere, unless the sphere was rotated to keep the powder uniformly distributed. Besides, here again aluminum was needed—for the spherical containment vessel.

Despite these frustrations, Halban, Joliot, and Kowarski managed to keep a scientific program going at Ivry, measuring, among other things, neutron-diffusion rates through paraffin and delayed neutrons after fission. The latter measurements brought them, unbeknownst, tantalizingly close to an *experimentum crucis* of bomb production. Noticing that some radioactivity persisted in a lump of uranium long after removal of the central neutron source, the idea occurred to them that a new element might be formed through resonance (non-fission) capture of neutrons in uranium, and that such an element might itself be fissionable. Alas, preoccupied with reactor experiments as they were, they did not pursue the matter. Nor did they conceive of the possibility that such material might constitute an alternative route to a bomb [8-18]. In any case, this research was rudely interrupted on the morning of 16 May.

No sooner had Dautry called Joliot that fair morning in May, than Joliot got on the phone with Henri Moureu, subdirector of the chemistry section of the Collège de France laboratory and a trusted friend. 'Henri, come immediately', he instructed. Moureu lost no time, and the two sat down for a talk. The heavy water must not fall into German hands; it must be evacuated to some safe place in the center of France. 'I give you the job', said Joliot simply. 'Absolute secrecy. You have a free hand' [8-19]. Moureu checked around, and found that the branch of the Banque de France at Clermont-Ferrand, 200 miles to the south, was willing to store the mysterious containers in its vaults. With crowds already beginning to flee Paris for the south, time was of the essence. Moureu had the 26 aluminum cans loaded onto a laboratory truck the same evening, and he himself left for the south near midnight.

On Dautry's suggestion, Moureu was joined by Halban as he left Paris. Their mission papers noted simply that Moureu was authorized to carry arms. They pulled up at Clermont-Ferrand at 5 a.m. on 17 May, and Product Z, the code for their secret cargo, was manhandled into the musky bank vaults. Moureu returned to Paris forthwith. Halban continued to nearby Monts Dore, a sleepy little town except for the fact that members of the French cabinet were installing themselves there just then. Halban's accent and general demeanor attracted the suspicion of the major of the town, a former general, and the physicist narrowly avoided arrest [8-20]. Cleared after some fast explaining, he set about doing what he had journeyed south for in the first place, organizing an improvised laboratory in a villa on Rue Étienne Dollet back in Clermont-Ferrand. Optimistically believing that a victorious military counter-offensive would halt the advancing German forces to the north, as had happened under Marshal Pétain in 1914, it was taken for granted that Joliot's team could continue its research in this quiet retreat. The University of Strasbourg had similar plans to relocate to Clermont-Ferrand, which would actually become an important center of the French Resistance.

In fact, the military situation was worsening by the hour, and the local Banque National manager grew worried, wanting Product Z out of his cellars. Again Moureu stepped in, and then the Minister himself. The stock was moved, first to the women's prison in Monts Dore, and then to the Central Prison in Riom. There the cans were secreted in an isolated cell reserved for dangerous criminals. The Riom prison would later acquire fame as the place of internment of Daladier and Reynaud when they were brought before the Supreme Court of Justice at Riom near the end of 1941, accused by the Vichy Government of having inadequately prepared the country for war [8-21].

In Paris, too, time was running out. On 10 June, the French Government fled the city amidst the flood of refugees clogging the highways leading south. At the Collège de France, Joliot and Moureu, virtually the only ones left, were preparing to leave, Paris having been declared an open city. On 11 June, the last load of scientific equipment—spectroscopes, ionization chambers, galvanometers, even lead bricks—left by truck for Clermont-Ferrand. Technical documents not needed for further research, but needing to be kept out of enemy hands, were burned. Unfortunately, to no avail; detailed progress reports submitted periodically to the Ministry of Armament offices were found by the Germans only days later in a railway wagon at La Charité-sur-Loire.

On 12 June German forces were across the Marne at Chateau-Thièrry, 50 miles away, with the skies darkened by smoke rising from blazing petrol stocks along the Seine all the way to Rouen. Joliot and Irène left for Clermont-Ferrand in their Peugeot 402, bringing along some gold and platinum, and Marie Curie's gram of radium from the women of America. Behind them followed Moureu in his own Peugeot. Two days later the Germans goose-stepped down the Champs Élysées as swastika flags were hoisted atop the Arc de Triomphe and Eiffel Tower.

At Clermont-Ferrand, the Joliot-Curies joined Halban and Kowarski, the latter having arrived on about 6 June [8-22]. No sooner had they pulled in, than Irène, quite exhausted, lay down on the floor and went to sleep [8-23]. The makeshift laboratory in villa Clair Logis was up and running; at least there was running water, and some equipment had been put in working order by army engineers. The day after their reunion, 16 June, was a Sunday, brilliantly clear and deserving of a small lunch marking resumption of researches. Just as the team sat down, a Simca drove up. Out stepped Allier, who drew Joliot aside. After a whispered conversation, Joliot called Moureu outside. The Allied armies were in full retreat, he explained. French collapse was imminent, and the Government had ordered the heavy water to Bordeaux and thence to England.

Later that evening Joliot went over the situation with Halban and Kowarski. Allier's orders from Dautry called for all three to accompany the heavy water to England, and Joliot concurred that his two colleagues should do so, placing themselves at the service of the Allies. Though it is believed that he had not reached a final decision as to what he himself should do, in all likelihood he never doubted his own role in events ahead, as custodian of science and organizer of resistance in occupied France [8-24]. In the course of his conversation with his two teammates that evening, he forecast European military and geo-political developments with remarkable foresight: breakup of the Nazi–Soviet Pact; Hitler's attack on the Soviet Union, and with it his ultimate undoing; speedy armistice in France, followed by a long period of occupation and rising French resistance under new leadership.

Events in Paris happened faster than Joliot could have predicted. The very same evening, Paul Reynaud and his cabinet resigned, and the 84-year-old Marshal Pétain, his successor, asked Hitler for an armistice. The next day, 17 June, another, defiant French voice was broadcast from London: that of General de Gaulle, stressing that the war was not decided by the battle of France, yet imploring his countrymen to continue the fight. As for Raoul Dautry, fleeing with the remnants of the Third Republic to Bordeaux, he would retreat to Provence were he holed up for the duration of the war. But for the moment his office held sway, as far as Joliot was concerned.

Early on 17 June, Allier and Halban showed up at the prison in Riom. The governor balked at handing over the heavy water, but acquiesced when Allier drew his revolver. Prisoners serving life-sentences carried the containers out of the cell to the waiting vehicle. Halban and Kowarski then speeded westward towards Bordeaux with the load, along with wives, children, and baggage, fortified with a multi-stamped pass to forestall repetition of Halban's troubles in Monts Dore. All the while they crossed countless roads radiating north–south from Paris, clogged with evacuees from the capital. They arrived at midnight, greeted by

Jean Bichelonne, Dautry's chief of staff at the Ministry of Armament's temporary office in a private school. Unlike that of his boss, Bichelonne's wartime career would end on an ignominious note [8-25]. But for now he tore a sheet out of a schoolboy's copybook lying around and scribbled an order for the party to board the *Broompark*, a Scottish coal-carrying steamer tied up down the estuary from the Bordeaux docks. Finding the collier with some difficulty, they were welcomed at the gangplank by an unlikely character whom, according to the orders handed them by Bichelonne, was to escort them and their mixed baggage across the Channel to England.

Moustached, shirt-sleeved with arms covered with tattoos, revolver in shoulder-holster and swinging a loaded hunting crop, their guide was the legendary 33-year-old twentieth Earl of Suffolk and Berkshire. The Earl had commandeered the *Broompark* for the intended mission. He was the British science attaché in Paris, determined largely on his own volition to bring out of France whatever could be salvaged: machine tools, industrial diamonds from Belgium, scientists, and engineers. He was fully aware of the importance of the 26 cans of heavy water trundled aboard, as well he should be: in the background of the ongoing exodus was Frank Foley of the SIS, also in town, as usual providing quiet guidance at the eleventh hour evacuation [8-26]. The Earl had the cans lashed to a raft in the hope of saving them should the ship be torpedoed—the odds of which he privately judged to be about even.

The next day, 18 June, Joliot and Moureu pulled into Bordeaux late in the morning. Irène and the children had been left behind in Clairvivre. Irène, ill with tuberculosis, waited to be admitted to a sanatorium there, while Hélène and Pierre were going on to l'Arcouest. All three would remain in France until Frédéric went underground late in the war [8-27]. At the Prefecture building, meanwhile, Joliot and Moureu ran into the Earl of Suffolk, who seized Joliot and pressed him to accompany the boat-party to England. Suffolk would personally look after Irène and the children. Joliot said little, and slipped out in search for Halban and Kowarski, whom Suffolk ensured him were safely aboard the *Broompark* with the heavy water. Joliot wanted to have a last meeting with his colleagues before returning to Clermont-Ferrand. But the ship was nowhere to be found; due to an air raid earlier in the day, it had been moved to a safer berth south on the River Gironde.

What if Joliot had found the vessel and gone aboard? In that eventuality, the speculation goes, it seems likely that he would have been detained by the captain and brought along to England with Halban and Kowarski. Due to Joliot's international reputation, he could well have ended up in the United States, with French scientists playing a relatively greater role than actually was the case in the wartime atomic bomb project [8-28].

Instead, Joliot sat down and talked the matter over with Moureu, his close friend and confidant. What now? In point of fact, there was little question of what to do. Halban and Kowarski were seemingly off on their mission, as agreed. France and French science needed him more than ever. Above all, there was Irène and the children. 'Definitely, I must stay', he declared. Joliot and Moureu returned to Irène at Clairvivre, and saw to it that she was admitted to the sanatorium. Moureu returned to Paris, and Joliot to Clermont-Ferrand. Frank Foley left Bordeaux just as German advance motorized units entered town, and headed south for the Spanish Pyrenees.

On board the *Broompark*, Halban and Kowarski had asked Suffolk for permission to go ashore for a last talk with Joliot. He told them it was up to the captain, who was sure to refuse, as they were preparing to sail. The crew, whom Suffolk had plied with champagne while loading went on, was apparently in good enough form on the afternoon of the 18th to weigh anchor. The ship headed for the Bay of Royan, where the same day a neighboring ship had been sunk by a magnetic mine. (Joliot, interrogated later in Paris, convinced the authorities that it was the *Broompark* that went down.) Twenty-four hours later, *Broompark* finally sailed for Southampton, and on 21 June she anchored in Falmouth Roads after an uneventful crossing.

The French party disembarked and were put on a train for London, with their precious cargo and still accompanied by Suffolk brandishing his revolver. In London, after a heated exchange with some civil servants, they were installed at the Mayfair on Halban's insistence. The 26 cans, unknown to them, were also off-loaded and brought to Wormwood Scrubs Prison. From there they went to, of all places, Windsor Castle in care of Owen Morshead, the librarian.

Though under lock and key once more, the last had not been heard of the French–Norwegian heavy water. Before the year was over, it would be put to use in an experiment at Cambridge that would vex the budding Allied uranium effort, just as an experiment in Heidelberg would inject a note of uncertainty in a corresponding effort barely under way in Nazi Germany.

Chapter 9

GERMAN ARMY ORDNANCE TAKES CHARGE

9.1. The Uranium Club; a tritium episode

At the end of April 1939, just as Joliot and his collaborators were shelving plans for a uranium bomb in favor of a power-producing 'engine', German physicists reached a contrary decision. On 29 April, the Reich Ministry of Culture (*Kulturministerium*) called a confidential meeting of nuclear physicists at the Ministry's headquarters on Unter den Linden in Berlin. It was presided over by Professor Abraham Esau, a political activist recently appointed President of the Physikalisch-Technische Reichsanstalt (PTR), as a reward for his pro-Nazi stance. Wilhelm Dames, Education Minister Bernhard Rust's scientific expert, held forth on the potentials of a nuclear bomb, and Esau chimed in on the importance of banning uranium exports. One piece of business accomplished at the meeting was coining of the term *Uranverein*, or 'Uranium Club', for those dragooned by Esau into showing up [9-1].

Quite like participants in the nuclear chain-reaction studies initiated under the aegis of MAUD one year later, members of the *Uranverein* would remain scattered among half a dozen laboratories, including Berlin, Hamburg, Heidelberg, and Leipzig. The German division was one of principle, however, forced by differences over technical approaches, ultimate goals, and ideological doctrines.

Among those in attendance at the closed-door meeting in Berlin was Josef Mattauch, the Austrian physicist whom Otto Hahn had hired as Lise Meitner's successor at Berlin-Dahlem. He was sitting in for his boss who, though first on Esau's list, had conveniently arranged to be in Sweden on a lecture engagement just then. The next day Mattauch shared his impressions of the meeting with Paul Rosbaud, who passed the disturbing news to Robert Hutton, a metallurgist from Cambridge, 'at a safe spot on a seat in the Mall' during one of his last visits to London [9-2]. Hutton, in turn, forwarded the information to Cockcroft,

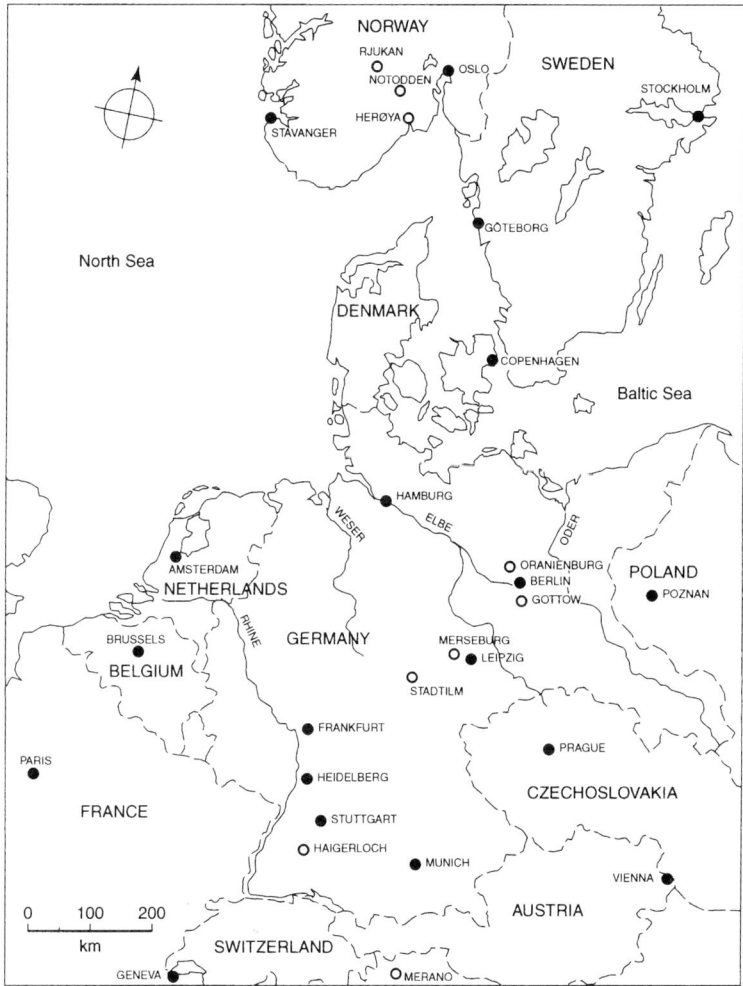

Figure 9.1. *Main research centers in the German uranium effort. Open circles indicate laboratories and industrial centers.*

ensuring that he was kept abreast of scientific developments in Germany, in keeping with Rosbaud's promise to Cockcroft the month before.

The week before Esau's meeting on Unter den Linden, a personal dialogue had taken place in the office of Paul Harteck at the University of Hamburg—a dialogue which would affect the fortunes of Esau before the year was over. Harteck had succeeded Otto Stern when the latter left for the United States as Hitler came to power. His assistant Wilhelm Groth, and a few others, called on him to talk him into proposing some research project that might prevent them from being drafted into a war that seemed imminent [9-3]. The upshot was a letter from Harteck

131

Figure 9.2. *Paul Harteck, April 1934, in Grantchester, UK. Courtesy AIP Emilio Segrè Visual Archives, Bainbridge Collection.*

and Groth to Erich Schumann, head of the weapons research branch of German Army Ordnance (*Heereswaffenamt*).

> We take the liberty of calling to your attention the newest developments in nuclear physics, which, in our opinion, will probably make it possible to produce an explosive many orders of magnitude more powerful than the conventional ones.... That country which first makes use of it has an unsurpassable advantage over the others. [9-4]

Harteck should know. A physical chemist by training who got his start in nuclear physics under Rutherford at Cambridge, he has been rated by Irving as 'the driving force behind much of the most far-sighted research on the German atomic project during the War years' [9-5]. As such, he warrants our attention, more so for his principal role in the soon unfolding heavy-water chronicle in Germany and occupied Norway.

Paul Harteck was the son of a high Austrian civil servant, born in Vienna in 1902 [9-6]. He was schooled in Vienna's excellent Schotten Gymnasium, run by the Benedictines, entering the University of Vienna in 1921. In only two years, he had finished all his course work, moving on to the University of Berlin where he obtained his PhD in physical chemistry under Max Bodenstein in 1926. Serving initially as assistant to

Arnold Eucken at Breslau University during 1926–1928, he then joined Fritz Haber at the KWI for Physical Chemistry. At both the University and at the KWI, he was exposed to a stream of scientific personalities from Einstein, Planck, and Schrödinger to Szilard, Michael Polyani, and Karl Bonhoeffer; 'It was obvious to me that a new era in science had begun' [9-7].

Harteck took leave from his latter assistantship in 1933 to serve a Rockefeller Fellowship under Rutherford at the Cavendish Laboratory [9-8]. The highlight of his year in Cambridge was clearly a paper he coauthored with Rutherford himself and Mark Oliphant dealing, in effect, with the first experimental demonstration of thermonuclear fusion [9-9]. In that experiment, deuterium compounds were bombarded with ions of heavy hydrogen (deuterons, or 'deutons' as Rutherford still insisted on calling them), with the emission of high-speed protons and neutrons. Also observed flying off the target were hydrogen and helium nuclei of mass three. Two distinct reactions were apparently involved. In one, the target and projectile deuterons fused to produce another isotope of hydrogen ('tritium' or 'triterium' as they first called it), consisting of one proton and two neutrons, with a simultaneous ejection of a proton and the release of 4.0 MeV of energy, according to

$$_1H^2 + {}_1H^2 \rightarrow {}_1H^3 + {}_1P^1 \qquad \text{or} \qquad H^2(d, p)H^3$$

for short. In the second d–d reaction, the two deuterons fused to produce a helium-3 nucleus (two protons and one neutron) plus a neutron and 3.2 MeV of energy. That is,

$$_1H^2 + {}_1H^2 \rightarrow {}_2He^3 + {}_0n^1 \qquad \text{or} \qquad H^2(d, n)He^3.$$

While the tritium and helium-3 nucleus 'appeared to be stable for the short time required for their detection, the question of their permanence [required] further consideration' [9-10]. He^3 is, in fact, stable but very rare in nature; H^3 is a highly unstable isotope.

Just how rare the isotope H^3 (tritium) is, compared to H^2 (deuterium), is illustrated by an experimental attempt during 1935–1936 to isolate it in the form of tritium oxide (T_2O) by electrolysis at Norsk Hydro, in much the same way deuterium oxide, heavy water or D_2O, is concentrated. The experiment was initiated by two nearly simultaneous suggestions to Norsk Hydro management in the spring of 1934, from Rutherford and from Harold Urey at Columbia—both regular heavy-water customers of Hydro. As the tritium episode has been overlooked in all the interest lavished on deuterium, it warrants a brief digression at this point in our chronicle.

In his letter to Norsk Hydro on 19 March, Rutherford argues that 'from analogy [with deuterium]', an isotope of hydrogen of mass 3 ought to be strongly concentrated in Hydro's electrolytic process for preparing heavy water [9-11]. 'As the discoverers of this isotope', Rutherford and his Cambridge colleagues are keen on laying their hands on a sufficient quantity of the material to determine its mass accurately in Aston's spectrograph, and to investigate the effects of the H^3 ion in their ongoing transmutation experiments. They suggest continuing electrolysis of, say, 1 liter of 98–99% heavy water with the expectation of obtaining a small final sample (a few cc) sufficiently concentrated in H^3 to allow its detection and separation by special laboratory methods. The importance of the experiment was underscored by Urey in his own letter to Hydro's Director Sverre Brænne, soon after returning from a visit with Rutherford at Cambridge [9-12].

General Director Aubert agreed not only to Hydro undertaking the proposed concentration experiment, but generously offered any forthcoming small sample of H^3-enriched material free of cost [9-13]. Rutherford also discussed the experiment with Leif Tronstad during the latter's visit to London in May 1935, and urged him to go over it in greater detail with Oliphant and the brothers Ladislaus and Adalbert Farkas at Cambridge [9-14]. Assuming a deuterium–tritium separation factor of two and Gould and Bleakney's estimate of 1 part tritium in 200 000 parts heavy water, reconcentration of 40 liters of heavy water to 1 ml by the method of Brun and Tronstad should do the trick.

The painfully slow experiment was duly performed at Rjukan by Brun, assisted by Fritz Bugge, a Hydro technician, using as starting material 43.4 kilos of 99.2% pure heavy water produced electrolytically from 13 000 tons of ordinary water. Between April 1935 and March 1936, this sample was electrolyzed to a final volume of 11 cc, and the concentrated material was delivered through ICI to Cambridge for analysis by Aston. Behold! No trace of tritium was observed, 'a disappointment for both of us', as Rutherford put it in his report to Hydro [9-15].

The negative result is not surprising; the natural abundance of tritium is less than 10^{-10}% (compared to 0.02% for deuterium). Even so, the valiant effort may not have been in vain after all. Visiting Norway in 1948, Urey reported that the sample in question had all been forgotten on a laboratory shelf during the war years, but was subsequently re-analyzed with a newer, more accurate method. It was found to be radioactive—a telltale indication of tritium, which is indeed radioactive with a half-life of 12 years [9-16].

Returning to the fusion experiment of Harteck, Oliphant, and Rutherford, the key to that experiment was a target of concentrated heavy water prepared by Harteck at Rutherford's behest. Having 'better luck than [Rutherford] anticipated' [9-17], he designed a series of very small (5 liter) electrolytic cells for the purpose. An electric current was passed through them for weeks on end, reducing many gallons of water, spiked with a small amount of NaOH, to about 30 gram of very pure (up to 98%) heavy water [9-18]. In this he was considerably assisted by Gilbert Lewis of Berkeley, who prior to then had shipped several precious batches of heavy water to Cambridge, and who supplied him with a detailed description of his own electrolytic procedure [9-19]. Harteck also participated in the Royal Society round-table discussion on heavy hydrogen that year—the one chaired by Rutherford. There he impressed the gathering of experts with several elegantly simple demonstrations of the difference between normal and heavy water [9-20]

Rutherford's scientific life was by then occupied almost entirely by transmutation experiments with deuteron projectiles ('diplons' as Rutherford still insisted) accelerated in a low-energy machine thrown together by Oliphant, replacing the cumbersome high-tension rectifier installations of Cockcroft and Ernest Walton. Rutherford, well into his sixties, had assumed a working style characterized by Wilson as a 'thundering nuisance': trembling hands, dribbling pipe-ash over apparatus and notebooks, shouting (and apologizing) to assistants and colleagues alike. Harteck recalls him as 'somewhat impulsive' but still 'a man of tremendous vitality and strength' [9-21].

Upon returning from his short but productive stay at Cambridge [9-22], Harteck had been appointed Professor and Director in the Institute for Physical Chemistry at the University of Hamburg, a position he still held five years later when Groth came knocking on his door, and Esau's confidential meeting was called to order. The consensus reached at Esau's meeting was to proceed forthwith on experiments with uranium oxide, U_3O_8. Esau lost no time in initiating procurement of 'Preparation 38', as they now dubbed the secret substance, on behalf of the Reich Ministry of Economics, heading off the Air Ministry which sought the same substance for luminous paints for instrument dials [9-23]. In so doing, he was instructed to inform Schumann, who would have to draft the voucher. After much difficulty in tracking him down, he learned to his astonishment that Army Ordnance had upstaged the Ministry the very week after Hitler's Reichstag speech. It ordered his own Reichsanstalt to cease meddling in uranium research at once! Unbeknown to Esau, Army Ordnance, then much higher in the hierarchy of science policy in Nazi Germany than the Reich Research Council in charge of all university research [9-24], had set about organizing its own nuclear project when Germany went to war.

For all his haughty attitude and Nazi Party loyalty, Esau's followers represented scientists of greater clout than those behind Schumann, and Esau's overthrow diminished the weight of academic scientists in the German uranium effort, at least for a long time. Samuel Goudsmit, in his account of the *Alsos* Mission in the wake of the Allied armies in Europe after the Normandy invasion, reports surprise and jealousy by German 'charlatans' like Schumann and company at how American academic scientists had worked so effectively with the United States Army in the Manhattan Project [9-25].

The Army's initiative was, in fact, prompted by Harteck. Harteck was not new to the Army. Since 1937 he had been advising Army Ordnance on chemical explosives, among other things. Though he did not receive a reply to his and Groth's letter of 24 April until August, the letter had not been ignored by military authorities. Immediately upon its receipt, it had been passed on up the chain to Schumann, who knew Harteck, and who in turn forwarded it to Kurt Diebner, the physicist who headed the section for nuclear physics and explosives in Army Ordnance.

Diebner was somewhat of a loner among the leaders of the German nuclear effort. He had obtained his doctorate in 1931 under Gerhard Hoffmann at the University of Halle. He then joined the PTR for a period, before being appointed against his wishes to a post doing research on explosives in Army Ordnance. A nuclear physicist, he repeatedly badgered Schumann to get him into a more appropriate area of research. After Schumann finally established a nuclear physics bureau more or less to appease him, Diebner was reproached on more than one occasion from higher up for wasting his time on 'atomic poppycock' [9-26]. Being an experimentalist, moreover, he was at odds with theorists, most specifically Heisenberg, who had criticized Diebner's work for his *Habilitation* and would retain a grudge against him in the years to come [9-27]. Even so, the admittedly difficult Diebner acquired a considerable reputation, and would soon rise high in the German nuclear pecking order, this in contrast to his boss. If he was the first to begin work on the uranium bomb problem, Schumann would also be the first to abandon it, losing interest in uranium research in early 1942, and transferring it to the *Reichsforschungsrat*. It helped that Diebner was also a Party member. He appears in the Farm Hall character sketches as

> outwardly very friendly but [having] an unpleasant personality and is not to be trusted. He is disliked by all the [other members of the Uranium Club] except Bagge. [9-28]

Be that as it may, Diebner prudently shared the Harteck–Groth letter with Hans Geiger, whose opinion counted and whose encouragement, readily offered, was all he and Schumann needed to get going. He also called on Erich Bagge who was on close terms with Heisenberg, under whom Bagge had obtained his doctorate only the year before. Whatever Diebner's own differences with Heisenberg, the doyen of German theorists, Heisenberg's support for the budding Army enterprise counted even more than Geiger's. Bagge's contact was one thing; his technical ability no less important. As a fresh recruit to the Uranium Club, Bagge's developing friendship with Diebner, despite his own background as a theorist, would 'lay him [too] open to suspicion' [9-29] but not detract from his contributions to the uranium project under both

Diebner and Heisenberg, nor from his advancement up the German academic ladder after the war.

Diebner's first task for Bagge was drawing up an agenda for the first of several 'Schumann conferences' on how nuclear physics might be exploited for military purposes. It was held on Hardenberg Strasse in Berlin on 16 September, attended (in effect under Army orders) by, among others, Bagge, Geiger, Georg Stetter, Hoffmann, Mattauch, and Flügge; even Hahn was summoned to attend. Absent, however, was Schumann himself; or, as Hahn put it curtly in his private diary for Saturday 16 September: 'Schumann-discussion. Nuclear physicists present, Schumann not' [9-30]. A descendant of the composer, he was himself a composer of military music and may have been away in that connection. Also absent, less surprisingly, was Esau, who only learned by chance and to his dismay that many of the attendees were those who had enthusiastically rallied behind his own initiative only months before. H Basche, Diebner's immediate superior in Army Ordnance, opened the meeting by outlining *Abwehr* evidence of uranium research under way abroad. Should the War Office support a similar effort? After a lively discussion, recalls Bagge, Bothe and Geiger ended the session in no uncertain terms. Bothe rose. 'Gentlemen', he declared, 'it *must* be done'. To which Geiger added, 'If there is the slightest chance that it is possible—it must be done' [9-31].

Harteck, too, participated in the conference that September, the first on nuclear physics in Germany, and was even asked by Basche to take overall responsibility for a German project on uranium fission. He declined, turning the usual objection around: the direction of such an undertaking was the job of a committee, not an individual. Besides, he added, he was a chemist, not a physicist [9-32]. Nevertheless, a nuclear research office was soon in full swing under Diebner in Army Ordnance. By the time war was declared in September of 1939, Germany alone among the feuding world powers had a military office exclusively devoted to wartime applications of nuclear energy [9-33].

No sooner had Diebner's office opened for business, than Army Ordnance requisitioned the building of the KWI for Physics at Berlin-Dahlem, also known as Max Planck Haus, as the seat for the new initiative. Peter Debye, the Dutch physicist then in charge of the Institute, was sent packing to America on a guest professorship, replaced by Diebner as the new 'administrative head'. (The title was a compromise to satisfy those who, like Heisenberg, did not think much of Diebner as a physicist.) While the KWI thus became, in principle, the center for the Nuclear Physics Research Group, in practice much of the research would be carried out under a few key scientists at university institutes scattered elsewhere, due to obstinate physicists refusing to relinquish their own institutional turf. In particular, uranium isotope separation studies were entrusted to Harteck in Hamburg; measurements of nuclear constants were the responsibility mainly of Walther Bothe at the KWI for Medicine at Heidelberg. Heisenberg, while remaining for the time being a professor at Leipzig, was requested to work on the theory of chain reactions.

Uranium-235, as Bohr had already shown and Heisenberg was quick to appreciate, was the isotope fissioning under thermal neutron capture. In pure enough form, the isotope appeared capable of exploding with unknown power. However, for the generation of nuclear power, even natural uranium could be used in an *Uranbrenner* or pile, if combined with a suitable moderator. If nothing else, such a pile could produce highly radioactive material that 'could have been thrown around' with nasty military implications [9-34]. A pile appeared also capable of producing transuranic elements, with even nastier implications than uranium. Fermi, we recall, had mistakenly thought he was chasing down transuranics in 1934. More recently, Joliot and company had skirted the same issue, and it would come to the fore in Germany as well. By 1940, it was generally realized, also in Germany, that if uranium-238 forms uranium-239 by neutron resonance capture, the latter isotope could transmute via beta emission to element 93, eka-rhenium [9-35]. This was soon shown to be the case at Berkeley, and in due course in Germany as well [9-36]. Element 93, neptunium as it became known, and its daughter product, element 94, eka-osmium or plutonium, assuming it too could be produced, ought to be fully as fissionable as uranium-235, according to the all-important Bohr–Wheeler theory. There appeared to be no end of nuclear explosives; the more pressing question was, could they be produced in time for use in the war under way in a situation which, in the German view, favored only short-term technical developments? [9-37].

Transuranics notwithstanding, a uranium machine seemed to be the way to go, in Berlin as in Paris, if for somewhat cross purposes. As for a moderator, there were essentially two candidates, as we have seen: carbon and heavy water. (Pure water might still work, but only with separated U^{235}, and that became more of a remote proposition in Germany as time passed by.) Of the two, carbon was clearly the easier material to come by. It fell on Bothe in Heidelberg to study its neutron-absorbing and moderating properties in detail. In so doing, curiously, the 'physicist's physicist', whose beryllium radiation had proved so important for Chadwick and Joliot, committed an unfortunate error, casting doubt on the suitability of carbon as a moderator. In the event, the Germans opted for heavy water instead, as we shall see, but, curiously again, not on account of the Heidelberg measurements.

9.2. A serious error

We last encountered Walther Bothe and his student Herbert Becker at the Reichsanstalt under Geiger *c*. 1930. Bothe's study of the penetrating radiation from beryllium under α-bombardment led to James Chadwick's identification of the neutron in 1932. The new German regime had not been kind to Bothe. He had accepted the professorship of physics at Giesen in 1931, and the same chair at Heidelberg the next year. There he was demoted from the directorship of the Institute of Physics in 1933 by none other than the aging Geheimrat Phillip Lenard, Nazi physicist personified, for not belonging to the Party and his lackluster attitude towards Aryan politics generally. Bothe's struggle with the authorities so unnerved him that he was forced to spend a long period in a Badenweiler sanatorium [9-38]. When he was sufficiently recovered in 1934, Planck saw to it that he was appointed to a position at the KWI for Medicine, also in Heidelberg and an institution where party politics were less in evidence. Among other things, he began the construction of a cyclotron which would sap even more of his energy for over half a decade.

Amidst his preoccupations with getting the first contracts for his cyclotron issued, a more urgent technical problem fell on Bothe in early 1941. The previous fall, his own neutron-absorption measurements in graphite had yielded an anomalously short diffusion length for thermal neutrons—some 61 cm, compared to over 70 cm expected theoretically. Presumably the purity of the graphite, nominally a very pure form of carbon, was at fault. Nonchalant, Bothe had Army Ordnance issue an order to the Siemens Company for a supply of electro-graphite, the purest grade of carbon commercially available. Together with Peter Jensen, he assembled a 110 cm sphere of the material, submerged in a tank of water with a standard radon–beryllium, fast-neutron source at the center—a geometry suggested by Heisenberg and depicted in figure 9.3. Neutron detectors were small, strategically located dysprosium-covered strips, similar to those used by Halban and Kowarski. The density distribution of thermal neutrons was mapped out inside the sphere, and from it the neutron absorption obtained anew. Behold! The new diffusion length was even shorter, 35 cm, yielding a neutron absorption cross section more than twice the value obtained by Anderson and Fermi at Columbia University—a value unbeknownst to Bothe, to be sure [9-39]. Hence, concluded Bothe, unless substantially purer U^{235} was forthcoming, pure graphite was out of the question as a moderator of neutrons in a chain-reacting pile.

The problem with Bothe's experiment, it was only learned from subsequent German measurements at Göttingen, lay not in the absorption measurement *per se*, but in Bothe's estimate of the impurity content of the electro-graphite. In order to determine the degree of purity of the Siemens sample, he reduced a section of the graphite to ashes in a platinum furnace and measured its neutron absorption in a separate calibration run. Unfortunately, in so doing he assumed without justification that any neutron-absorbing impurities originally present would remain intact during the combustion. Wilhelm Hanle at

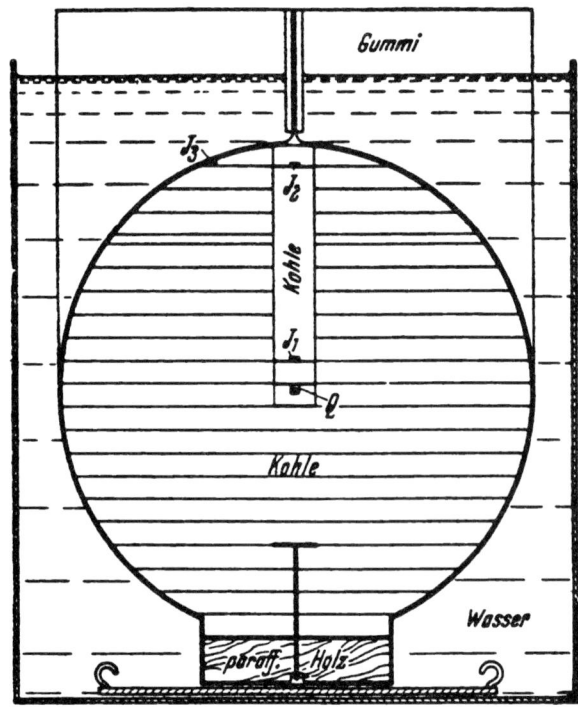

Figure 9.3. *Bothe's apparatus for measuring the absorption of neutrons in graphite. A 110 cm diameter sphere is mounted atop a paraffin-soaked wooden stand in a cylindrical tank of water. Access to the radon–beryllium neutron source Q at the center of the sphere is via a vertical channel closed with tightly-fitting graphite blocks. At J$_1$, J$_2$, and J$_3$ are located neutron probes in the form of small dysprosium-coated strips. Bothe W and Jensen P 1944 ZP* **122** *751.*

Göttingen showed not only that such impurities are largely lost during combustion, but also that even the Siemens electro-graphite contained boron and cadmium, both ravenous neutron absorbers [9-40].

To indulge in pure speculation, it is tempting to believe that Bothe may also have been preoccupied with thoughts of a certain lady in the middle of this. In the summer of 1939, he had visited the United States where, together with Heisenberg, he was scheduled to attend a cosmic-ray conference at the University of Chicago. During the passage to New York aboard the liner *Hamburg*, he ran into Ingeborg Moerschner, thirteen years his junior, who was on her way to San Francisco to work for the German Consul General, a former adjutant to Adolf Hitler and, incidentally, a spy. Ingeborg and Walther hit it off very well. The two toured the World's Fair in New York, and Bothe caught up with her again in San Francisco, where he also called on Ernest Lawrence at nearby Berkeley, in whose laboratory his student Gentner was learning the cyclotron art just then. Bothe returned to Heidelberg in a highly infatuated state, and went back to work on his own cyclotron and took charge of nuclear constants. On the first anniversary of his and Ingeborg's meeting, he wrote her that 'I have been speaking of physics the entire day, while thinking only of you' [9-41].

Bothe was not informed of the Göttingen results, perhaps because Hanle was outside the Army power project and a former associate of Esau, whose uranium initiative had been torpedoed by the War Office. Hanle's own measurements would show that carbon, properly prepared, would in fact work perfectly well as a moderator, but at a cost of production in industrial quantities ruled prohibitive by Army Ordnance.

Bothe's problem confounded Halban and Kowarski at about the same time, now installed at the Cavendish Laboratory in Cambridge. The two scientists had not been idle since they arrived in London in care of the Earl of Suffolk on 22 June 1940. The same evening they had an informal meeting with Cockcroft, who put them in touch with MAUD Committee officials. Before the Committee could decide just what to do with the visitors, the two were requested to prepare a progress report on the latest work under Joliot [9-42]. They were also asked to call on Peierls, Frisch, and Oliphant in Birmingham, Thomson in London, and Chadwick at Liverpool, giving each of them a personal briefing on the French work. The Committee then met in a closed session, during which mixed opinions were exchanged on the wisdom of incorporating Joliot's colleagues into the British effort. It was not yet clear in British circles that there was any direct connection between a controlled chain reaction of slow neutrons and a nuclear bomb. There was also the awkward matter of existing patents on most of the French work [9-43]. Nevertheless, in the end the MAUD Committee decided to accept the French scientists's offer of assistance, and lend them laboratory space at the Cavendish, where the Norwegian heavy water had already arrived.

Halban and Kowarski were soon installed at Cambridge, dining at the High Table in Trinity College and marveling at academic discussions on the European situation as the Battle of Britain raged in the skies overhead. They lost no time preparing to repeat the Ivry experiment, despite British misgivings, with a rotating aluminum sphere filled with a slurry of uranium oxide and heavy water. However, for a truly meaningful experiment, far more heavy water was needed than the 185 kg on hand from Norsk Hydro, and the British had their hands full with the problem of separating uranium-235. Carbon was still not ruled out as a viable moderator alternative. In October they placed an order for a thousand cubes of graphite *in lieu* of those left behind in France; more were ordered early in 1941. Alas, the British supplier, Imperial Chemical Industries, was no more able to guarantee the absence of impurities than Siemens, with predictable results. In April of 1941, Halban reported a value for the neutron-absorption cross section for carbon only slightly below Bothe's, once again rendering a carbon-moderated reactor in doubt. Like the German War Office, the British Government was unwilling to gamble on the industrial production of tons of graphite of the requisite quality. Fortunately for the British, they had a way out, however. They could leave pursual of the carbon-moderator problem to Enrico Fermi and co-workers in the United States.

Bothe's report to Army Ordnance on his results, dated 20 January 1941, apparently never came to Harteck's attention, judging by his post-war insistence that 'there was no publication' and his claim that he 'just did not know exactly what Bothe did' [9-44]. This is all the more curious since Harteck, as he goes on to point out in his interview with Ermenc at Rensselaer Polytechnic Institute in 1967, had himself been preparing a parallel experiment, a little earlier, with dry ice (frozen carbon dioxide) as a moderator [9-45]. For this purpose he had obtained a massive block of dry ice from I G Farben free of charge, including rail transport from Farben's Leuna ammonia works at Merseburg, courtesy of Paul Herold, the firm's Nazi director of research [9-46]. Farben, sensing a potential commercial opening, could afford to be generous just then. Carbon dioxide, which lasts for about a week, was used chiefly for food preservation, but was not needed that early in the year, mid-April. Obtaining uranium oxide proved less simple. For one thing, scientists at the KWI for Physics in Berlin under Heisenberg, an adviser at the KWI, were also planning experiments with uranium. However, the Auer-Gessellschaft of Berlin, the well known German chemical concern and principal supplier of uranium, only had about 150 kg of uranium oxide in stock just then, but expected to have much more on hand before long. Heisenberg himself wrote Harteck, asking whether he could perhaps postpone his experiment in favor of one in Berlin, since Harteck would presumably need plenty of time to prepare for his own experiment. On the other hand, Heisenberg added on a generous note, if Harteck was really in a hurry, they would gladly oblige [9-47]. The decision was, at any rate, up to Diebner at the War Office.

Time *was* of the essence for Harteck. Dry ice was on its way, melting ever so slowly, with no prospects for any more significant amounts. All preparations for the experiment had been completed. In the event, not quite 200 kg of 'Preparation 38' arrived, the absolute minimum Harteck had specified for a meaningful test. Along with it came 15 tons of carbon dioxide, again less than he could have used [9-48]. As Harteck privately predicted, his hastily assembled pile, seven feet tall and six feet square, gave no hint of any neutron multiplication [9-49].

Heisenberg, in defending his subsequent criticism of Harteck's experiment, felt that Harteck was rushing prematurely into a large-scale experiment without further ado—using willy-nilly 'thirty tons of carbon dioxide, twenty tons of very dirty uranium oxide, things like that'. Such an experiment takes lots of time, he noted. 'You then discover that you have too much dirt in the uranium oxide and have to send the thirty tons to the chemical company' [9-50].

Bothe, for his part, was not out of the woods with regard to nuclear constants, it turned out soon after German troops marched into Paris.

9.3. Joliot's guests

With the departure of the *Broompark* from Bordeaux, and the signing of the armistice on 22 June, Joliot had gone to Clairvivre to join his family and await further events. In Paris, meanwhile, the everyday city routine was at first scarcely disturbed by the occupying troops, who were on their best behavior. The Germans even opened some of the hastily vacated laboratories of the Collège de France, as well as the Laboratoire Curie, seeking to ingratiate themselves with what famous scientists were still around. However, on 15 July the Military Governor of Paris ordered Joliot's laboratory at the Institute de Chimie Nucléaire closed again, with entry only possible with a special permit. Two weeks later Joliot wrote his mother, still living in Paris, that he had received instructions to return to his laboratory, which had been partially occupied by German scientists. 'Don't worry about that', he added. 'I think I know one of them very well' [9-51].

The scientist he was referring to was Wolfgang Gentner, the physicist who had resided with the Joliot-Curies during 1932–1935. It was he who had vouched for Joliot's Geiger counter late in an evening of 1934, and in so doing he had played an indirect role in the discovery of artificial radioactivity by his mentors. Since leaving Paris he had maintained occasional contact with Joliot. Shortly after the outbreak of war in 1939, Gentner, a determined anti-Nazi, had been drafted by German Army Ordnance, and soon found himself in Paris, along with Walther Bothe, his boss as of late and whom he had been assisting in the assembly of the Heidelberg cyclotron. In Paris they joined up with Erich Schumann, now General Schumann in addition to Herr Professor, and Kurt Diebner, who both had descended on Joliot's laboratory immediately after the occupation of the city [9-52].

On 3 August, the four met with Joliot in his laboratory. Schumann held forth in glowing terms, in front of the assembled staff, on Joliot's important work, and expressed his wish for fruitful collaboration between German and French scientists in the future. The Germans then withdrew with Joliot to the privacy of his office, and the conversational tone changed abruptly. Joliot was interrogated in no uncertain terms, with Gentner, unhappily, serving as translator. Where, Schumann demanded, was the heavy water that Allier had smuggled out of Oslo? The French military had put it on board a British ship, Joliot replied; he believed it was a ship known to have been sunk in the Bay of Royan. Where was the uranium and the uranium oxide? It was withdrawn by the Ministry of Armaments, but Joliot did not know its present whereabouts. But surely, insisted Schumann, 'you of all people must know!'. Not so, replied Joliot [9-53]. Next, Schumann wanted to know the condition of Joliot's cyclotron. Cyclotrons were a sensitive topic for the Germans. Bothe's machine at Heidelberg was still not completed and unlikely to be so anytime soon, with military priorities preempting the necessary human and material resources. A cyclotron under the unlikely auspices of the German Post Office at Miersdorf was even more behind schedule. Joliot cautiously explained that his own machine was still being installed, and expected to be completed 'in due course'. In fact, the Paris cyclotron had been completed for some time, but was down for repairs due to a faulty oscillating system and general lack of maintenance as of late. The troublesome radiofrequency system had never worked properly since it had been built on a shoestring budget by Culmann et Compagnie, a French furnace maker [9-54].

As the interrogation came to a close, ending to the apparent satisfaction of the Germans, Gentner managed to draw Joliot aside and whispered to him, 'Can we meet somewhere this evening?'. Joliot nodded, naming a café on the Boulevard Saint Michel. There they met in the back room at 6 o'clock that evening. Gentner told Joliot that the original German intention had been to seize the cyclotron as war booty and ship it to Germany, for use as an accelerator or simply for its valuable spare parts. Removing it proved impossible, since the machine, with its massive electromagnet, was securely entombed in the vault of a sub-basement under the Collège de France. Instead, it was decided to recommission the machine in its present location with the aid of a German team of skilled technicians. Gentner would most likely be put in charge of the team, because of his cyclotron experience at Berkeley and Heidelberg, but he would only agree to do so if Joliot concurred and if it was of mutual benefit to both parties, French and German. Joliot assured him that he did; he would much prefer to have his experienced friend involved than some perfect stranger.

Gentner also asked Joliot in confidence whether his statements about the heavy water and uranium during the interrogation were true. Army Ordnance was convinced that both materials had been spirited to Algiers, and Gentner feared that he would find himself on a wil-o'-the-wisp chase for it there. Joliot insisted that what he had said was correct, as far as he knew. (In the event, Gentner argued successfully with his War Office superiors that it would be futile to track down pure uranium, due to its low level of radiation.)

Following his clandestine meeting with Joliot, Gentner informed Army Ordnance, by then quartered in an office in Paris, that he was prepared to take on the assignment in Joliot's laboratory, having already secured Joliot's cooperation. On Gentner's urging, Joliot received a written understanding from the authorities that he himself would stay on as director of the laboratory, that he would be kept fully informed and remain in overall control of the work done there, and that the only research undertaken with the cyclotron would be fundamental, non-military in nature, emphasizing biological studies. Irène, still in charge of the Radium Institute, agreed to all of this as well.

That settled, Joliot went back to work, and Gentner took up his new duties. The head of the Paris office of Army Ordnance, a Professor Bender, proved most helpful in every way. (Incidentally, Bender, a reserve Major, was in civilian life a professor of music. Musical forte seems to have been a strong card in the upper ranks of Army Ordnance; aside from his composing skills, virtually all of Schumann's work in physics concerned piano acoustics.) The young German technicians under Gentner bent willingly to their tasks. Their duty in Paris gave them a measure of freedom and security not enjoyed by their fellow conscripts back home; they were fulfilling their military duties in a far more satisfactory location than, say, the eastern front. One by one the machine components were taken apart, overhauled, and reassembled. The high-frequency transmission system, in particular, was rebuilt with parts obtained from Germany and Oerlikon of Zürich (who built the cyclotron magnet, among other components).

No sooner was the cyclotron restored to working order than Joliot was contacted by Bothe, Gentner's pre-war boss, requesting beam time on the machine. On the whole, German scientists were respecting the gentlemen's agreement about using the installation only for non-military research. Bothe turned out to be less obliging, insisting on using it for uranium cross-section measurements. Curiously, whenever he showed up for a run, the cyclotron failed. The problem, unknown to Bothe, was the French chief mechanic, who operated the machine jointly with one of the Germans in residence. Once the machine delivered beam to Bothe's specifications, the Frenchman slipped out of the control room. Invariably, soon after he excused himself for some necessary errand, the machine overheated and an insulator usually cracked, traced to a tap in the cooling water system being shut off. The partially irradiated uranium samples sent back to Germany for analysis yielded nothing much of interest— certainly no indication of transuranic elements, as observed at Berkeley in similar experiments. Bothe, nominally in charge of the run, took the blame for the aborted experiment. Gentner overlooked the maneuver, and the rest of his German colleagues never suspected sabotage [9-55].

The occupying powers had not quite given up on the Norwegian heavy water. In December of 1940 Allier, who was then in charge of a subcommittee within the Vichy regime responsible for dealing with German demands for war materials, was visited by a German general. The general bade him look into just what did become of the heavy water which, recalled Allier after the war, he explained to him 'was essential for the conduct of the war' and 'had been seized in Norway by a French officer whose name was not mentioned' [9-56]. Maintaining his composure, Allier brought the request to a high-level meeting with Marshal Pétain and some of his ministers. They instructed him to report back to the general's office that the material had been shipped out at Bordeaux and had most likely been lost when the ship went down. That proved to be the last German inquiry on the subject, and Allier's last wartime dealings with it as well. He spent the occupation years largely undisturbed, at least as far as the heavy water went [9-57]. So did Dautry, as noted earlier.

To be sure, it would not be the last word from the redoubtable Jacques Allier. He would have much to say about heavy water, about uranium, and about Joliot's nuclear-energy program in the early post-war years. Norsk Hydro, too, would again benefit from his wise counsel.

Meanwhile, Gentner proved invaluable for Joliot in more ways than one. The cyclotron continued to function well on the whole, while he turned a blind eye to occasional clandestine activities. Following Paul Langevin's arrest by the Germans in October of 1940, Gentner persuaded the authorities to grant him a partial reprieve, and Langevin was permitted to go into exile in the town of Troyes [9-58]. Gentner was able to see to that Joliot's home at Antony, which had been commandeered, was returned to him. When Joliot himself was arrested in June 1941, Gentner advised the local SS that an arrest of employees of the Collège de France could only be undertaken with the permission of the High Command in Berlin. Moreover, Gentner warned the SS officer, the work under Joliot had the highest priority, being a secret assignment classified as decisive for the outcome of the war. Joliot was promptly released. Gentner, too, arranged for Irène to attend a Swiss sanatorium at Leysin for a period, channeling all mail to and from her family through Heidelberg. His superiors gradually lost patience with Gentner's style of supervision and excessively friendly attitude toward Joliot. In the Spring of 1942 it came to a head, when he was denounced by a German compatriot guest at the Collège de France, and he was ordered back to Heidelberg. There, profiting from his experience in Paris, and Berkeley much earlier, he finally coaxed a beam out of Bothe's cyclotron in late 1943.

Gentner's replacement in Paris was Wolfgang Riezler from Bonn, who continued the supervisory work at the Collège de France largely in Gentner's style. With the aid of Walther Gerlach, who became Chairman for Nuclear Physics in the Reich Research Council, Esau's old post, Riezler ensured that the cyclotron remained untouched when the Germans withdrew from Paris. As for Gentner, the French remembered his stance in wartime Paris and made him an officer of the Légion d'Honneur.

Chapter 10

HEAVY WATER TAKES CENTER STAGE

10.1. Pressure on Norsk Hydro mounts

The economics of high-purity graphite on an industrial scale, rather than neutron cross section experiments *per se*, effectively eliminated carbon as a moderator in a German *Uranbrenner*. There was still deuterium oxide, alias heavy water, and there was no dearth of experts on heavy water among German scientists. There was Harteck, for one. Another was Karl-Friedrich Bonhoeffer, an early expert on the physical chemistry of heavy hydrogen; he had been among the first to alert Norsk Hydro on the potential importance of producing heavy water as a sideline. Yet another authority was Karl Wirtz, who had been Bonhoeffer's assistant at Leipzig before moving on to the KWI für Physik at Berlin-Dahlem in 1937. Wirtz had studied the physical properties of deuterium, particularly its specific gravity. All the same, when war broke out in 1939, hardly more than 10 liters of heavy water was to be found in all of Germany. There was only one firm producing heavy water on a commercial scale: Norsk Hydro. What then, we may ask, was the situation at Norsk Hydro in newly occupied Norway?

Soon after the fall of southern Norway in May of 1940, Norsk Hydro was again open for business, if not quite as usual; the last-minute fighting at Rjukan had affected the firm's facilities less than might be supposed. However, production at the Rjukan–Vemork complex, including that of heavy water, was at first severely curtailed over what it had been at the outbreak of hostilities. The reason was not so much the onset of wartime conditions as the shortage of water for powering the plants, as a result of a serious dry spell. The capacity of Lake Møs, the reservoir feeding the Vemork penstocks, was nearly down to zero. At the same time, snow was late in melting in south-central Norway that year. When production was finally restored, the German authorities at first paid Hydro no more attention than other Norwegian concerns of comparable size and importance. By and large, Hydro's administration was left alone to manage its own affairs; chiefly, it would seem, because the principal stockholders and capital resided in France, and France, too, was occupied and under German control. Instead, the Germans opted to look after their own interests in Hydro through I G Farbenindustrie which was, after all, an important partner in the giant conglomerate. This cozy arrangement remained, however, predicated on the attitude of Hydro's central management *vis-à-vis* that of the occupying power [10-1].

Early in the German campaign in Norway, it became clear that Norwegian hydroelectric power was high among German priorities in Scandinavia: in the short run for military purposes; in the long run for supplementing Germany's own natural resources. Above all, the German armament industry was counting on expanded Norwegian hydropower for assuring the supply of aluminum and magnesium for aircraft production. This had been repeatedly emphasized by Josef Terboven ever since he had been appointed head of the *Reichskommissariat* for *Festung Norwegen* by Hitler himself—the center of power throughout the German occupation (to the continued annoyance of the Quisling government). The initiative for a German-serving, light-metal industry was taken in 1941, with the establishment of *Nordisk Lettmetall AS* as an extension forcibly grafted onto Norsk Hydro by the German authorities. The new concern was located at Herøya on the southeastern coast, constructed by the Todt Organization with the objective of producing 12 000 tons of aluminum and 10 000 tons of magnesium. In the event, the plant was destroyed by Allied bombers in 1943, at considerable loss of Norwegian lives, it might be added. Thereby ended one phase of German involvement with Norsk Hydro in wartime Norway, in a remarkably close parallel with the demise of the heavy-water industry of more immediate interest here [10-2].

In January of 1940, four months before the German invasion of Norway, Kurt Diebner called a meeting in his office at 10 Hardenberg Strasse in Berlin with Werner Heisenberg. Also present were Karl Wirtz and his mentor, Karl-Friedrich Bonhoeffer. Heisenberg's report to Army Ordnance dated 6 December had become the blueprint for a German chain-reaction initiative [10-3]; in it, Heisenberg had recommended adopting Harteck's proposal for a layered arrangement of uranium and heavy water for avoiding resonance absorption in a pile of unseparated uranium [10-4]. Diebner asked Heisenberg whether he favored building a plant for the full-scale production of heavy water right away in Germany. Heisenberg said no; he preferred to start with a test of neutron absorption in a small sample of heavy water. All it would take was a few liters of liquid. Diebner agreed to look into procuring about 10 liters from Norsk Hydro.

Harteck himself wrote Heisenberg on 15 January, also stressing the importance of the production of heavy water, which had been vociferously impressed on him by his colleague, Hans Suess. At first, Harteck had been completely taken aback by Suess's suggestion that heavy water might be in the running for moderating a chain-reacting pile. Who, if anybody, Harteck now wanted to know, was looking into the problem of mass production of heavy water in Germany? Heisenberg wrote back, advising Harteck to talk it over with his Leipzig colleague, Bonhoeffer [10-5]. Harteck did just that, asking Bonhoeffer if he would be interested in jointly pursuing the production of heavy water by the catalytic exchange between hydrogen and water vapor—a process quite different from the electrolytic process normally used, and one explored by Harteck and Suess some time earlier. Bonhoeffer, in turn, passed the buck to I G Farben Ammoniakwerk at Merseburg, who expressed guarded interest in Harteck's suggestion [10-6].

Here, then, we have the reason why a representative of I G Farben had called on the firm's affiliate, Norsk Hydro, in January of 1940, as discussed in chapter 7. Speaking on behalf of the German Government, Farben attempted to place a substantial order for heavy water, more than Hydro could deliver and a request which raised suspicion among the firm's management in Oslo. It seems that Diebner's original promise to Heisenberg of 10 liters had been inflated by the War Office, perhaps put off by Harteck's letter to Army Ordnance on the 24th suggesting that *tons*, not liters of heavy water would be needed sooner or later [10-7]. The War Office was also miffed by Harteck's direct communication with Heisenberg on the matter, rather than through standard Army Ordnance channels, judging by its curt reply to Harteck. Army Ordnance was taking care of the matter, they assured Harteck [10-8].

The successful invasion of Norway in April of 1940 reopened the heavy-water question in a new light, with Germany now in control of Norwegian affairs, as Harteck reminded Bonhoeffer within days after the invasion [10-9]. However, existing stocks of heavy water had gone to France and, presumably, thence to the bottom of the Bay of Royan. Moreover, initial inquiries with the military economic liaison office in Oslo had established, as we already know, that for the time being Hydro could only supply its German customers 'minute quantities of SH.200' [10-10]. The total production for the year ending 1 July 1940 had been 20 kg, and the typical output for a 24 hour period by August amounted to 0.3 kg [10-11]. Karl Wirtz was instructed by Army Ordnance to proceed to Norway without delay to personally look into the troublesome situation at Vemork.

Wirtz traveled by train via Copenhagen, across the Sound to Malmø and on through Norrkoping to Stockholm, pausing in the Swedish capital (as did everybody on similar errands) to enjoy a city not at war. Thence, he journeyed on to Oslo, stopping over there at the German military economic liaison office and Norsk Hydro headquarters. He also used the occasion in Oslo to buy a wedding ring for his fiancée back home, Ottoni von Zicgner. From there, he was driven to Rjukan and Vemork, where he was able to inspect firsthand the impressive production facilities—in particular, the high-concentration plant occupying 62 square meters of basement space in the eight-story electrolysis building on the rock shelf at Vemork, 600 feet above the Maane River. Much of what he saw was unsettling; he had not appreciated the sobering conditions under which his Norwegian colleagues worked [10-12]. He also met with Jomar Brun, still the chief engineer at the Vemork plant. Wirtz had known Brun by reputation, though they had never met before. They had several conversations; on at least one occasion Aubert himself joined in [10-13].

In order to meet the German demands, it was agreed that Hydro would carry out a plant expansion along the lines Brun had suggested in his confidential plan to Director Stephansen early in February. This expansion was indeed accomplished during the course of the fall of 1940 [10-14]. The new facility consisted of a cascade of Pechkranz electrolyzers and cells, six stages in all, followed by a high-concentration stage expanded to nine specially constructed cells, each rated for 500 A. Alas, it took longer than expected to bring the installation up to speed, with the heavy-water output remaining well below full capacity. For one thing, water from Lake Møs remained in short supply for powering the Vemork generators. Part of the trouble, excess foam in the electrolyzers, was brought on by Brun himself, as we shall see. At any rate, Hydro files show that the total production between July 1940 and July 1941 was a mere 288 kg: to be sure, a factor of ten over the previous year, but not commensurate with the heightened plant capacity.

Late December of 1940 found Brun in Trondheim, where he discussed the heavy-water situation with Tronstad. After Christmas, Tronstad reciprocated with a visit to Vemork. General Director Aubert had remained tight-lipped about what he had learned from Allier back in March, and Tronstad and Brun were still in the dark as to the reason for the German heavy-water demands. Rumors flew, especially within Hydro, that it had to do with some form of poison gas—something Tronstad and Brun, both excellent chemists, were inclined to dismiss [10-15].

However, we are getting ahead of events. Returning to Germany from Vemork, Wirtz had been temporarily sidetracked in the building of the 'Virus House' at Berlin-Dahlem in the fall of 1940. There, a paraffin-moderated, uranium-oxide pile, quite similar to Harteck's earlier one at Hamburg, proved no better at neutron multiplication than Harteck's. Neither did a third attempt by Robert Döpel in Leipzig. Hamburg, Dahlem, and now Leipzig. Could there be any doubt? Neither light water, nor paraffin appeared capable of sustaining a chain reaction in natural uranium. That left heavy water. As if there were any lingering doubt, Döpel, assisted by his wife Klara, went a step further, using what little heavy water they could lay their hands on, about nine liters. Together with Heisenberg, they measured the diffusion length of thermal neutrons in heavy water. The diffusion length was still too short compared to its theoretical value, but only by a small amount [10-16]. Heavy water just might work as a moderator with natural uranium. Indeed, the first significant quantity of Norwegian heavy water would be put to good use in Döpel's Leipzig laboratory a year later, as we shall see.

By March of 1941, it was clear that heavy water output at Vemork was still not up to German expectations. That month, Brun was notified by Hydro management to take drastic steps to improve the production capacity. Nothing less than the full nine stages of the normal hydrogen-electrolysis plant were to be deployed in the pre-concentration process. The nine stages combined would produce water with its deuterium content enriched to about 13%. At this point, a beefed-up, high-concentration plant, 18 special cells of 700 A capacity, would take over from the main plant, with the heavy water further electrolytically concentrated to about 99.5% purity. In the event, putting all this into practice took time. Despite *Wehrmacht* demands for a yearly production of 1500 kg [10-17], by the end of 1941 the production during a typical 24 hour period was still only 1.5 kg.

In October, Harteck joined Wirtz in another visit with Brun. Accompanying them was Consul Schoepke, head of the German military economic liaison office in Oslo, who subsequently became, in Brun's words, the 'evil spirit' in the trumped-up collaboration with the Germans [10-18]. German by birth, he had spent years in Norway, most recently as the German Consul in Larvik, and in due course had become a Norwegian citizen. As such, he was arrested after the war, convicted in the Norwegian courts of treason, and served several years behind bars.

Brun, Harteck, Wirtz, and Schoepke had a long discussion, with Directors Stephansen from Oslo and Nilssen from Rjukan joining in. The three scientists met further alone at Grand Hotel in Rjukan [10-19]. Two conclusions were inescapable. First, electrolysis by itself, the process employed at Vemork for heavy-water production, was much too inefficient to meet likely German demands for the rare isotope, even with the full plant capacity brought to bear. Second, the economics of producing heavy water ruled out establishing a production plant in Germany, even if the necessary coal-fired electrical generators had been up to the task. A new production process was urgently needed to augment Hydro's standard production process. There were several possibilities: among them, catalytic exchange between hydrogen and water, repeated distillation at low pressures, and rectification of liquid hydrogen at normal pressure. All were thrashed out at Army Ordnance in a meeting called by Diebner and Bagge on 22 November 1941 [10-20]. But which to choose?

Harteck had the answer in the catalytic-exchange process devised some time ago by Hans Suess and himself. They had not pushed it at the time, when it appeared that Vemork's production alone would suffice to meet foreseeable demands for deuterium. The catch now was, the exchange process was hardly out of the experimental stage [10-21].

The electrolytic process, as explained earlier, takes advantage of the empirical fact that the residual water in electrolytic cells is richer in deuterium than the hydrogen given off during the electrolytic decomposition of water. The enriched residue is fed back into the system, reoxidized and electrolyzed repeatedly as it passes from one stage to the next. A shortcoming of the electrolytic separation as used at Vemork was that the unburned, deuterium-rich hydrogen was fed to the ammonia plant, and thus wasted as far as heavy-water production was concerned. Burning the hydrogen in oxygen, on the other hand, produced more deuterium-rich water, as was done in the final high-concentration stage of deuterium production; however, in so doing hydrogen was lost for ammonia production, which was, after all, Hydro's main business at Rjukan! The Harteck–Suess catalytic-exchange process avoided loss of hydrogen, regenerated the deuterium otherwise lost, and, on top of that, promised to be far cheaper than electrolysis. In it, a mixture of ordinary hydrogen and HD gas is passed through water in the presence of a platinum or palladium catalyst suspended as a colloidal solution. An exchange reaction takes place, transferring deuterium from the rich hydrogen to leaner water, given by $H_2O + HD \rightarrow HDO + H_2$ leading to an equilibrium concentration which is richer in heavy hydrogen in the gas mixture compared to what it was in the original water. The deuterium formed is partially removed by absorbing it in water, or scrubbing the gas, in the parlance of physical chemistry manipulations. (This scrubbing procedure, developed for deuterium by Harteck and Suess, was the key element in their catalytic-exchange process—the trick for saving the deuterium normally wasted in Hydro's electrolytic process.) The deuterium is further concentrated by passing it through a series of countercurrent towers with electrolysis of the water between successive towers [10-22].

Unfortunately, noble metals, such as palladium, were needed for a variety of catalytic synthesizers, including for that matter, ammonia synthesis for fertilizer production. Nickel was another catalyst, even though it was not as good as palladium. However, even nickel was in short supply, due to the war effort. By chance, tests in Harteck's laboratory had shown that even small amounts of nickel, suitably dispersed in a large volume of pumice stone, would work as a catalyst for the Harteck–Suess process. In practice, however, assembling the huge drums of pumice stone took time and effort, and was one reason why implementing catalytic exchange on a large scale would prove unexpectedly troublesome [10-23].

The upshot of the latest discussions at Vemork was a decision to wait and see; this despite an increase of the original German order received just then, raising the total order from 1500 to 5000 kg. For the time being, the ongoing plant expansion took all of Hydro's resources.

In September, Brun was again in Trondheim, where he met as usual with Tronstad, whom he realized was getting deeper into organized resistance work [10-24]. They were no closer to an explanation for the escalating German demands, but Tronstad told Brun to watch carefully for hints from the many German scientists who came and went at Vemork. Shortly after returning to Vemork, Brun received word that his friend had left Norway, leaving a message at NTH in Trondheim that he was off for a visit to Rjukan [10-25].

The month that Tronstad disappeared, October of 1941, Wirtz had, as noted, paid one of his regular visits to Vemork, accompanied by Lieutenant Colonel Tzschirner, Chief of *Wehrwirtschaftstab Norwegen*. On that occasion he asked Brun whether he was familiar with the confidential transport of heavy water to the Joliot-Curies shortly before the invasion of Norway. Brun replied that he remembered shipping a considerable quantity of heavy water to Oslo, but that was all. Tzschirner and Wirtz had a good laugh. They knew everything there was to know about that affair! [10-26].

By the end of 1941, despite the hurdles of expanding the Vemork plant, heavy-water production had increased tenfold to a respectable 100 kg/month, or I G Farben's original requested production rate [10-27]. By then, Hydro had delivered the first 361 kg of the original order placed by Army Ordnance. However, celebrations were premature. The production soon fell sharply, from a record of 140 kg in December to 100 kg in January, and to 91 kg in February [10-28]. On 15 January, Consul Schoepke and the military authorities in Oslo lost patience with the production figures. A letter over the signature of Colonel Tzschirner to Hydro headquarters ordered Brun to personally report to Kurt Diebner's office, *Forschungsabteilung Ia* on Hardenbergstrasse 10 in Berlin-Charlottenburg, on 20 January [10-29]. Both Colonel Tzschirner and Consul Schoepke came along with Brun for the trip. In Diebner's office they joined Diebner, Harteck, Wirtz, and Hans Jensen. Also present for the first meeting was *Oberregierungsrat* Dr H Basche, Diebner's immediate boss in Army Ordnance, and Herr Belitz of I G Farben [10-30].

The focus of the discussion, as intended, was how to implement the Harteck–Suess process for boosting production at Vemork by retrofitting stages 4, 5, and 6 of the electrolytic cascade, as indicated in figure 10.1, prepared for the meeting. By so doing, production was estimated to rise to ~8 kg for a 24-hour period. At the same time, it was planned to install heavy-water apparatus at Hydro's smaller Såheim electrolysis plant, about 1 mile east of Vemork (in Rjukan), as well as at Notodden. To this end, Brun also visited several German firms to look into equipment

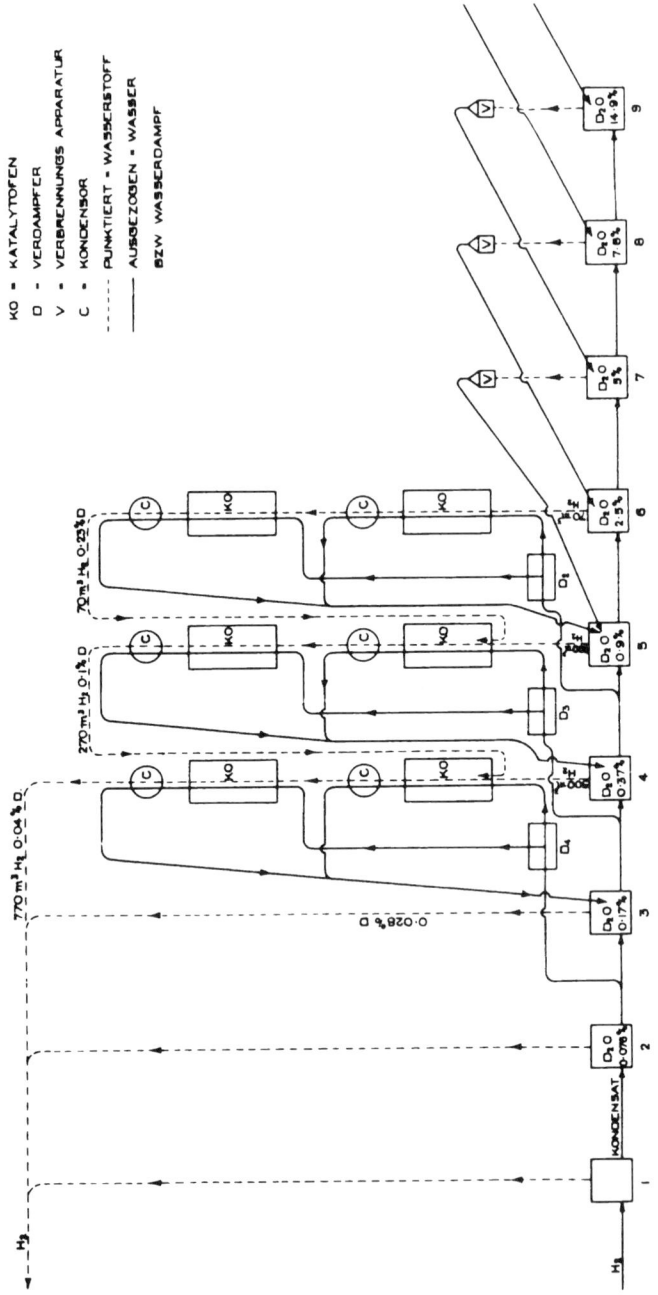

Figure 10.1. *German plans for a catalytic exchange installation at Vemork. Wirtz, G-198, 26–28 February 1942, figure 3.*

for augmenting such a drastic expansion. As always, I G Farben would remain the prime subcontractor.

Subsequent discussions, in which Harteck, Wirtz, and Jensen briefed Brun in detail on the catalytic-exchange process, were held in Wirtz's office in Max Planck Haus on KWI grounds in Dahlem—formerly Diebner's, until he moved to Hardenbergstrasse 10. Among other things, nickel was closely scrutinized as a potential catalyst, with Brun voicing skepticism about its practicality under conditions at Vemork [10-31]. A precondition for implementing the exchange process was *in situ* tests of suitable nickel catalysts at Vemork.

A little later, between meetings, Brun was invited by Wirtz to visit the actual laboratories of the KWI, though the 'Virus House' down the road, and the domineering high-tension hall, were off limits. Nearly everybody spoke freely, even about politics; only one subject was *not* open for discussion: the reason for all the German interest in heavy water. All he was told was that it had no military significance, though Brun knew enough of the developments in modern physics, coupled with what little he gleaned from the conversations, that nuclear physics was involved.

While touring the KWI, Brun was startled to see two glass carboys in a corner of Wirtz's laboratory where analysis work was done, containing about 130 kg of heavy water—representing well over a month of normal Vemork production, with the price tag of 130 000 Norwegian kroner. Though he was on close and friendly terms with his host, he could hardly refrain from lecturing Wirtz on the inadvisability of storing such a valuable commodity in fragile containers in the face of say, eventual air raids on Berlin. To which Wirtz retorted that the quantity of heavy water was as safe there as at Vemork—'something he eventually saw come to pass', adds Brun in his memoirs [10-32]. Brun also had a private conversation with Hans Jensen, who made no secret of his anti-Nazi views, as well as expressing his belief that Germany would ultimately lose the war. Dissatisfaction with the Hitler regime seemed to be the norm among most of the scientists, all of whom made a point of warning Brun to avoid discussing politics with certain of their colleagues, particularly Kurt Diebner.

Returning from Berlin, Brun was called in for a chat with Axel Aubert, who by then had stepped down as General Director of Norsk Hydro. As of 1942, he was Chairman of the Board instead. Aubert wished to know Brun's impression of the KWI. In telling him, Brun apologized for not learning the real reason for the German heavy-water effort. Brun sensed that Aubert was tempted to share with him what he himself knew, but Aubert said simply, 'My mouth is sealed with seven seals!' [10-33]. Aubert did allow, however, that, in his opinion, German experiments with heavy water posed no immediate danger to the Allied cause. It appears that Aubert, himself a competent chemist, had been assured from Austrian sources that the ongoing German uranium project, of which Brun, of course, was unaware, was on such a small scale as to represent negligible risk from a military point of view. In this, Aubert was supported by Bjarne Eriksen, who succeeded Aubert as General Director, and sought to reassure the Allies during a visit to Stockholm [10-34].

10.2.   SIS, SOE, and the Galtesund affair

Diebner and company were not the only ones kept abreast of the heavy-water situation in Norway by Jomar Brun. So was Leif Tronstad, his old colleague and codesigner of the original high-concentration plant at Vemork.   Upon arriving in England in the fall of 1941, Tronstad had, characteristically, sought active duty in the fledgling Norwegian armed forces training in the United Kingdom. Fortunately, one might say, his request was turned down, though he did talk his way through a short officer's training course in Scotland.   Instead, he ended up heading Section IV of the Norwegian High Command with the rank of Major, in charge of intelligence, espionage, and sabotage.   In this capacity, he bridged the gap between the two separate and distinct British organizations dealing with secret operations [10-35].   The SIS (Secret Intelligence Service) trained agents in pure military intelligence gathering before sending them back into their home countries.   It operated under Commander Eric Welsh, whom we have met on several occasions. The SOE (Special Operations Executive) was a more sinister organization.   It had been created personally by Winston Churchill in 1940, with the mandate 'Set Europe Ablaze!'.   Operating under Lord Selborne and Major-General Colin Gubbins of the Ministry of Economic Warfare, its mission was to train agents for clandestine warfare, ranging from subversion to sabotage. Not surprisingly, SIS and SOE were often at odds; sabotage actions invariably drew the immediate, merciless wrath of the Gestapo, and with it jeopardized the work of intelligence-gathering agents who avoided attention at all cost.

By Tronstad's time, the head of the Norwegian Section of the SOE was an Englishman, Colonel John Skinner Wilson, former Director of the Boy Scouts Association, and before then Deputy Commissioner of Police in Calcutta.   A seemingly rather shy and retiring person, he was the antithesis of the outgoing Eric Welsh.   In fact, both fit their respective roles to a tee, each in his own way. Wilson was outwardly pious, but highly ambitious; Welsh was a hard-drinking extrovert, but born of a singleness of purpose. At any rate, Tronstad got along very well with both Wilson and Welsh.   The mutual relationship between Tronstad, Wilson, and Welsh was just what was needed at this juncture in coming to grips with heavy water and German intentions concerning it. Intelligence-gathering came first, and none were more qualified than Tronstad in evaluating scraps of information coming from Norwegian agents about IMI, the British code word for heavy water (borrowed from 'repeat message' in Morse code). 'Juice' was Tronstad's preferred term. In the long run, active sabotage was beginning to appear unavoidable.

Fittingly, the center for British secret operations was at 64 and 81 Baker Street in London, above Baker Street subway station and not far from the legendary chambers of Sherlock Holmes. Links between Tronstad's intelligence section, in Kingston House South, and Norsk Hydro took several forms, including radio transmissions and couriers. One of the sources of intelligence was hidden microphones installed in places frequented by high-ranking Germans at both Rjukan and Notodden. On at least one occasion, they revealed German technical developments of the utmost importance, if not directly concerning heavy water [10-36]. However, effective communication between Tronstad and Jomar Brun called for a reliable radio operator having good access to the Hydro facilities—somebody not yet on hand at the start of 1942. That problem was solved by the daring undertaking of Einar Skinnarland, and fast action on part of the British.

Einar Skinnarland was a native of Rjukan, tall and slender, then in his early twenties. He had been mobilized as a Sergeant in the Corps of Royal Engineers of the Norwegian Army on 9 April 1940, and saw sporadic and futile fighting before returning to Rjukan. There he resumed his regular occupation, construction supervisor of the hydroelectric dam on Lake Møs, the highest of the series of dams feeding Vemork and its neighboring power plants along the Vestfjord Valley watershed. He soon became involved with the local resistance network in Telemark, which operated under the leadership of Milorg, the central military resistance organization headquartered in Oslo. Since he knew several of the engineers on the Vemork staff, was a moderately proficient radio operator and an excellent skier, and, on top of all that, spoke English fluently, Skinnarland was soon tapped by Colonel Rolf Palmstrøm, Milorg's communication group leader. Palmstrøm knew Skinnarland from before the war through his wife, also a native of Telemark. Would he be willing to undergo special SOE training in England and then be parachuted back into Norway as soon as possible, there to serve as a communications link between Jomar Brun, Tronstad, and Palmstrøm? If so, Palmstrøm suggested he leave for England via the 'Shetland Bus', which ferried agents to and from Shetland by Norwegian weather-beaten fishing boats recruited or hijacked for the purpose [10-37].

Skinnarland agreed on the spot, but opted for another way across the ocean. In the middle of March 1942, he and five companions seized at gunpoint the *Galtesund*, a Norwegian coastal steamer of 520 tons serving the coastal stretch between Oslo and Stavanger. They commandeered the vessel on the south coast near Flekkefjord, and piloted it across the North Sea to Aberdeen, Scotland. Skinnarland and two of his comrades, Lieutenants Odd Starheim and Andreas Fasting, both already commissioned officers in the Linge Company, a Norwegian unit attached to Special Forces, England, reported to Colonel Wilson on 19 March. After hurried commando training as a newly enlisted Sergeant in Company Linge, and briefing by Tronstad on his mission to Vemork, plus one practice parachute jump, Skinnarland was ready to go. He was dropped back onto the Hardanger Plateau on the shore of Lake Møs near Rjukan from an RAF Halifax bomber on 25 March. It was not quite two weeks since he had landed in Scotland. His 'leave' from his work at Rjukan had not yet expired, and his absence had gone quite unnoticed [10-38].

10.3. Mild sabotage; frank talk

Einar Skinnarland's return to Rjukan was well timed. Deuterium production was picking up. The month of his whirlwind escapade in absentia, fully 100 kg had been produced, a record to date. However, production plunged to zero during April. The immediate reason for the sudden halt, according to Hydro's monthly reports for Vemork production, was a need for overhauling the interior of a malfunctioning Vemork hydrogen gasometer [10-39]. This temporary interruption notwithstanding, net production was not gaining as fast as demanded by Army Ordnance. Wartime scarcity of materials, as in the case of noble metal catalysts, was partly to blame. Moreover, in spite of pressure from the German authorities to increase the heavy-water output as fast as possible, the Germans were weary of sharing results from their own secret research with scientists in occupied countries—to the extent that Brun and his colleagues never did receive copies of the formal Harteck–Suess report on the all-important catalytic-exchange process [10-40].

There was another reason for sluggish production: plant sabotage. Brun, increasingly troubled by the German push, had been secretly adding castor oil to the electrolyte, causing heavy foaming in the system. At the start, he viewed his little pastime mainly as a test of effective foaming agents, should serious sabotage become necessary [10-41]. Unknown to Brun, however, other workers at the plant were of similar sentiment, in particular Chief Engineer Alf H Larsen and his assistant Gunnar Syverstad [10-42]. They were adding cod-liver oil to the high-concentration containers at the same time, despite the fact that obtaining cod-liver oil in wartime Norway was 'like pulling down the Moon' [10-43]. Hydro's monthly reports for heavy-water production echo the recurring problem: 'There have been a few interruptions in the plant, especially because of foaming' (July 1942); 'The concentration has, as usual, been disturbed . . . ' (September 1942) [10-44]. The frothing and foaming grew to an alarming extent, necessitating Brun to seek some counter-measure. In this he was less successful, and the foaming continued in varying degrees as long as he himself remained on the job in

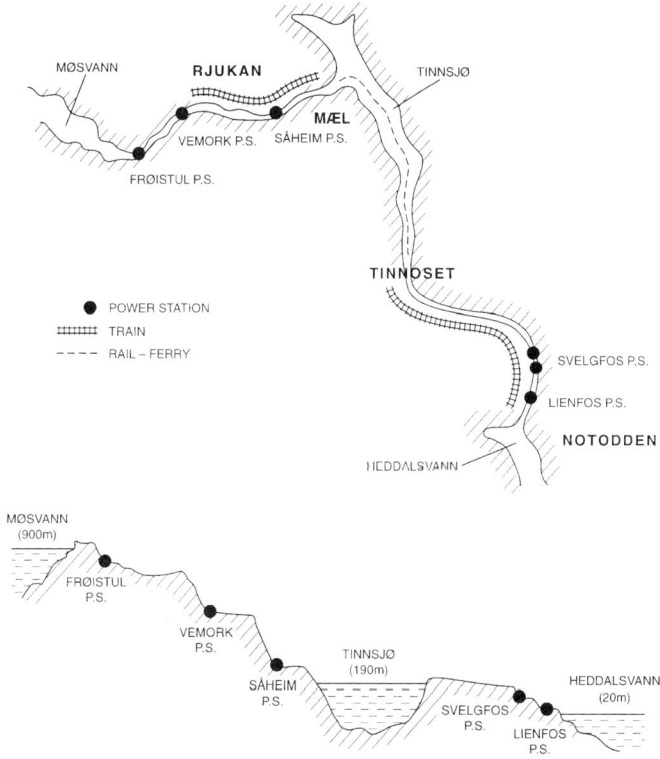

Figure 10.2. *Power stations along the Telemark water-course from Rjukan to Notodden. After Anker Olsen Kr 1955 Norsk Hydro Gjennom 50 År (Oslo: Norsk Hydro) p 225.*

Norway [10-45]. The German authorities never did discover the source of the problem, which was, of course, overshadowed by the events of 28 February 1943 and its dramatic aftermath.

Consul Schoepke had his suspicions, to be sure, blaming the erratic production on a 'certain passive resistance' [10-46]. He found a way around the problem. Partly on his initiative (prompted by Harteck), preparations for heavy-water production were also under way at nearby Såheim, as noted earlier, and a team was studying the deuterium content of the electrolyzers at the Notodden facility well to the south (figure 10.2). The Såheim power station (actually in Rjukan) is nearly as old as Vemork, dating from 1915. The original Såheim electrolysis plant, smaller than the one at Vemork, went into operation in 1929. It was expanded in 1940, and was now again in the process of upgrading in 1942. It was to be retrofitted with a number of new Swiss Pechkranz electrolyzers, plus two more from Bamag-Megiun in Berlin [10-47].

The main thrust of the heavy-water plant expansion, however, was at Vemork. *Wehrwirtschaftstab* had ordered the Vemork plant upgraded

159

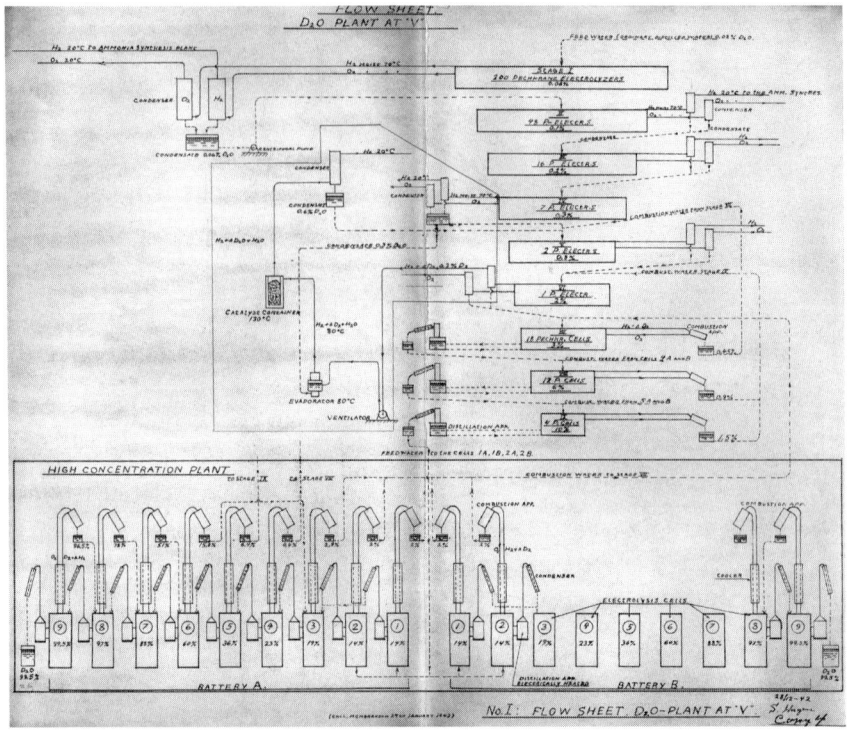

Figure 10.3. *Flow sheet for heavy-water plant at Vemork, as drawn for the British authorities by Jomar Brun in London, dated 28 December 1942. Copy obtained from Norsk Industriarbeidermuseum, Vemork, courtesy Norsk Hjemmefrontmuseum, Oslo.*

from 9 to 18 high-concentration cells, arranged as shown in the flow sheet, figure 10.3. (This figure, and figure 10.4, as well as another figure not shown here, were prepared by Brun for the British authorities after his escape to England in the fall of 1942, along with a comprehensive report detailing the methods for producing heavy water at Vemork [10-48].) Preceding the high-concentration plant were nine successively smaller stages of Pechkranz electrolyzers, ranging from stage I with 200 multi-cell electrolyzers, to stage IX with four individual P-cells. In the early stages, the condensate was used as feed water for each successive stage. At stage VII, however, the makeup water for stage VIII was obtained by distillation of the electrolyte. Simultaneously, regeneration of the deuterium content of the hydrogen gas was accomplished by combustion apparatus resembling large Bunsen burners. In the high-concentration plant, two rows of electrolyzers, each containing 9 series-connected cells of 700 A capacity, were connected in two parallel circuits, forming 'Battery A' and 'Battery B', as shown at the bottom of the flow sheet. The vessel of each cell (figures 10.4 and 10.5) measured

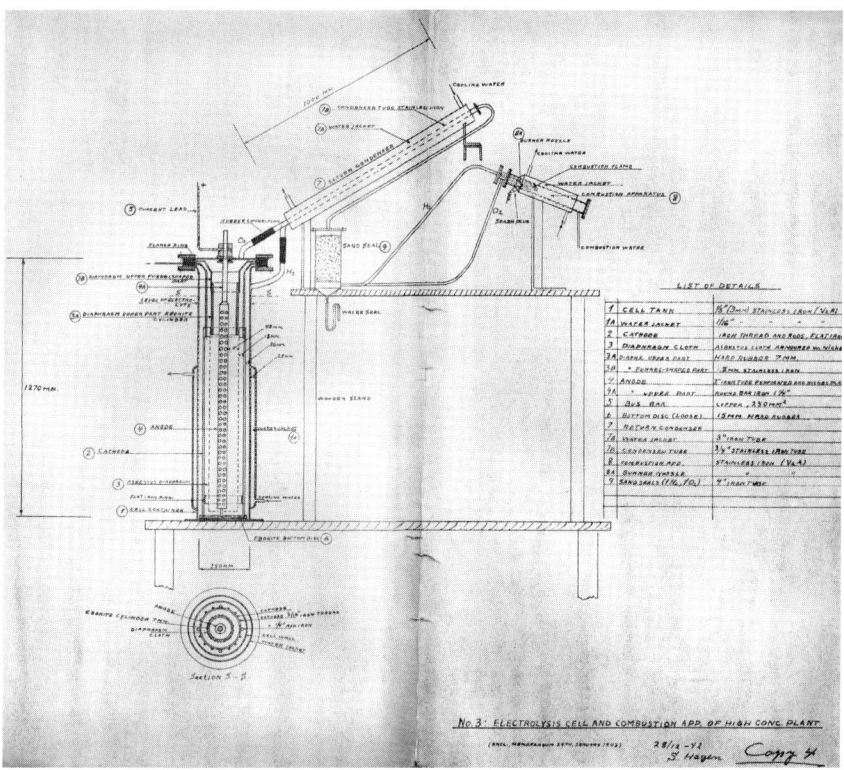

Figure 10.4. *Electrolysis cell and combustion apparatus in heavy-water plant at Vemork, as drawn for the British authorities by Jomar Brun in London, dated 28 December 1942. Copy obtained from Norsk Industriarbeidermuseum, Vemork, courtesy Norsk Hjemmefrontmuseum, Oslo.*

1270 mm in height by 250 mm in diameter, and contained 50 liters of potash lye. Equipped with lead pipes, condenser tubes, sand seals, water seals, rubber connections, and flanges, a cell was a complicated affair. In fact, it consisted mainly of an anode in the form of a perforated steel tube mounted on the axis, surrounded by a concentric, cylindrical cathode, also of mild steel. Between the anode and cathode was a cylindrical asbestos diaphragm, and the wall of the containment vessel incorporated a water-cooled jacket. Every cell contained, besides the water-cooled combustion unit (shown in figure 10.4), reflux gas coolers and an electrically heated distillation unit (figure 10.3). As the 'juice' flowed from cell to cell in each battery, the deuterium concentration increased from 14% in cell 1 to 99.5% in cell 9.

The plant expansion at Vemork had been under way since February of 1942, and despite the various hangups, by June Hydro was tapping heavy water from the new, additional row of 9 high-concentration cells

161

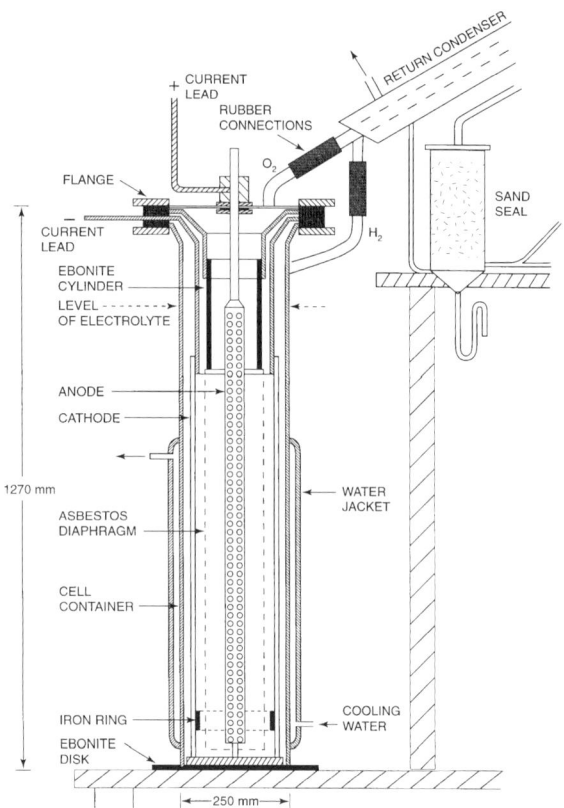

Figure 10.5. *A high-concentration cell at Vemork, based on Jomar Brun's original drawing (figure 10.4).*

[10-49]. On 28 July, Rjukan Director N Stephansen reported to Berlin that 'we are now again up in production and if the present water supply can be maintained the production will be about 125–130 kg per month'— something the production figures for July–August confirmed [10-50]. In a lecture before the KWI at Berlin-Dahlem on 24 February, Wirtz lauded the Vemork operation as a 'masterwork by Norwegian engineers and scientists' [10-51], capable of 1.5 tons per year of D_2O with the expanded conventional facilities, and perhaps 5 tons per year with the Harteck–Suess modification. Both electrolytic and catalytic processes are succinctly described in Wirtz's lecture.

The surge in activity among the Telemark facilities that spring and summer was accompanied by considerable traffic of scientists and VIPs to and from the plants. Shortly after Brun's return to Vemork from Germany, Hans Suess arrived there to conduct, with Brun, the agreed-upon tests of various catalysts that could be spared from the list of strategic materials. Preliminary tests carried out by Suess in Hamburg, mainly on nickel catalysts used in the fat-hardening industry, were already in Hydro hands [10-52]. They were needed, as noted, for implementation in a dual-temperature variant of the catalytic-exchange process intended to augment the regular electrolysis system. Assisting Suess and Brun in these experiments was the Hydro engineer Alf H Larsen, who would succeed Brun as head of the Vemork operation, and take the heat during the forthcoming Allied commando and bomber raids in Telemark. The catalyzer tests got off to a shaky start, but suitably dispersing nickel in pumice stone finally brought results. Brun and Suess subsequently prepared three reports to the War Office on the ongoing efforts to increase the heavy-water output generally, and catalyst results in particular [10-53].

In May the Vemork staff was 'honored' by a visit of *Reichskommissar* Terboven himself, along with his entourage, and Brun showed them around. 'It was obvious they lacked any comprehension of the production effort', adds Brun [10-54].

In June, Suess returned to Vemork, and it was definitely decided to go ahead and retrofit stage 6 with a nickel-based catalytic converter. The next month, Wirtz and Friedrich Berkei of Army Ordnance's Gottow laboratory also arrived, along with three additional Hydro engineers, among them Egil Hvoslef Eide, for a conference on steps to boost heavy-water production still further. In addition to reviewing the retrofitting of the production line at Vemork, they discussed the Såheim plant in nearby Rjukan, as well as prospects for concentrating heavy water on German soil. Nicolai Stephansen, the local director of Norsk Hydro, also attended most of the discussions [10-55].

Suess has left a record of a candid conversation he had with Brun during his latest stay at Vemork. According to him, after considerable soul-searching he decided to risk discussing German plans with his Norwegian friend, fully aware that Brun was in contact with the British. There might also have been microphones in Brun's office—something, as we know, was indeed the case at Rjukan and Notodden. For that reason, the conversation took place on one of the balconies of the main eight-story Vemork building. There was even a chance that the Germans, then poised before Sevastopol on the Eastern Front, would win the war, after all, and Suess could see himself tried for treason in a German court of law.

The thrust of the German program, Suess explained, was a uranium pile for power production, largely along the lines spelled out in a patent by Joliot-Curie. As to heavy water contributing to a German victory, Suess reassured his unhappy friend, he was convinced it would take many years. He guessed 'a minimum of five, before anything practical could result from the research which used heavy water'. 'So this is all Zukunftsmusik?' asked Brun. Just so, replied Suess. In his opinion, those who believed in a quick victory surely were hoping for its peaceful application after the war, and those who expected a long war were thinking that they must have some knowledge of all the possibilities that might result from such research. [10-56]

Whether those 'possibilities' included a nuclear bomb, Suess was careful not to say. Brun, for his part, told Suess some years later that he had been confident he was telling the truth to the best of his knowledge, and Brun himself would do his best to safeguard his friend if he found himself in position to pass on the confidential information. However, he harbored some suspicion that Suess was the victim of wishful thinking, which may have been one reason why Brun's subsequent report created little stir on Baker Street.

Not long after Suess's latest meeting with Brun, Hans Jensen also had a frank talk with Norwegian colleagues, as he had in his private chat with Brun in Berlin. Jensen was a co-worker of both Harteck and Suess. He worked with them both on heavy-water problems, and with Suess on basic nuclear structure studies (the 'shell model' of the nucleus, for which Jensen shared the Nobel Prize in physics much later with Maria Goeppert Mayer). Jensen's mission to Scandinavia in the summer of 1942 was twofold. In Copenhagen, he paid a visit to Bohr's Institute on behalf of Heisenberg, his mentor. There he spoke candidly with Bohr and his assistant, Christian Møller, about his heavy-water research [10-57]. From Copenhagen he continued to Oslo, where he attended a colloquium with Norwegian scientists. Most of those present were from the University of Oslo, including the physicists Harald Wergeland and Egil Hylleraas, and Tom Barth, the geophysicist.

Jensen, the featured speaker, purportedly warned the group not to take notes during his informal talk. However, Wergeland recalled that Barth signaled him to ignore Jensen's admonition. Afterwards, Wergeland passed what notes he did take on to another colleague, Brunulf Ottar. In Ottar's recollection, Jensen had dwelt mainly on the fact that, in Heisenberg's opinion, Germany would be unable to produce the bomb [10-58]. A day or so after the colloquium, one of Ottar's co-workers in XU, the Norwegian Resistance intelligence organization [10-59], told him to see Wergeland immediately; Wergeland had some important papers that had to be transmitted by courier to England without delay. Ottar stopped by Wergeland at his home; he thinks Jensen was there as well. That Wergeland's report did indeed reach England is attested to in the official British history of intelligence [10-60].

Karl Wirtz, too, was in Norway just then, as noted. He knew Wergeland well, since they had studied together under Heisenberg at Leipzig in the 1930s. Among their early common interests was heavy water; Wergeland began his career as a chemical engineer studying deuterium hydride under Tronstad at NTH in Trondheim, before turning to nuclear physics [10-61]. On his first trip to Norway in 1940 Wirtz had visited Wergeland in Oslo, offering apologies for German brutalities in Norway. He had been at pains to reassure Wergeland that he was not a passionate Nazi, and urged more scientific collegiality between Norwegian and German scientists. Now in his reunion with Wirtz, Wergeland, in turn, hinted that he could arrange safe passage for Wirtz through Sweden to England, without explicitly revealing his own involvement in the Norwegian Resistance network. Wirtz declined; like Heisenberg, he felt obliged to remain in Germany and look after German science in difficult times.

A similar offer was in the cards for Suess in 1944. Norwegian SOE agents were instructed by Tronstad to inform Suess, should he visit Norway again, that passage to England was assured if he stopped by in the Norwegian legation while passing through Stockholm. A code word only familiar to Suess and Jomar Brun was intended to assure Suess of the authenticity of the offer. Unfortunately, Suess did not visit Norway again while the war lasted. After the war, he told Brun he would gladly have accepted the offer [10-62]. His intimate knowledge of German developments could have significantly aided the Allies in arriving at a realistic assessment of the true state of the German wartime uranium effort.

Wergeland, we might add, was soon summoned to England, along with Lise Meitner, to join the atomic-energy program then in full swing there. Wergeland declared his willingness to serve the Allied cause abroad, but, on moral grounds, not as a practicing nuclear physicist. In the event, he remained in Oslo as an XU agent, keeping watch on German scientists [10-63]. Meitner, too, declined on similar grounds to participate actively in bomb development work. She undoubtedly contributed useful information to English intelligence, however; possibly directly through Rosbaud to SIS, but more likely through a different channel to London involving Njål Hole [10-64]. Hole was a young nuclear physicist, trained at NTH in Trondheim, where he obtained his doctorate in 1938. He was 'ex-filtrated' from Norway in September 1941, at the same time Tronstad left Trondheim. Tronstad saw to it that Hole received an appointment in Siegbahn's Institute for the duration of the war. There he interacted with Meitner on a daily basis, and, using the code-name Epsilon, reported regularly to London through XU in Stockholm on German scientists passing through Sweden and what he could learn from them [10-65].

10.4. Exit Jomar Brun

'Our juice is very important', began a message from Tronstad to Brun, delivered via courier in early spring of 1942, shortly after Skinnarland's return to Norway. The courier brought greetings from 'Mikkel' (the Fox, alias Tronstad), and bade 'Master' (Brun) send by return courier complete information about German plans for the Vemork installation, technical specifications of same, and a preliminary plan for the construction of a similar heavy-water plant in Great Britain or the UK [10-66]. Brun's report was entitled 'High-concentration plant for enrichment of glucose'. In it, he provided a detailed list of apparatus for bringing a comparable high-concentration plant on line; essentially an inventory of all Vemork equipment. He estimated approximately 3 months for its construction, at a cost of roughly 100 000 Norwegian kroner. Brun included a report on his recent visit to the KWI in Berlin-Dahlem, and on German plans for augmenting Vemork production with one or more catalytic-exchange stages.

The courier returned in May, and this time Brun sent back sketches of the full complement of Vemork apparatus, along with news of German plans for a major plant expansion at the Hydro facility. As earlier, all reports destined for the office of FO4, the section within the Norwegian High Command dealing with the Resistance movement in Norway—that is, with Milorg—were microphotographed by Dr L D'Arcy Shepherd, a trusted friend at Rjukan. They were sealed in the form of undeveloped film in small glass ampules; the ampules, in turn, were inserted into 'unopened' toothpaste tubes. In retrospect, Brun considered this mode of shipment 'one of the most foolish things he undertook', in light of the popularity of the method—a common smuggling technique all too well known to friends and foes alike [10-67]. In the event, all shipments reached their destination without incident.

The latest courier also brought a query from Mikkel as to the feasibility of bringing a significant quantity of 'juice' to some location near Vemork, from where it might be conveyed to Sweden and thence to the UK by reliable couriers. Mikkel also wished to know whether Master himself was willing to join him in Britain. Master's reply, sprinkled with enigmatic language, read partly as follows:

> The shipment of juice is presently virtually impossible, since all plants here and elsewhere are guarded by the foreigners. Attempts to interrupt production are, consequently, ruled out as well. Apparently flak-batteries are soon to be installed at several of the plants . . . , also around here. Hopefully a visit from your end is not contemplated. At any rate, be sure to warn us in advance. As far as I can tell, the juice poses no dangers, at least not for several years. We are anxious for information from Mikkel concerning what the juice is used for. That is, if for other than atomic physics.

> Concerning relocation of Master, I take it to be under consideration for a somewhat later time frame, which is fine, since he prefers to put it off as long as possible. In the event of a hasty departure becoming forced by events, it would still have to be arranged in advance, and in strict confidence.

Master [10-68]

In a subsequent transmission, Brun added that if his own departure for England was seriously considered, he would insist on bringing out his wife as well. The only chance of smuggling out juice seemed to be by landing a small British plane, say a Lysander, on some frozen lake in the mountains above Vemork, perhaps Lake Såheimsbotn. Returning to technical matters, Brun's message informed London that steps have been taken

> ... to implement a method, suggested by the scientists from abroad, for the exchange over a nickel catalyst. It does not yet appear too viable, but then it is too early to judge. Combined production of juice now amounts to about 100 kg per month. Plans are also under way to produce juice at the two other plants within Master's firm, but the outlook for accomplishing that is more remote. Significant production can only be expected $1\frac{1}{2}$ years hence, at the earliest. At that point the nearby plant should be producing around 500 kg per year, the more distant plant 200 kg per year. By that time production in the main plant will most likely have reached 1500–2000 kg per year, implying a combined production of not more than 2700 kg per year...

All is well with Mrs. Fox and the two cubs.

Master [10-69]

In August Tronstad received additional sketches and photographs, this time of the newly expanded high-concentration plant at Vemork, shipped by the usual 'export' route via Sweden. This material would play a central role in the planning of operation 'Gunnerside' in the wake of the disastrous operation 'Freshman' presently being hatched in Britain without Brun's knowledge. Among the August sketches were some of stage 6 in the cascade of cells retrofitted with a catalytic exchange unit. That unit became operational on the 14th of the very next month, September of 1942, incorporating 'nickel catalyst number 5233' [10-70].

On the evening of 22 October, the telephone rang in the Brun residence in Vemork [10-71]. The caller, identifying himself as a student named Berg, said that he had arrived at Rjukan with a message from 'Mikkel' and needed to see Brun in person. 'I'll see you here in an hour', answered Brun, assured by the code name that Mr Berg could be trusted. The Vemork plant and its little community, perched as it was on a rock ledge on the south side of the ravine formed by the Maane River, was only accessible from the main road to and from Rjukan by a suspension bridge crossing the ravine. The bridge was guarded by German soldiers. Brun phoned the guard house, identifying himself and explaining that he was expecting a student visitor from the University of Oslo, whom they could safely let pass.

When Berg arrived, having walked the $2\frac{1}{2}$ miles from Rjukan, it turned out he was Fredrik Bachke, a member of the Special Mission

Command of the Resistance. He had direct orders from Major General Wilhelm Hansteen of the Norwegian High Command in London to see that Brun left for London without delay to fill an important post [10-72]. Brun's wife, too, was authorized to come along. Despite the late hour, the shock of learning of her husband's involvement in clandestine affairs, and worry over parents and relatives to be left behind, Fru Brun, née Tomy Johanne Svingjom, rose to the occasion. She served up a rare meal of canned fruit and a leg of mutton, washed down with a good bottle of wine, and Berg spent the night with them. The next morning Brun accompanied him back across the suspension bridge, also to meet the school bus. The guard gave them no trouble, knowing Herr Brun, and since they were joined on the other side by a crowd of bubbling school children which lent an air of gaiety to the occasion.

The Bruns had all of two days to prepare for their departure. Confiding in factory foreman Thor Abrahamsen, Brun collected a packet of technical drawings, documents, and photographs weighing several kilos. All the items were subsequently microphotographed by contacts in Oslo, who saw to it that the material reached FO4 in London. Brun also packed 2 kg of heavy water 'expropriated' from the Vemork production line, plus poison ampules obtained from a friendly Rjukan apothecary, in case they were caught fleeing the country.

> My wife and I left Vemork on October 24, 1942 [recalls Brun]. I had explained to my superior, Director Nilssen at Rjukan, that I needed to make a trip to Oslo to visit an eye specialist... Thanks to the thorough preparations of Tronstad and Welsh, it had been arranged for escorts to guide us through the woods to Gravberget in north Finnskog [near the Swedish border due east of Elverum in south-central Norway]. As we arrived at a river close to the frontier, our guide explained that he dared not risk accompanying us any further. Two bridges crossed the river, and he felt the upper one to be the safest. Not finding it, we crossed the lower one instead. That was fortunate. After we had crossed over the border into Sweden, we learned that the Germans had staged a raid at the uppermost bridge just as we were about to cross. It was the only time, all during the war, when we were actually in real danger. [10-73]

After a delay of several days in Stockholm, Brun and his wife departed Bromma Airfield for Scotland on 9 November, and arrived in London two days later. There they were installed in the De Vere Hotel in South Kensington. The same evening Commander Welsh and Major Tronstad came calling [10-74]. They immediately brought Brun to a conference with Reginald V Jones of the Air Intelligence Section at SIS and F Charles Frank, his assistant. Both Jones and Frank were particularly interested in Brun's impressions from his meetings in Max Planck Haus at the KWI earlier that year.

Only after meeting with Tronstad was Brun informed of operation 'Freshman', the operation conceived by SOE for putting the Vemork heavy-water plant out of business. In fact, the operation was far along in the planning stage by the time Brun learned of it. Brun's departure from Vemork was carefully timed by Tronstad to preclude his presence there when the installation was attacked by British commandos. The very day the Bruns took off from Bromma, 18 October, an advance party of Norwegians, already parachuted onto the Hardanger Plateau, made their first wireless contact with London, assisted by a storage battery provided by Einar Skinnarland's brother, Torstein Skinnarland. Though the planning for 'Freshman' was virtually complete, Brun, who now went under the pseudonym Sverre Hagen, was able to provide Tronstad and the operation's planning staff with the final, necessary information on the placement of apparatus in the basement of the electrolysis building, and the best route into the plant.

Brun also learned from Tronstad that General Director Aubert had been very bitter upon being informed of his departure from Hydro without his knowledge. After the war, with Aubert dead, Aubert's friend Professor Jens Bache-Wiig told Brun that Aubert characterized Brun's unauthorized modus operandi as a 'dagger in his back'. 'Needless to say', adds Brun, informing Aubert 'would have been impossible. Besides, the courier who came to fetch us was quite explicit on the need for absolute secrecy' [10-75].

As an added note of interest, Brun's wife would also contribute to the Allied cause. Soon after their arrival in London, she journeyed to Dumfries in Scotland to attend a military course. Following that, she went through a training course for corporals at the Norwegian War College, and then went to work at FO4's offices in London.

Chapter 11

AMERICA JOINS THE QUEST

11.1. Stirrings in the new world

While scientists on the Continent were becoming preoccupied with heavy water as the key to a chain-reacting pile, the rare isotope was largely ignored in the United States, at least well into 1941. Yet the breakthrough so assiduously sought in Paris and at Ivry, in Hamburg, Berlin and Leipzig, and in London and at Cambridge, was achieved in Chicago without the benefit of heavy water, scarcely a month after Jomar Brun fled Norway with more bad news about deuterium in German hands. In just one year, the Americans and their emigré colleagues had leapfrogged the Europeans from a placidly academic attitude toward nuclear physics during 1940–1941 to the brink of a concerted push for a nuclear bomb in December of 1941—the month America, too, went to war. These developments, presaging a scientific and technological undertaking of staggering proportions, are not of primary concern in our chronicle. All the same, a cursory account of the scientific mobilization of nuclear resources in the United States is needed at this juncture, if only to place certain of the Continental milestones in perspective. In particular, the American effort would, in typical American fashion, produce heavy water measured in tons, not kilos, simply as a backup for graphite, by several interesting variations on the production schemes relied upon at Vemork.

We last touched on American developments in the closely contested race between the Fermi and Joliot camps to observe neutrons from uranium fission. Realizing that natural uranium and water both consumed neutrons faster than they were produced in fission, Fermi, like Joliot, cast about for a less ravenous neutron moderator than plain water. As fate would have it, at about the time Bothe reached his unfortunate conclusion about carbon at Heidelberg, measurements at Columbia and at Princeton indicated positive results with graphite, *provided it was extremely pure*. Gambling on the success of a graphite-moderated, power-producing pile [11-1], Fermi was slower in abandoning his gut feeling that 'there was little likelihood of an atomic bomb, little proof that we were not pursuing a chimera' [11-2]. By March of 1939 he was nevertheless sufficiently concerned to broach the US Navy on the matter. And that autumn the irrepressible Szilard, aided by his Hungarian colleagues Eugene Wigner and Edward Teller, prompted Einstein into writing his celebrated letter to President Roosevelt.

Alas, Einstein's letter led simply to the establishment of a committee. The Advisory Committee on Uranium, under the chairmanship of Lyman J Briggs of the National Bureau of Standards, in turn spawned several more committees. Indeed, the initial American fission frenzy in early 1939 had subsided as quickly as it came about. Save for the chain-reaction studies at Columbia, centrifuge studies by Alfred O Nier at Minnesota (confirming conclusively that uranium-235 was the isotope that fissioned with slow neutrons), and a push by Lawrence and company at Berkeley that would identify both neptunium (element 93) and plutonium (94), 1940 came and went in the United States with nuclear studies conducted in a leisurely fashion—mainly focused on power production via slow neutrons (as in France). Not so in Britain, recalled Arthur Compton.

> The British approach was from the beginning... focused specifically on a definite weapon whose construction was roughly outlined. Though the Americans were aware of this weapon as a possibility, it was more than a year later that it became for us the focus of attention. In 1940, it was still difficult for us in America to concentrate our thoughts on war, while for the British it was their prime concern. [11-3]

Committees are good and well, provided they lead to concrete results. What turned the tide in the United States, rescuing nuclear physicists from stagnation, was such a committee—to be sure, a British committee. The British move to strengthen research ties with the United States began in an early meeting of the MAUD Committee. Professor A V Hill was about to cross the Atlantic on a scientific fact-gathering mission, and G P Thomson asked him to report back on anything of interest. The imminent fall of France underscored the urgency of closer Anglo-American cooperation. Professor R H Fowler, son-in-law of the late Rutherford and a scientific heavyweight himself, was leaving for Ottawa to establish scientific liaison between research efforts on mutual military research problems, and was empowered to share British developments with their Canadian and American colleagues. In the event, his mission was rendered somewhat passé by the Tizard Mission in August of 1940.

Splendidly organized by Henry Tizard, the avowed skeptic of nuclear energy for military purposes, his mission was, in fact, mainly concerned with sharing air defense secrets and Naval bases in the West Indies with the Americans, in return for lend-lease destroyers. (The Norden bombsight, long the most desirable American military secret device in British thinking, was no longer considered of vital importance to Britain.) The centerpiece of the British offerings was demonstrated before American scientists and military officials in Washington: the cavity magnetron valve Otto Frisch had skirted under Oliphant at Birmingham. The 'most important item in the whole Lease-Lend' [11-4], the magnetron, carried around in a wooden box with its lid fastened by thumbscrews, played an important role in fostering trans-Atlantic atomic cooperation by alerting the Americans to the advanced state of much British wartime research. It also broke the ice over lingering American doubts about British willingness to put all their cards on the table. Shortly afterwards, James B Conant, President of Harvard, arrived in London for the purpose of organizing an office for furthering the exchange of information, and before long a trickle of American scientists passing through grew to a steady stream.

More important for the exchange of specifically nuclear information was the fact that, from the early fall of 1940, Fowler received minutes of the MAUD Committee for transmission to the US authorities. At least one of these, albeit reporting on French developments, proved of great interest. For its part, an offshoot of the Briggs Committee churned out fission reports of a general nature, but still focused on power generation, not bombs. The real turning point came in July 1941, when Thomson personally handed copies of the MAUD Report to Vannevar Bush and James Conant. Here were specific plans for producing a bomb based on fast, not slow, neutrons in uranium-235 separated by gaseous diffusion on a massive scale [11-5]. It brought at the same time a sobering reminder that, as of late, much of the expertise of Nazi Germany's Kaiser Wilhelm Institutes in Berlin-Dahlem—the birthplace of nuclear fission almost three years earlier—had been set aside for uranium research.

Vannevar Bush, then President of the Carnegie Institution of Washington, DC, was by now the central mover in marshalling American forces of science for war. Shortly after the outbreak of war in Europe, Bush had taken it upon himself to act as a sounding board for the US scientific community in convincing the Roosevelt Administration of the need for action, by setting up a national science organization. He soon had the ear of Roosevelt's closest advisor, Harry Hopkins, and then Roosevelt himself. By June of 1940 he seized programmatic responsibility for uranium research, by reorganizing and expanding the Uranium Committee into a scientific body of substance in the form of the National Defense Research Committee—a body even authorized to allocate research funds. With the MAUD Report in hand, Bush had a long conversation with the President and Vice President Henry A Wallace, briefing them on the British findings. Roosevelt instructed him to set the wheels in motion to ascertain whether a practical bomb was really in the cards, and at what cost. He was to regard all decisions as the purview of a Top Policy Group consisting of the President, Vice-President, Secretary of War Henry L Stimson, Army Chief of Staff General George C Marshall, Bush, and Conant, but the technical effort was not to proceed beyond research and development without further presidential authorization.

On 6 November, one month and a day before Pearl Harbor, a committee under Arthur Compton of the University of Chicago confirmed the basic conclusion of the MAUD Committee on the practicability of a transportable fission bomb at the cost of $50 to 100 million in isotope separation. On 19 January 1942, Roosevelt responded to Bush's latest cover letter accompanying the Compton Committee Report, now as Commander-in-Chief of a nation at war with Japan, Germany, and Italy. 'V.B. OK—returned—I think you had best keep this in your own safe FDR' [11-6]. The nation had his approval for the production of an atomic bomb.

11.2. 'Graphite versus deuterium' once more

As early as July of 1939, Szilard had urged Fermi to throw caution to the wind and start a large-scale experiment with carbon as a moderator without further ado. 'If we waited for the absorption measurements we would lose three months...' [11-7]. The trouble was, Szilard reckoned that upward of 50 tons of carbon (and 5 tons of uranium) was needed simply 'as a start', yet he himself disposed of no funds whatever. Fermi did, but was weary of a full-scale commitment simply on the basis of 'the obviously optimistic hypothesis that carbon should have no absorption at all for neutrons' [11-8].

Besides, what was really wanted was heavy water, but that was an even more remote proposition; Szilard was not sure whether it was 'physically possible' to obtain even the smallest amounts [11-9]. Fermi dismissed heavy water simply as 'too expensive' [11-10]. What saved the day was the allocation, in February 1940, of a paltry $6000 in Army funds from the fledgling Briggs Uranium Committee for raw materials for the Columbia experiments, including four tons of graphite—the first outlay in funds for the American uranium project.

With a small store of graphite in hand, mainly thanks to Szilard's untiring energy, Fermi and Anderson got back to work. On the seventh floor of Columbia's Pupin Laboratories, they began assembling an experiment for determining the neutron distribution in a stack of the coal-black material, similar in design to that of Halban and colleagues half a year earlier. Columns of graphite, typically four feet on the side and ten feet high, were stacked up high enough to make it necessary for Fermi, a short man, to precariously climb on top of the slippery mass of extruded bars. A radium–beryllium neutron source was located at the bottom, and the diffusion of neutrons up and out of the column was studied with neutron-sensitive rhodium foil detectors [11-11]. Fermi enjoyed two advantages over Halban the previous fall. He was much the better theorist, able to account for the finite size of his column in estimating the neutron-absorption cross section. Also, his graphite was purer than Halban's, again thanks to Szilard's perseverance. By May, he and his team had determined an absorption cross section for the graphite of about half of Halban's latest value [11-12]. This, then, was the physical basis for Fermi's confidence in a graphite-moderated pile (see p 170).

The fall of France shocked the United States and Britain into closer cooperation, as noted. In October of 1940, John Cockcroft, who had taken command of the Tizard Mission when Sir Henry returned early to Britain, attended a meeting of the President's Advisory Committee on Uranium at the National Bureau of Standards in Washington, DC. There he spoke not only of the ongoing, widespread nuclear research in England, but also of the work by the French earlier at the Collège de France. Though the uranium problem was less urgent than radar along the English shores, who knew what the future might bring, and he felt the need to impress on his American colleagues the importance of nuclear research [11-13]. A little later a copy of Halban and Kowarski's own report to the MAUD Committee on their researches under Joliot at Paris and Ivry reached the US as well. In mid-November Cockcroft also visited Columbia University, where he conferred with Fermi whose attention was diverted only with difficulty from his stacks of graphite blocks interspersed with lumps of uranium metal. The last time they had met was at Bologna in 1937.

A month later, December of 1940, Halban and Kowarski's heavy-water experiment at Cambridge achieved a qualified success, at least in their own opinion. A repetition of their experiment so hastily abandoned at Ivry, it featured a 60 cm rotating sphere containing a slurry of uranium oxide powder and 120 kg of the Norwegian heavy water (leaving 65 kg in reserve), surrounded by a liquid hydrocarbon. A neutron source was placed in the center, and measurements of the neutron intensity were taken at various distances from the center with dysprosium detectors. The result, a very slight increase of neutrons (if not even as much as in the tantalizing results in Leipzig by Heisenberg and the Döpels), was enough to convince Cockcroft [11-14]. Returning to Britain from his mission to America, he wrote Fowler, the British liaison officer in North America, almost at once.

> Halban has obtained strong evidence that the D_2O_1U [deuterium oxide–uranium] slow neutron reaction will go…. He is now proceeding to try carbon. Can you now press Urey to develop his ideas for producing D_2 in quantity? [11-15]

Harold Urey, preoccupied as he was at Columbia University with developing gaseous diffusion for separation of uranium-235, had not lost track of heavy water. He had already been alerted to 'the heavy-water problem' attending chain reaction studies by his colleague Hugh

S Taylor, the British physical chemist long settled at Princeton University, late in the previous fall. (Taylor, born in Lancashire, England had been a professor at Princeton since 1922. He was not a newcomer to heavy water, being one of the first to prepare it in quantity by electrolysis as early as 1933.) The occasion had been a meeting in Cambridge, MA, where various methods for concentrating deuterium had been discussed briefly. On Cockcroft's urging, Fowler passed the news of Halban's apparent success to Urey on about 1 February 1941. In reviewing the origins of the wartime American heavy-water project two years later, Urey recalled in a letter to Conant that 'Dr. Fowler called our attention to Halban's work, in which the conclusion was reached that the chain reaction was definitely possible with heavy water, using probably a homogeneous system, but they believed certainly so for a heterogeneous system' [11-16]. Briggs, too, likewise alerted by Fowler to the news from Cambridge, took the Halban results seriously. One month after hearing from Fowler, he wrote E O Lawrence on the subject. Lawrence was the west-coast mainstay of his Uranium Committee; only months earlier Briggs had enlisted his aid in determining the thermal-neutron capture cross section of deuterium—a matter 'important enough to justify a good deal of serious effort' [11-17]. 'Prof. Fowler has advised me that he told you about Halban's experiments with heavy water', he now wrote. 'In view of the possible importance of this information, I believe it very desirable that it should be kept strictly confidential' [11-18].

Urey began at once looking into methods for mass-producing heavy water. He soon concluded that the diffusion of hydrogen gas as a means of separating heavy hydrogen was not the way to go, being 'definitely discouraged by the Kellogg Company' which controlled patents for such separation processes. The distillation of water was another possibility, but on closer examination 'looked very expensive and quite beyond the reach of any program which was envisioned by the Committee headed by Dr. Briggs' [11-19]. The catalytic exchange between hydrogen and water seemed much more promising, and seemingly amenable for support by the Briggs Committee. Briggs himself concurred, but with at least one reservation. 'If we keep jumping around we are not going to get anywhere', he warned in reply to Urey's request. 'I think that the heavy water project should be looked into carefully, but I should be sorry to see it interfere in any way with what you now have on the books' [11-20]. What Urey 'had on the books' was, as noted, uranium-isotope separation, the quickest route to a weapon in the opinion of most [11-21].

While Urey was shopping around for suitable methods for separating deuterium, the Americans asked their British colleagues to review the Halban–Kowarski experiments on which so much seemed to hinge. James Chadwick, a cautious experimenter if ever there was one, went over the Cambridge team's laboratory notebooks with great care. In May, he concluded that 'in my opinion the results can be accepted with confidence and I think their claim that the fission chain is potentially divergent is fully justified' [11-22]. A committee under Compton to review the American fission work was less convinced, finding the Cambridge results 'encouraging' but far from clear-cut; a repetition of the experiment with perhaps a ton of heavy water, produced in a dedicated pilot production plant, was advisable in their view.

The basic problem with the Cambridge experiment was the scant number of neutrons diffusing out of the uranium oxide sphere, a parameter which Halban and Kowarski labeled Q_{ext}. Their experiment yielded a value $Q_{ext} = 1.06 \pm 0.02$. In view of the large uncertainty, the system was only *potentially* divergent—all the authors claimed, in fact. This was not good enough for the Americans; a clear-cut indication of divergency would require a value for Q_{ext} 'comfortably' above unity, though a sensible margin was anybody's guess. Moreover, the amount of heavy water required was enormously sensitive to the precise value for the neutron parameter. Only more heavy water would settle the issue.

Urey talked the problem over with Pegram and Fermi. As cautious as Chadwick, Fermi agreed that heavy water was bound to be useful, if nothing else as a coolant for a graphite reactor, but would not admit that deuterium could be a better moderator than graphite until he saw exhaustive proof to that effect [11-23]. That was all the encouragement Urey needed. Fortified as well by the Compton committee's conclusion that heavy water held the hope of being 'the most promising method now on the horizon', he returned to Briggs who held the purse strings. Briggs, though lukewarm about heavy water, could not very well ignore the committee's advice to set up a pilot plant. He agreed to award the Standard Oil Company a contract for $50 000 to set up a pilot plant in Louisiana. The plant would study the liquid–gas phase exchange reaction under high pressure, while Urey and Taylor turned to the further development of a catalyst [11-24]. Later in the year, Pegram and Urey toured Britain, lugging steel helmets and gas masks wherever they went on the insistence of the State Department. Among others, they met with Halban and Kowarski at Cambridge. Urey, the discoverer of heavy water, had never beheld more than a test tube worth of the rare liquid. He was all the more astonished when Kowarski handed him a two-gallon can of D_2O! At any rate, Urey and Pegram's account of what they learned in England was, arguably, even more important than the MAUD Report in persuading their countrymen that a nuclear weapon was a practical possibility.

Halban paid a return visit to the United States in early 1942, braving the U-boat-infested North Atlantic by ship, as a weak heart precluded his flying at high altitudes. He arrived amidst a suggestion put forth by the Tube Alloys Directorate, which shortly before had supplanted the MAUD Committee for handling all British fission work, that he be put in charge of a heavy-water research program under British control in North

America. Halban got along fine with Urey, whom he met again during his American visit. Urey thought the Tube Alloy suggestion a good one [11-25]. The notion of Halban continuing his work on heavy water in the United States or Canada had grown out of the ongoing interaction between Imperial Chemical Industries (ICI) and the defunct MAUD Committee; the British had sounded out Urey on the matter during his own visit the previous fall. However, Halban did not fare so well with some of the other American leaders he met. His connections with ICI were suspect, as was his Germanic background, and his unwillingness to join the heavy-water effort simply as a worker under American direction ran counter to American policy—certainly to Conant's. If the British insisted, let Halban go to Canada, where there would be no question of difficulties of 'extra territoriality' [11-26].

Moreover, by then the Americans were starting to forge ahead on their own initiative, concluding that they could not learn much more from the British—a conclusion totally at odds with the assumption on which Halban's visit was predicated. The British, too, were becoming keenly aware that their American allies were pulling ahead of their own effort. The American project, conceived only months before, was already utilizing both university and industrial resources on a scale far greater than would ever be possible in wartime Britain [11-27].

With Fermi's increasingly promising results, first at Columbia and then at Chicago, the decision was finally made by Compton to proceed full-speed with graphite, despite lingering skepticism of Conant. On 1 April 1942, Briggs, Compton, and Conant spent a day thrashing out the relative merits of graphite and heavy water, and the die was cast.

> This was shortly after Halban had met with the S-1 [Uranium] Committee... and had argued his point of view. Taking up the cudgels for heavy water, I then argued strongly that we should go slow on the Chicago development and concentrate our efforts on the alternative method which would involve heavy water with the idea of producing element '94' by that route. Dr. Compton convinced me that under the schedule then set up for heavy water from Trail (which as a matter of fact turned out to be much too optimistic) he stood a chance of getting through by this method many months before one could hope for success along the heavy water route. Dr. Briggs and I, therefore, were converted to Dr. Compton's view. [11-28]

So graphite it was—a momentous decision reached just as German Army Ordnance was pressuring Norsk Hydro for stepped-up heavy-water production at its Telemark plants, and Leif Tronstad was beseeching Jomar Brun via courier for information about German heavy-water progress and how Britain, too, might benefit from ongoing activities at Vemork.

11.3. North American heavy water in abundance

The name Trail, in Conant's 7 April letter to Urey, embodies the initial phase of the belated, yet enormous heavy-water effort within the Manhattan Project, as the American atomic bomb project became known for short. The city of Trail, population 10 000, is nested in the Columbia River Valley, six miles east of the 'Gold City' of Rossland, in the Boundary Country of British Columbia barely across the US–Canada border from the State of Washington. Trail began as a Columbia River stern-wheeler port in the early 1890s, serving the miners as they followed the lure of gold, silver, and copper in the region. In 1896 a hydro-electric smelter was constructed at Trail, which grew into Consolidated Mining and Smelting Company of Canada Ltd (Cominco), one of the largest lead and zinc smelters in the world.

When Urey and Taylor began looking into a successor to the pilot plant in Louisiana for large-scale production of heavy water by the hydrogen–water exchange process, they realized full well the advantage of using hydrogen produced by electrolysis, since hydrogen thus prepared has a head start in possessing a slight enrichment of the heavy isotope. A British subject, Taylor also knew that Cominco had an existing hydro-powered, electrolytic-hydrogen plant for ammonia production—the closest counterpart to Norsk Hydro's facilities and the largest producer of electrolytic hydrogen in North America. Before the end of 1941, he visited Trail and convinced Cominco management that it would be practical to operate heavy-water equipment as a loop in the existing ammonia plant installation [11-29].

Jomar Brun, too, was aware of the possible importance of Cominco for wartime production of heavy water. In one of his courier dispatches from Vemork to London during 1942, he reminded the Allies that 'one place potentially suitable for production is Consolidated Mining & Smelting Co., Trail, Canada, where there exists an electrolysis plant of roughly 20 000 kW capacity' [11-30].

Though ultimately the Trail effort would produce ten times the heavy water resulting from the German effort in Norway [11-31], the American project would prove hardly more beneficial for its intended wartime purposes than would the German, if for different reasons. However, though heavy water had been reduced in importance at an early stage to America's 'second line of defense', it was prudently retained as a potential backup for graphite as a neutron moderator. In mid-1942 contracts were let with Consolidated Mining & Smelting for a heavy-water plant at Trail. Construction began in September of 1942, the very month Norsk Hydro's expansion at Vemork under German pressure was completed.

The method used to isolate heavy water at Cominco was more sophisticated than at Vemork, being carried out in two separate operations involving a primary and a secondary plant [11-32]. The primary plant concentration utilized a combination of catalytic exchange and electrolysis. The deuterium concentration was brought to about 2.37% purity in a cascade of stages of decreasing size, each stage featuring a catalytic exchange tower and a battery of electrolytic cells. Both platinum and nickel catalysts were used, neither rare metal being as desperately short in supply as in Germany's Army Ordnance, though platinum on activated charcoal was found to be the best; not only did platinum work well, but also it was easily recovered by burning off the charcoal. The secondary plant used electrolysis alone to bring the concentration from 2.37% to 99.8% purity.

Platinum on charcoal, a catalyst developed under Hugh Taylor at Princeton, was, incidentally, independently found to be an excellent heavy-water catalyst by a team including Jomar Brun at Imperial Chemical Industries at Billingham in northeastern England, where Brun was assigned after escaping from Norway [11-33]. This catalyst turned out to be far superior to Harteck's nickel catalyst, and went under the code name 'Russian oxide' [11-34].

The Trail plant thus incorporated features of the original Norwegian electrolytic process and a considerably improved variation on the subsequent exchange process of Harteck and Suess, with the final concentration about the same [11-35]. Individual units at Trail were subcontracted to various organizations, among them E B Badger & Sons, The Harshaw Chemical Co., Stone & Webster, Stuart Oxygen of San Francisco (an American firm specializing in electrolytic concentration of deuterium), and several universities, including Minnesota, Northwestern, and Princeton. Standard Oil Development Company was responsible for the basic design of the entire heavy-water plant under the direction of E V Murphree and F T Barr, and Cominco naturally took responsibility for the adaptation of their hydrogen plant to the heavy-water operation.

With the mobilization of the American uranium project into an all-out effort in December of 1941, Compton was placed in charge of the chain-reaction work (though Conant retained overall charge of the American project, including Compton's reactor team, through the summer of 1942). A professor of physics at the University of Chicago, Compton decided to consolidate all the widely scattered participants in this technical area, including the Columbia group, into a central organization at Chicago dubbed the Metallurgical Laboratory. Fermi himself relocated without enthusiasm. The Columbia team had prospered with that mixture of theory and experiment that suited him to a tee; the Metallurgical Laboratory loomed as an ungainly, bureaucratic organization in which the Columbia team ran the risk of being swallowed up without identity. Moreover, his travels out of New York were hampered by his classification as an enemy alien—a classification struck down by President Roosevelt on Columbus Day 1942. Urey, too, had his misgivings, but saw the wisdom of Compton's decision. 'I am very sorry to see Fermi leave us', he wrote Compton. 'He is somewhat unhappy about the matter but I think he will do everything he can to help the problem anyway. In any case, it seems to be a very good idea to get the group together in one place' [11-36].

By May 1942 the Columbia team was settled in Chicago, assembling a succession of exponential experiments—so named after an exponent entering into the mathematical expression for the neutron-diffusion process. Each assembly was a lattice structure of cans of uranium oxide or lumps of metallic uranium spread throughout a mass of graphite; both the uranium and graphite purity increased as time went by, and the neutron reproduction factor rose correspondingly, steadily approaching unity. By October enough uranium and graphite was on hand to start building a full-fledged, chain-reacting pile beneath the stands of Chicago's disused Stagg Field, under Zinn and Anderson's direction and Fermi's overall supervision. The pile grew to a flattened ellipsoid, measuring 766 cm in equatorial diameter and housing 6 tons of uranium oxide organized in lumps distributed throughout 250 tons of graphite. Strips of cadmium strategically inserted here and there served to keep the reaction under control, with buckets of cadmium salt solution hefted nearby in case of unforeseen catastrophe. Since there was some doubt whether the dimensions as planned would be sufficient for the system to reach criticality, the whole structure was erected inside a huge tent of balloon-cloth fabric that could, if needed, be evacuated in order to eliminate parasitic-neutron absorption by atmospheric nitrogen [11-37].

The memorable day dawned on 2 December 1942. At 2:20 p.m. Fermi instructed George Weil to 'take control rod 21 out another twelve inches' [11-38]. Fermi nodded to Compton, 'this is going to do it'. The cursor on his slide rule flew back and forth. The moment was at hand. The pile was critical, displaying a reproduction factor k of 1.0006, with the neutron intensity doubling every two minutes. It ran for 28 minutes at a power level prudently not exceeding half a watt.

The success of CP-1, as the first Chicago pile was called, confirmed the decision of General Groves, who had taken charge of the Manhattan District of the Corps of Engineers, newly formed that summer from the Office of Scientific Research and Development under Vannevar Bush, to go with graphite *in lieu* of deuterium. In the event, construction of the heavy-water plant at Trail was beset by one difficulty after another: low priority (Grove's decision as well), conflict with the engineering demands for synthetic-rubber production, and frequent breakdowns. The discreetly phrased monthly production reports from Cominco reveal a mixed bag of optimism and pessimism, echoing perhaps the ups and downs of Norsk Hydro's monthly reports half a year earlier [11-39].

December 14, 1942:
400 pounds of platinum catalyst now on hand, awaiting tests. The present program calls for the completion of the whole project by April 1, 1943.

January 12, 1943:
Catalyst contamination; Hugh S. Taylor is *en route* to Trail to look into the problem

March 11, 1943:
In view of delay in production of Product no. 80, negotiations have begun for the purchase of an additional batch of Product no. 43. Castings for condensers and reboilers are faulty.

May 6, 1943:
Castings still unsatisfactory, nor is erratic catalyst behavior yet understood.

To the consternation of Conant, the completion date for the Trail facility slipped slowly but surely, as he reminded Urey in no uncertain terms.

It is disappointing to all of us that the work at Trail has proved to be so slow, and it is very fortunate that we did not abandon the carbon route as some advocated a year ago with the assumption that the Trail program would live up to the optimistic hopes of those who were concerned with it. [11-40]

Eger V Murphree, Urey's contact with Standard Oil, could only agree with Conant's sentiments in his comments on Conant's harsh letter to Urey, though on 1 July Murphree reassured Conant that 'the first-stage towers at the Trail plant were started up June 14' and 'passed all tests' [11-41]. Though Urey himself was disappointed in the long equilibrium time proving inherent in the production process [11-42], by mid-June the facility had produced enough heavy water for Fermi to make some measurements [11-43]. These were the first thermal neutron capture cross-section measurements on deuterium since E O Lawrence's measurements on Briggs' request in the winter of 1940. Newly back from a visit to the Metallurgical Laboratory, Urey wrote to Conant on 21 June:

Fermi's results with 15 kilos of heavy water from Trail, containing about 99.7% D_2O, with the nature of the impurities not closely determined, indicate that the absorption cross-section of D_2O is extremely small.... While Fermi's experiments are not entirely satisfactory to him because the absorption on only 15 kilograms of D_2O is too small to measure, it does now seem certain that a homogeneous heavy water pile will operate; in fact, it may well be that Halban's original estimates are nearly correct. [11-44]

Ergo, deuterium absorbed virtually no neutrons, and was a perfectly good moderator. Alas, it was too late; the success of Fermi's graphite reactor CP-1 was history.

The exchange between Urey and Conant in the late spring of 1943 must have rankled Conant no end. It also emphasizes the shifting priority of heavy water in American thinking, nagging doubts over the 'Halban results', and the pros and cons of bringing Halban to American shores, along with the 185 kilos of heavy water spirited out of Norway so long ago.

Urey to Conant, April 2:
In the past week or ten days I have been trying to look more carefully at the whole problem of heavy water and its place in our general program. My conclusion at the present time is that it is probable that we should have foreseen that the use of heavy water would be the easiest, quickest, and cheapest method of accomplishing the production of element 94.

Fundamentally... I believe the reason for a non-vigorous program of heavy water during the first half of 1942, was due... to doubts on the part of the [Briggs] committee and its advisers, as to the full validity of Halban's experiments. I also believe that an important hindrance to a vigorous program lay in the fact that separate groups were charged with research, with the production, and with the use of heavy water; thus neither Mr. Murphree nor I took an active interest in the use of heavy water for the production of power and element 94, and the relative feasibility of engineering procedures as compared with graphite pile.

Shortly after Halban's results were reported to us by Dr. Fowler, the suggestion was made, probably originally by the British, that Halban should come to this country to work on the use of heavy water. This was a proposal of which I approved.... It was not possible, however, to bring Halban and his supply of heavy water to this country during 1942.... It is my opinion that if Halban could have been brought to the country during the past two years, that [his] experiments... would have been repeated, using much better techniques, and that we should have had, long before this, all that could have been gotten from the amount of heavy water in his possession. [11-45]

Conant to Urey, April 7:
From the time of my association with the S-1 [Uranium] project... I never heard anyone suggest that a repetition of Dr. Halban's experiment with the amount of heavy water he had on hand would be of any significant value. [11-46]

Urey to Conant, April 16:

In regard to the usefulness of Halban's heavy water, it seems to me that nearly every time I have discussed the question of Halban's coming to this country, someone had raised the question if it were not possible to get Halban's heavy water without having him come to the United States.... Some of us agree that Halban as individual would be useful, though I think that the general consensus of opinion was that there were others in the United States who could do the work equally well but that without the heavy water, no one could do the work. [11-47]

Urey to Conant, June 21:

Had the results, which we have now [from Fermi's measurements on D_2O] been secured earlier [with Halban's heavy water], we might have been stimulated to greater speed in the production of heavy water, and we might have had even more experiments as early as last December, to enable us to design heavy water piles with confidence.

The time that might have been saved by the use of this heavy water has now been irretrievably lost. There is nothing more to be gained by the use of their heavy water, for our own supplies will be adequate for even more extensive experiments soon. [11-48]

Conant to Urey, June 29:

I find myself in complete disagreement with your letter of 21 June. I find that your statement... is based on misinformation and erroneous analysis of the various factors involved.... I never assumed that Halban's results were in error.... All the decisions which were made... seem to me to have been made on the assumption that the results were essentially as presented then and, as you now say, the new work of Fermi indicates. This being the case I do not see how our having confirmed what we assumed to be the facts could have changed our program. I hope very much that the British will agree to re-establish interchange with us in the near future.... [11-49]

Alas, the Anglo–American exchange of information bogged down in a state of mutual suspicion in 1942, as noted, though in June of that year Roosevelt and Churchill decided that the bomb should be built in the United States, backed by all the resources that both sides could spare. In the event, the British wound up distinctly junior partners in the alliance.

As for the Trail plant, it was finally coaxed into producing about 40 lb/day of 99.8% D_2O [11-50]. Smarting from the frustrating experience in getting production up to speed, the American project leaders now turned to the du Pont Company instead. Du Pont was already shaping up as a full partner in General Grove's organization, as the prime contractor for chemical separation at the Clinton Labs in Tennessee and for plutonium production at the Hanford Engineering Works, as well as in a lesser role at the Metallurgical Laboratory. In their feasibility report on reactor designs in November 1942, du Pont management rated heavy-water-moderated piles very high. Grove responded positively, seeing du Pont's interest as a way to circumvent the difficulties at Trail. Du Pont was asked to forego sophisticated schemes based on catalytic exchange or electrolysis, and proceed with deuterium production by the crude but more direct method of fractional distillation of steam, based on the slight difference in the boiling points of heavy and ordinary water, augmented by electrolysis for a final concentration stage. Though more expensive than the Trail process, the distillation method promised faster production by utilizing excess steam capacity at several du Pont ordnance works. This was duly undertaken at three separate ordnance works that the company was building at Morgantown, West Virginia, Childersburg, Alabama, and Newport, Indiana [11-51]. These plants operated over a two-year period on a budget chosen as the lowest of three alternative options offered by du Pont, and were producing about 2400 lb of 99.8% D_2O per month (half the planned output) when shut down in late 1945 [11-52]. They produced a total of about 23 tons at a cost of $28 million— really a drop in the bucket compared to the nearly two billion dollars spent for America's fission project as a whole.

By 1944 there was, consequently, more than enough heavy water for the 6.5 tons needed to fill the world's first heavy-water reactor CP-3, first named P-9 after the American code word for heavy water. CP-3 was built at Argonne, Illinois, ostensibly as a pilot reactor backup should the Hanford graphite production reactors fail; it was as much a ploy by General Groves to calm troubled waters among unhappy Chicago physicists. They were uneasy with du Pont's plans for the Hanford reactors on technical grounds, antagonistic over du Pont's management style, and even more convinced that heavy water was the way to go in the wake of Fermi's CP-2 tests on heavy water from Trail.

CP-3 consisted of a cylindrical aluminum tank, six feet in diameter, filled with 6.5 tons of heavy water and surrounded by a graphite reflector approximately 2 feet thick [11-53]. Through the top of the tank were suspended 136 natural uranium rods sheathed with aluminum. The reactor reached criticality on 15 May 1944.

To be sure, CP-3's plutonium output was modest: some 100 milligrams per day by the fall of 1944, by which time the first of the three Hanford production reactors, pile B, was operational. CP-3 nevertheless had its comeuppance. Pile B went critical near midnight on 26 September 1944, and ran for about an hour at a record power level. The reactivity then began dropping steadily, and within a few hours the pile was dead. It was coaxed back to life, ran for another twelve hours, then faded again. Not by accident, Princeton theorist John Wheeler was on hand, as was Fermi who had overseen the startup of the pile, just like CP-1. They immediately suspected fission product poisoning—a non-neutron-absorbing fission product decaying to a neutron-absorbing daughter product starving the chain reaction and shutting down the pile. The culprit was almost certainly xenon, a daughter of iodine with 150 times the appetite of cadmium for thermal neutrons. A call went out to Samuel Allison at Chicago—Compton's right-hand man at the Metallurgical Laboratory. Allison passed the word to Walter Zinn at Argonne, who had just shut down CP-3 for the day. Zinn got the heavy-water pile going again, ran it at full power of 300 kW for twelve hours, and confirmed the xenon effect without difficulty.

Halban's fortunes would wax and wane, as we shall see in due course. His and Kowarski's small but famous stock of heavy water, smuggled out of Norway and then out of France, had yet another voyage through troubled waters in store. It wound up in Montreal, two years after their experiment in Cambridge, its purity undoubtedly suffering from much use and abuse.

Chapter 12

ACTION VEMORK

12.1. Germany's uranium machines: off to a promising start

December 1941 marked the beginning of Germany's reversals in its offensive on the eastern front. In the gray dawn of 5 December, fresh Siberian troops charged the *Wehrmacht* lines amidst ground-level bursts of snow and red-hot shell fragments in the forests northwest of Moscow. On the 14th the Siberians punctured the line near Klin; about the same time Red Army cavalry struck over the frozen rivers south of the Capital. The Soviet counteroffensive was under way; Germany was now on the defensive. Adolf Hitler reacted swiftly. The week before Christmas, he took over as operational commander of the Army.

The very day the Siberians launched their attack, Erich Schumann, a high official in Army Ordnance, wrote the leading scientists of the Uranium Club, warning them that continued support of nuclear fission research could be justified only if benefits could be expected in the foreseeable future. The Army was concerned with winning the war; in that context, and in light of the changing military fortunes, nuclear power, the focus of the German uranium project, was largely irrelevant.

Reflecting the new priorities, and despite the insistence of Army Ordnance scientists (among them Kurt Diebner) that nuclear power must nevertheless continue to be pursued in the interest of the German economy in general and for military applications in particular (e.g., as power plants for submarine propulsion, as Joliot had also speculated on much earlier), the Army command was leaning toward relinquishing control of nuclear power altogether [12-1]. In early February 1942, Schumann's boss, General Emil Leeb, called a meeting with Schumann and Albert Vögler, the President of the Kaiser Wilhelm Society. Leeb suggested the Kaiser Wilhelm Society was the right vehicle for guiding the research henceforth—a suggestion heartily endorsed by Vögler.

It did not quite go the way General Leeb suggested; the Reich Research Council snatched the prize away from the Kaiser Wilhelm Society in a power play, as we shall see. That maneuver notwithstanding, we have here an interesting development in the winter of 1941–1942, concerning the uranium projects in Germany and the United States, respectively. In Germany, the course of events forced the conclusion that nuclear fission, at least as a route to super-explosives, was irrelevant to the prosecution of the war, and the Army abdicated control in favor of other agencies—in effect, handed nuclear physics back to the community of academic scientists, whence the project sprang in the first place. In the United States (as in Britain), the authorities came to quite the opposite conclusion. A nuclear bomb appeared practicable, as it was assumed the Germans had concluded as well, and control of nuclear research was entrusted to the Army from thence onward [12-2].

Shortly before Army Ordnance relinquished control to the Kaiser Wilhelm Society, Schumann scheduled a high-level research conference for the end of February at the KWI for Physics, the objective being a forum for the assessment of overall progress in the various branches of the uranium project at this critical juncture. In the end, the agenda for the three-day conference, held in KWI's Harnack Building on 26–28 February, listed twenty-five papers covering isotope separation, heavy-water production, pile configurations, and measurements of nuclear constants. For our purpose, it suffices to note that the papers on heavy water included a report by Karl Wirtz on heavy-water production in Norway [12-3], a paper by Hans Suess on catalytic exchange [12-4], and one by Karl Clusius and Kurt Starke covering laboratory experiments on the distillation of liquid hydrogen—a deuterium separation process rejected as too complex by Urey and colleagues, but actually carried to the pilot-plant stage near Munich in wartime Germany [12-5].

While Schumann's research conference was still in the planning stage, the Reich Research Council decided to hold a concurrent, popular series of lectures for a select audience of Government officials, senior party members, military commanders, and industrial spokesmen, co-sponsored with the Army. These lectures were intended as a showcase for nuclear physics and its potential applications, and were scheduled for half a day in the 'House of German Research' in Berlin-Steglitz on 26 February, so that participating scientists could make it to the purely technical sessions across town later in the day. Due to a major secretarial snafu, many of the influential invitees to the popular session were sent an invitation to the research conference instead. Among those receiving the wrong invitation were Heinrich Himmler, Field-Marshal Wilhelm Keitel, and Grand-Admiral Erich Raeder, all of whom declined 'with regrets' [12-6]. In fact, the popular lectures [12-7] proved so popular that the Minister of Education, Bernhard Rust, was emboldened, possibly buoyed by the absence of the chiefs of the armed forces, to wrest control of the nuclear-power project from the Kaiser Wilhelm Society and award it instead to his own Reich Research Council.

With Army Ordnance no longer in charge of nuclear power, the KWI for Physics, which had been requisitioned by the Army for war work, reverted to the aegis of the Kaiser Wilhelm Society. Since Peter Debye had been sacked and was no longer around after the Army took over [12-8], it became necessary to find a new director. A tug-of-war arose between the proponents of Bothe as the best experimental physicist available (Army Ordnance, the Kaiser Wilhelm Society) and Heisenberg, the dean of theorists (the Reich Research Council). After rather complex behind-the-scenes maneuvering, the post went to Heisenberg, reflecting the Research Council's new-found clout. His appointment also forced Kurt Diebner, the Army's provisional head, out of the institute—a result of murkier animosity between Diebner and Heisenberg's inner circle, chiefly himself, Carl-Friedrich von Weizsäcker, and Karl Wirtz [12-9].

Diebner's forced disappearance would have technical ramifications in due course, but for the moment the takeover by the Reich Research Council had little immediate impact on the German project. For that matter, the project's technical momentum could hardly be arrested. The burgeoning research is amply attested to in the presentations of the two back-to-back sessions in Berlin, and perhaps well illustrated by a recapitulation at this point of the various model pile experiments under way in Germany since the fall of 1940.

The first subcritical pile experiments (after Harteck's dry ice experiment in the summer of 1940) were, as may be recalled, directed by Karl Wirtz in the 'Virus House' on the grounds of the Institute of Biology and Virus Research down the street from the KWI for Physics in Berlin-Dahlem. The conspicuous feature of the Virus House was a circular, brick-lined pit, 2 m deep, which housed the early pile assemblies. B-I, finished in December 1940, was a cylindrical array, 1.4 m tall and about as large in diameter, of layers of U_3O_8 separated by thinner layers of paraffin wax as the moderator [12-10]. Immersed in the water-filled pit, the water acted as a neutron reflector and shield, for bouncing stray peripheral neutrons back into the core of the pile. A neutron source was lowered down a tube into the pile center. Not surprising to its designers, considering the small quantities of uranium and its moderator available, results were negative; in fact, the neutron count was reduced, not increased. A second attempt a few weeks later, B-II, did no better.

Robert Döpel's concurrent effort in Leipzig was at first equally discouraging. L-I featured a spherical arrangement of uranium oxide and paraffin wax, again immersed in a tank with water as the reflecting mantle, and supplied with a neutron source in the center of the sphere [12-11]. It, too, gave no hint of neutron multiplication. The question now was, would a uranium 'machine' (as the Germans were wont to call them) show any better performance with heavy water as the moderator, albeit not nearly as much as the 5 tons calculated to do the trick by Heisenberg, Döpel's theoretical partner in the Leipzig experiments. The answer was yes and no, after an experiment was completed with the Norwegian heavy water that was received half a year later. The first Norwegian heavy water was delivered to Döpel's laboratory late in the fall of 1941, augmenting the 10 kilos or so already on hand.

Since October of 1941, Norsk Hydro had delivered 850 kg of 'juice'—only 17% of the amount judged sufficient by Heisenberg and Fermi alike to ignite a self-sustaining chain reaction at zero power. The recipients were a closely guarded secret; all Jomar Brun at Vemork could inform Tronstad and company in England as late as the summer of 1942 was that the shipments went to either of two addresses: 'KWI for Physics, Boltzmannstrasse 20, Berlin-Dahlem', or 'Schering (Quinine factory), Leipzig'. In fact, the shipments went as follows. All heavy water was transhipped to Germany in aluminum 'bottles', each containing five liters of heavy water. Four bottles were packed in a wooden box, and went by train from Vemork to the Oslo quay office of Norsk Hydro. There the boxes were taken in hand by *Wehrwirtschaftstab Norwegen*, and the forwarding address was changed to that of an office of the *Oberkommando der Wehrmacht* under Lt Co Dr Basche in Berlin: *Forschungsabteilung I*, Hardenberg Strasse 10, Berlin-Charlottenburg. From Charlottenburg the military authorities forwarded the boxes to Berlin-Dahlem, or to Leipzig, where the heavy water was stored at the Schering plant [12-12].

L-II, Döpel's latest pile experiment, still gave no detectable evidence of an increase in the neutron count. However, the quantity of heavy water deployed was only 164 kg, along with 142 kg of uranium oxide. L-II was thus about the same in size (75 cm diameter) and weight as the heavy-water pile of Halban and Kowarski at Cambridge, one year earlier. Fermi's first 'potentially divergent' graphite pile would only be ready in the late spring of 1942 [12-13]. More to the point, if allowance was made for neutron absorption in the aluminum structure separating the concentric layers of L-II, it appeared that a positive neutron multiplication coefficient was close at hand.

Heavy water was not the only pile improvement in sight by then. Powdered uranium *metal* was now becoming available from German industry; to be sure, it too was in exceedingly short demand at first. The first uranium machine furnished with uranium metal was B-III at Berlin-Dahlem, featuring horizontal layers of uranium and paraffin in a spherical container. Though smaller than the cylindrical layer models B-I and B-II, B-III produced slightly more neutrons. However, its thunder was stolen at the Berlin research conference by Leipzig's latest contestant, L-III—the first pile to employ both heavy water, still only 164 kg, and uranium metal (108 kg). With one-quarter as much uranium as B-III, it, too, came close to a positive neutron count [12-14].

The showstopper was again at Leipzig, later that spring. L-IV, depicted in figure 12.1, housed three-quarters of a ton of uranium metal powder, in addition to the 164 kg of D_2O, and that was enough for a

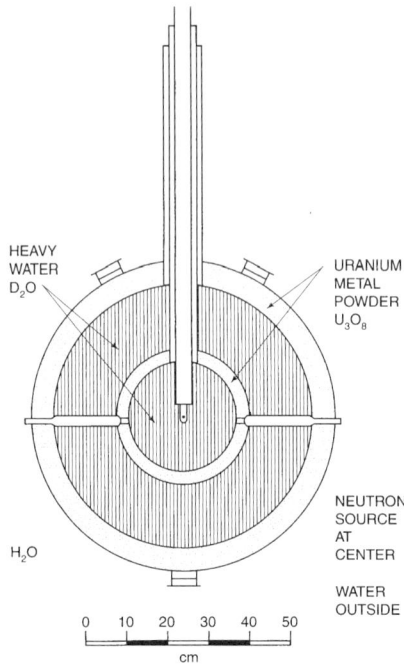

Figure 12.1. *L-IV, the promising Leipzig pile. Spherical shells of uranium metal powder, enclosing two regions of heavy-water moderator, are surrounded by a mantle of light water; a neutron source is mounted in the center. Based on information in Irving D 1967 The German Atomic Bomb (New York: Simon and Schuster).*

decisive jump in performance. This time there was no doubt that more neutrons, fully 13% more, were escaping the spherical uranium surface than were injected by the Ra–Be neutron source into the center of the pile. L-IV had the distinction of being the first model pile in the world exhibiting a positive neutron production coefficient, beating Fermi's subcritical pile by the narrowest of margins [12-15]. The latest Leipzig results were clear-cut, confirming a by then widespread theoretical rule of thumb: about 5 tons of heavy water and twice as much metallic uranium should suffice for a self-sustaining chain reaction.

The success of L-IV was marred by a serious accident, fortunately at the end of the experiment, which destroyed the pile; this incident precluded further iterations of this promising series of pile experiments, say by adding heavy water or by adopting cast-metal uranium instead of uranium powder—essentially the source of the mishap. Döpel and his lone technician, noting bubbles emerging from the immersed sphere, had hoisted it out of the pit and were opening it up to inspect the moisture content of the uranium—the machine having been under water for several weeks by then. Upon removing one of the aluminum covers, there was a dull thud, and a jet of flame shot out of the machine. The pile was doused with water, and gradually the flames subsided. Döpel had the precious heavy water in the inner sphere pumped out, and the machine was lowered back into the water pit. At this point, Heisenberg stopped by, on his way to hold a seminar. All seemed well enough, so he departed for his appointment. However, soon bubbles began rising from the bath again, then steam hissed forth; L-IV's uranium was clearly burning, heating up at an alarming rate. Heisenberg was called back. He and Döpel watched as the outer sphere buckled, then sprang for the door as the sphere burst, showering the laboratory with burning uranium. The local fire brigade arrived posthaste, and soon had the blaze in the building under control, but the fire within the sphere continued for a day and a half, until it petered out in a 'gurgling swamp' of burned-out fragments.

In the aftermath, Heisenberg, unhurt like Döpel, was congratulated by colleagues, tongue-in-cheek no doubt, for a successful uranium bomb. News of the chemical explosion reached British Intelligence soon after the event. More ominous rumors of the accidental explosion of a uranium bomb eventually reached the United States as well [12-16].

While the latest Leipzig experiment was still in progress, Heisenberg, along with Hahn, Diebner, Harteck, Wirtz, Jensen, Bothe, von Weizsäcker, Bagge, and several other scientists high in the uranium project, met in Berlin with Albert Speer, Hitler's Minister of Armaments and War Production, and his technical staff and advisors. Also present were Albert Vögler of the Kaiser Wilhelm Society, Generals Leeb and Schumann of Army Ordnance, General Friedrich Fromm, Leeb's superior, Field-Marshal Erhard Milch representing the Air Force, and Admiral Carl Witzeil of the Navy. This meeting, held in Harnack Haus on 4 June, was probably the most important gathering during the course of the German uranium project; in effect, it determined its fate.

In his presentation as spokesman for the assembled scientists, and mindful that the military heads had missed the briefing session at Berlin-Steglitz earlier in the spring, Heisenberg got right to the point of military applications of nuclear physics, and launched into an explanation of how to make an atomic bomb. In the question-and-answer exchange, Heisenberg took pains to warn, however, that a bomb was an economic impossibility in Germany for the time being. Speer pressed him for how much money he might need to carry on; Heisenberg, with von Weizsäcker chiming in, mentioned some ridiculously low sum. Giving him the benefit of the doubt, Speer offered to raise the amount to a few million marks. With the RAF bombing campaign beginning in earnest, he threw in an allowance for an underground bunker for a full-size uranium pile in the grounds of the KWI for Physics. General Milch, skeptical, asked some more questions about atomic bomb effects, shook his head at Speer, and left the Helmholtz Lecture Room. (Two weeks later Milch authorized mass-production of the V-1 flying bomb, part of a multi-hundred million mark program.) Speer closed the meeting, privately convinced the atomic bomb could no longer influence the course of the war [12-17]. Formally, some construction money was ensured for a project of limited scope, confined to the development of an uranium machine, pure and simple. The project was still alive, if only barely.

On 23 June, the day L-IV began to act up, Speer discussed the uranium project and the 4 June meeting with Adolf Hitler in Berlin—item 16 on a long agenda. Noted Speer simply afterwards: 'Reported briefly to the Führer on the meeting concerning atomic fission, and on the assistance we have rendered' [12-18]. That was that.

With some exceptions [12-19], a sense of complacency, or lack of urgency, was permeating Germany's uranium effort in and out of Berlin. (Speer is reputed to have expressed it as there not being 'much music in the thing' [12-20].) In London and Washington the Allies were watching the same developments with mounting apprehension. Many factors contributed to this growing unease on their part, as we shall touch on again. Among them were a lack of confidence in reassuring reports coming from Paul Rosbaud at the center of events in Berlin and Jomar Brun no less in touch with things in occupied Norway. Guarded disclosures by several senior German scientists, notably Suess and Jensen, to their Norwegian colleagues, were viewed with plain disbelief. Another factor was German secrecy, so effective that it may be said to have backfired, giving the Allies the misguided impression of important technical progress in German bomb research.

Be that as it may, six months after Speer's fateful meeting with the scientists in Berlin-Dahlem's Harnack Building, the Allies struck. Rjukan, the last bastion of defense in the desperate fighting of 1940 in southern Norway, had made it to the top of the British Air Ministry's Target List.

12.2. Freshman: an unqualified disaster

The bleak sun over Scotland was westering on 19 November 1942 as two Halifax bombers rose from Skitten airdrome near Wick, each towing a Horsa glider crammed with British Royal Engineers (sappers) and their equipment—guns, food, and plastic explosives [12-21]. On reaching cruising altitude, the bombers and their charges headed east into the gathering twilight over the North Sea. They were bound for the barren Hardanger Plateau north of Rjukan, 400 miles away in occupied Norway, whence Jomar Brun had fled about one month earlier. There, too, a party of Norwegian parachutists, code-named *Grouse* ('Rype' in Norwegian), was waiting in a base camp on the Plateau. The reception party was equipped with a Eureka ground-to-air homing device and torches (flashlights) for guiding the bombers to the landing zone on the Skoland Marsh, south of Lake Møs, where Einar Skinnarland had jumped last March (figure 10.2). Operation *Freshman* was under way.

Plans for *Freshman* had been set in motion five months earlier, as the result of a rising influx of coded reports from agents in 'Festung Europa', pointing to secret German weapons under development, from Peenemünde to Berlin-Dahlem. Thus news of a chemical explosion in Leipzig rattled SOE offices on Baker Street soon after the event. The news reached Leif Tronstad in London through Roscher Lund, the uncle of Bjørn Rørholt, his old compatriot in the Trondheim woods; the source of the information was none other than Paul Rosbaud [12-22]. Offsetting the sinister implications of the explosion, Rosbaud learned also of Speer's momentous decision at Harnack House that June, over coffee in a Ku'damm café with some of the scientists involved [12-23]. He brought the news in person during one of his many visits to Oslo, for transmission to Eric Welsh via Harald Wergeland and the XU network. By chance, his visit was shortly before Hans Jensen shared essentially the same piece of information with Wergeland's colloquium at the University of Oslo. As a result, the SIS in London was forewarned from at least two reliable sources that the German atomic program was less of an imminent threat than had generally been supposed.

That British authorities were nevertheless not placated about German intentions may, in part, be attributed to a difference in attitude between SIS and SOE. The task of SIS, Secret Intelligence Service, was pure intelligence gathering without ruffling enemy feathers; SOE, by its nature, was *gung ho* for action, be it sabotage or other forms of subversive warfare, and tended to put a sinister spin on anything learned from intelligence sources. Action with regard to the German uranium project was undoubtedly in order. The Germans had a two-year head start in nuclear research. Moreover, it was abundantly clear to the British, as it was to the Germans, that element 94 was not only likely to be fully as effective as uranium-235 as a super-explosive, but could be produced in a heavy-water-moderated pile of natural uranium. As of mid-1942, however, a self-sustaining chain reaction had yet to be achieved anywhere, and Germany was in control of the only heavy-water plant in existence, reporting a record monthly production figure for July. As we know, Tronstad was in regular communication with Jomar Brun and other engineers at Vemork, and the escalating German demands for production of 'juice' could hardly be ignored.

If anything, the Americans were more concerned than the British. At the Metallurgical Laboratory, the perennially worrisome Leo Szilard held forth about the German threat, fortified by a message of doom relayed by Fritz Houtermans from the Continent through a refugee who was leaving for America; a report on the threat, blown out of all proportions, eventually made its way to the British via Conant [12-24]. At any rate, in the spring of 1942 Tronstad had been approached by American officers about bombing Vemork, which he strenuously objected to. The high-concentration plant was tucked away in the basement of an eight-story structure of armored concrete. Extreme flying range from Britain, high mountains, and dubious weather conditions in late summer and autumn did not bode well for a successful bomber strike at night—the only British bombing practice at the time. (General Ira Eaker's American B-17 crews were training for daylight bombing, but were not yet operational.) Stray bombs were as likely to hit the liquid-ammonia storage tanks, endangering the whole Rjukan community.

Plans for action against Vemork were revived in July, when the British War Cabinet put Combined Operations Command under Vice-Admiral Lord Louis Mountbatten on notice to prepare for a ground-level attack sufficiently strong to put the heavy-water plant out of business. Combined Operations, in turn, approached SOE for advice; SOE allowed that it had, in fact, an advance team of specially trained Norwegian agents standing by in Britain to render assistance in just such an undertaking! The upshot was the first Allied glider operation of WWII. Thirty Royal Engineers of the First British Airborne Division underwent training for an operation code-named *Freshman*, scheduled for the November moon of 1942 [12-25]. Two Horsa gliders would each carry 17 men: one officer in command, 14 other ranks, and two pilots. SOE's advance party of four Norwegians [12-26] was to be dropped by parachute onto the Hardanger Plateau several weeks before the attack. Unlike the Royal Engineers, they were not to be in uniform and would not actively participate in the actual attack on the plant. Their job was to reconnoiter the plant and its environs, provide weather reports for the November moon period, guide the aircraft to the landing site on the marshy edge of Lake Møs, and lead the sappers to the Vemork target.

The main airborne force would make landfall at Egersund on Norway's southwest coast and fly directly across the mountains to the landing site. For the final attack, the sappers would advance in a formed body on the target along a windswept road on the north side of the Maane River Gorge, and cross the suspension bridge linking the hamlet Våer to Vemork across the gorge. This would entail eliminating two German sentries on the bridge. Alternatively, depending on the final recommendation of their *Grouse* guides, they could make their way down the 600 foot descent to the plant along the ladders flanking the penstocks leading down into the Vemork power station. That approach, however, entailed passing a German flak battery on the Vemork heights. After destroying, in order of priority, existing stocks of heavy water, the electrolytic production apparatus, and eight of the ten DC generators [12-27], the force would split into smaller groups and head for the Swedish border, several hundred miles away under winter conditions— by far the riskiest part of the mission.

In the event, the advance party did not have an easy time of it. After two aborted attempts, Sub-Lieutenant Jens Anton Poulsson and his three countrymen, heavily burdened with equipment, jumped on 18 October under a full 'bomber's moon', albeit 45 km from the intended drop zone, with additional equipment parachuted after them over a widely scattered area of the inhospitable countryside. It took them two days to collect it all, and then a violent snow storm broke over the party. When it finally cleared they found that the paraffin for their primus stoves had been lost, necessitating a change in the intended route to their objective to search for alternative fuel and additional huts. (Wood is very scarce on the Vidda, except for scraggy juniper bushes.)

After a forced march of two weeks under intolerable skiing conditions, they reached their operational base near Skoland Marsh on 6 November. Just as wireless contact with SOE London was established, the battery of the transmitter went dead. A fresh battery was obtained from Einar Skinnarland's older brother Torstein, manager of the nearby Møs Dam and someone active in the underground, like his brother. (For the sake of security, Einar Skinnarland, despite his prominent involvement in clandestine Telemark affairs, had not been informed of *Grouse* and the upcoming operation.) On 9 November, *Grouse* finally made contact with SOE headquarters, where concern over their fate had everybody on edge. Ten days later, 19 November, London signaled that *Freshman* was poised for takeoff at sunset that afternoon.

In Britain, meanwhile, the planning for *Freshman* was nearing completion by the time Jomar Brun arrived from Norway via Sweden, in the nick of time to supply vital details about the heavy-water plant and its *Wehrmacht* defenses [12-28] (and himself safely out of the way, as Tronstad had factored into the mission of spiriting Brun out of Norway). The troops had undergone rigorous physical training in the mountains of North Wales, and technical training with visits to English and Scottish hydroelectric and condenser plants similar to Vemork. The air crews, highly skilled to be sure, had exclusively flown Wellington and Whitley bombers up to that point, and had no experience towing gliders; consequently they, too, underwent their own training, including briefings by *Grouse* members before that team left for Norway. On arrival of the main force at Skitten airfield, the field was sealed off; to further avoid suspicion a cover story, already in circulation, had been put out involving gliders and long marches in an endurance contest between British and US Army Engineers for a mythical 'Washington Cup'.

On the afternoon of 19 November, the *Grouse* team awaited tensely at Skoland Marsh. It had been a clear day over Telemark, but soon after dark low-hanging clouds closed in over the site. About the time of *Freshman*'s expected arrival, the drone of aircraft engines was heard in the distance. Haugland turned on the Eureka set, communicating, in principle, with a Rebecca receiving device in the bomber tugs, and landing lights were lit in a prearranged L-shaped pattern. The sounds overhead intensified, then faded. No gliders arrived. Back at the base airfield, a faint radio signal was received twenty minutes before midnight. Believed to come from Halifax B, the second aircraft to take off, it read simply: 'Released into sea'. A second signal, in plain language, was picked up fifteen minutes later from Halifax A: 'Glider ditched into sea'. Nothing more was heard from Halifax B; Halifax A returned to base some hours later, without it Horsa glider. The pilot, Sqn Ldr A B Wilkinson, RAF, reported circling in thickening overcast in the target area, until shortage of fuel compelled the bomber to head back for Scotland. Over or near the Norwegian coast, the towline iced up and parted.

Two days later, the following announcement was heard over German radio:

> On the night of November 19–20th two British bombers, each towing one glider, flew into Southern Norway. One bomber and both gliders were forced to land. The sabotage troops they were carrying were put to battle and wiped out to the last man. [12-29]

The full story of what happened to *Freshman* was only pieced together in 1945. Halifax A and its Horsa had taken off from Skitten at 5:50 in the evening, Halifax B twenty minutes later. It was drizzling lightly over Wick, and cloudy over the North Sea and the Skagerrak; however, moonlight, essential for the operation, was guardedly predicted over the target area, with at least one important forecaster voicing serious concern [12-30]. Soon after takeoff, the telephone links between both tugs and their gliders failed—a bad omen. Halifax A made landfall near 10 000 feet over Lista, between Egersund and Stavanger, as planned, by which time conditions over southern Norway were deteriorating. The Halifax–glider tandem searched twice in vain for the drop zone, then turned for home. When the tow-rope snapped, the glider crashed, not into the sea as the pilot had believed, but on a snow-capped mountain overlooking Lysefjord near Stavanger. Eight were killed, and four seriously injured. Upon capture, the injured were administered drugs by a German doctor, then strangled or shot, and their bodies dumped at sea. The survivors were transported to Grini concentration camp near Oslo, repeatedly interrogated, and shot at Trandum, a small wood near the Capital where many patriots were executed.

Halifax B also crossed the coast at Egersund, cleared the mountains and, like Halifax A, searched for the landing site. After turning back, either its tow-rope parted, too, or the pilot let go to gain height. The glider crashed on a mountain slope in the Helland region, 25 miles northeast of Egersund, and the tow-plane flew into a nearby mountain. All aboard the bomber were killed, but only three aboard the glider. The fourteen survivors, reached by German infantry troops at first light the next morning, were briefly interrogated at battalion headquarters at Slettebø, near Egersund, then shot the same evening in accordance with

Hitler's standing order (*Führerbefehl*) that 'crews of sabotage planes are to be shot at once by the troops detaining them' [12-31].

That Vemork had indeed been the target of a sabotage mission was abundantly clear from the glider wreckage: guns and demolition equipment, winter gear, Norwegian currency, even a silk map with Vemork encircled in blue ink. The uniforms of the assault troops proved equally damaging [12-32]. The Germans responded in force. On 3 December, not quite a day after Fermi's reactor went critical in Chicago, a virtual army of *Wehrmacht* troops, Gestapo, German and Quisling police, descended on the intended landing site and the Rjukan-Vemork area, led by Colonel-General Nikolaus von Falkenhorst and *Reichskommissar* Josef Terboven, heads of the German military and civilian high commands in Norway. A state of siege was declared in the area, with the Germans making a wide sweep over the Vidda and many townspeople arrested [12-33]. The Rjukan garrison was strengthened, and all approaches to the Vemork plant sown with mines.

The *Grouse* team survived the uproar. Along with receiving the bad news from London, they were instructed to stand by for another attempt during the December moon. When the large-scale German raid got under way, they retreated westward into the interior of the Vidda, finally reaching Poulsson's own cabin, 'Svensbu', where they took refuge in grueling winter weather to await further instructions. For the Allies in general, the situation was no less grave. The Vemork plant was intact. The Germans knew what the British were up to; not only could further raids be expected, but also the abortive strike must have confirmed German suspicions that the Allies were pressing ahead on the development of a nuclear weapon [12-34]. Not only that; Jomar Brun, in his technical report on the heavy-water situation in Norway, and on other possibilities for the production of heavy water in countries controlled by the Axis, warned that 'compared with these possibilities of the Axis countries the prospects for achieving a similar production within the Allied countries appear rather poor' [12-35].

12.3. Gunnerside: a qualified success

The *Freshman* disaster brought the question of bombing the Hydro facilities to the fore again. Reluctantly, Jomar Brun and Jan Herman Reimers, another of Tronstad's old physical chemistry students at Trondheim and now assigned to the Air Ministry, worked out a tentative plan for operation *Gerd*, with the objective of bombing the embankment of the Skarfoss Dam, the dam feeding both the Vemork and Såheim power stations—perhaps with airborne torpedoes. Alternatively, the motor-driven ventilating plant on the heights above Vemork might be put out of action with much the same result. Either action, they concluded, ran the risk of unacceptable loss of life [12-36].

That left what Tronstad had argued for in the first place: an attack on just the high-concentration plant itself by a small party of highly trained saboteurs—specifically ski commandos of the so-called Linge Company [12-37]. Colonel 'Jack' Wilson, the head of the Norwegian Section of SOE, went carefully over the idea with Tronstad, then broached it to Major-General Gubbins, Director of SOE; he, too, was persuaded after some misgivings. Within hours, the British War Cabinet gave its blessings, and Wilson and Tronstad boarded the overnight express for the SOE's Norwegian depot at Aviemore in the Cairngorms, the Scottish highlands. They were on their way to enlist Sub-Lieutenant Joachim Rønneberg, the obvious choice to lead Operation *Gunnerside*, as the mission was code-named.

Rønneberg was briefed on the general nature of the mission, to be undertaken jointly with the *Grouse* team still encamped on the Hardanger Plateau, renamed *Swallow* for the new mission. Also joining in would be Einar Skinnarland; he, too, was still at large on the plateau, despite his brother's arrest. As outlined in Operational Instructions No 1 [12-38], they would advance as a team in full uniform [12-39] on Vemork by a route different from that to be used by *Freshman*. Once there, reads paragraph 8 of the instructions,

> Gunnerside force will attack the storage and production plant at
> VEMORK with high explosives, so that present stocks and fluid
> in the course of production are destroyed. [12-40]

Afterwards, part of the group would remain in Norway to help organize home forces in the district, while the rest would escape to Sweden. If all went well, that is. Wilson and Tronstad impressed on Rønneberg the disastrous outcome of the Combined Operations raid shortly before. Unperturbed, Rønneberg's mind was only on selecting five good skiers to accompany him. The choice was easy: Sub-Lieutenants Knut Haukelid and Kasper Idland, Sergeant Fredrik Kayser, plus Corporals Hans Storhaug and Birger Strømsheim.

All six of the *Gunnerside* team had already been through the regular airborne infantry training course and specialized training in commando warfare at SOE's training camp at Glenmore in Scotland. Now, after further briefing by General Wilhelm Hansteen's Norwegian defense staff in London, they were shipped to Special Training School Number 17, where they would concentrate on the specific job at hand. The centerpiece of School 17, cleared of all other training personnel, was a wooden mock-up of the Vemork high-concentration plant, erected in a well guarded shack under the supervision of Jomar Brun, whose presence in Britain was not revealed to the men. They practiced ceaselessly laying dummy explosive charges among the apparatus, pored over aerial photographs of the plant and gorge, and fashioned two sets of plastic explosives and detonators, each set containing one charge for each of

Figure 12.2. *A few of the saboteurs in England, pictured with Leif Tronstad. Front row, left to right: Jens Anton Poulsson (in command of Grouse, later renamed Swallow), Leif Tronstad, and Joachim Rønneberg (in command of Gunnerside). Rear, left to right: Hans Storhaug (Gunnerside), Fredrik Kayser (Gunnerside), Kasper Idland (Gunnerside), Claus Helberg (Grouse/Swallow), and Birger Strømsheim (Gunnerside). Not pictured are Arne Kjelstrup and Knut Haugland (both of Grouse/Swallow), and Knut Haukelid (Gunnerside). Also missing are Knut Lier Hansen and Rolf Sørlie who, with Haukelid, took part in the subsequent Hydro ferry operation. Courtesy J A Poulsson.*

the 18 high-concentration cells by then in place in the basement of the electrolysis building.

The next and final stop for the team was Special Training School No 61 outside Cambridge, housed in a country manor all by itself in a large park—a staging spot for agents waiting to be air-dropped into occupied Europe on secret errands [12-41]. Here they whiled away time on site and in town, dividing their days between close-combat training and entertainment, marking time before the next moon.

A full moon was one thing; one discernible from the Hardanger Plateau in the winter of 1942–1943 was another matter. Fully three months the six soldiers cooled their heels, first at STS 61, and then on a solitary island in Scotland, while blizzards swept the plateau. To be sure, the months were even more trying for the 'swallows' braving the fierce elements in and around Poulsson's cabin. By December their supplies were practically gone; they were reduced to subsisting on a gruel of reindeer moss (lichen) and a few grains of oatmeal, the foul weather ruling out hunting.

On 23 January, at long last, *Gunnerside* was driven by lorry to a remote airfield, clad in British battle dress with white skiing suits on top, loaded down with parachutes and gear, including a poison capsule inserted in the emergency bandage. 'I cannot tell you why this mission is so important', Tronstad told them, 'but if you succeed, it will live in Norway's memory for a hundred years. Do your best, and good luck!'. With that, they clambered aboard a Halifax, and the four-engined bomber roared off the tarmac into the night. Some four hours later, the aircraft crisscrossed the Vidda, unable to spot the red lights laid out by Poulsson and his men. The pilot gave up, and set course back for Scotland.

The next full moon was 16 February. On the advice of *Swallow*, a new dropping point had been selected, and, for additional security, it was decided that *Gunnerside* should make a blind drop, and find its way unassisted to Poulsson's cabin. The drop went well, albeit not where intended, and in a rising gale. All the same, the two parties, toiling around in what became a blizzard of major proportions, met up about a week later. They pooled their resources, and when sufficiently recovered pressed on to the jumping-off point for the attack on Vemork, a cabin at Sandvatn where the Vidda dips into the Vestfjord Valley, some twenty miles from Rjukan.

Once at Sandvatn, Rønneberg assigned each man his duty in the forthcoming action. Haugland and Skinnarland had already left for a cabin at Skarbu, where they would remain in contact with London, and report on the results of the operation. Four, Kayser, Strømsheim and Idland with Rønneberg in command, would comprise the demolition party, and the remaining five, led by Haukelid, would make up the covering party. Meanwhile, Claus Helberg, a native of Rjukan, set off into town on a scouting mission. After the attack, he would bring news of the operation, success or failure, to the two radio operators still up on the Plateau. At that point, all of *Gunnerside* except Haukelid would head for Sweden, still in uniform. Poulsson and Helberg, in civilian clothing, would go to Oslo and await further orders; Haukelid and Arne Kjelstrup, also in civilian dress, would join Haugland and Skinnarland at Skarbu for possible future action in the region.

Down in Rjukan, Helberg got in touch with Vemork construction engineer Rolf Sørlie, a reliable underground man, and learned that the suspension bridge crossing the gorge was guarded by two sentries armed with Schmeisser submachine guns. The second way of reaching the plant on its rocky shelf, via the ladders along the penstocks leading down from the top of the mountain into the turbine hall, was also guarded by sentries and covered by booby traps and trip wires of every type. The astonishing good news concerned the third approach route, namely the railroad tracks running along a ledge hewn out of the mountainside. The tracks passed into the plant yard through an iron-barred gate in the high fence surrounding the entire plant. That gate, somehow, was unguarded.

Figure 12.3. *Vemork, as seen from the plateau to the north of the Vestfjord Valley. The electrolysis plant, on a rocky shelf overlooking the gorge of the Maane River, partly hides the power-station building behind it, fed by the penstocks leading down the mountain side. The suspension bridge across the gorge is visible at the lower left, as is the railway line, left center, leading out of the plant. Courtesy Norsk Industriarbeidermuseum, Vemork; negative No UF-129.*

The catch was that, to take advantage of this piece of good luck, the precipice leading up from the bottom of the gorge had to be scaled. This problem, too, Helberg had anticipated. He had personally found a few spots along the gorge where the rock wall apparently could be negotiated. Once inside the plant yard, they had to circumvent, if possible, a platoon of about 15 *Wehrmacht* infantry quartered in a barrack hut between the main electrolysis building and the turbine hall behind it. The guards on the bridge and penstocks were changed every two hours. Floodlights had been installed on the roof of the main building; by throwing a switch the entire plant area, including all access from Vemork to the hamlet of Våer across the gorge, could be bathed in light. There were many chances for things going wrong; with luck, however, the operation appeared feasible.

The final decision concerned extricating themselves after the attack—perhaps the trickiest part of the operation. Time would be of the essence, with the garrison in Rjukan only a few miles away. Both Haukelid, in command of the covering party, and Rønneberg argued that they flee by

Figure 12.4. *Vemork, viewed from the hamlet Våer across the Maane River gorge. The suspension bridge is quite apparent at the lower left. Courtesy Norsk Industriarbeidermuseum, Vemork; negative No UF-131.*

the fastest way out, the suspension bridge, even if it meant shooting their way across. Besides, climbing the gorge twice in the same night seemed much too exhausting. In the end, however, it was decided that killing German sentries posed too great a risk of reprisals against townspeople; the gorge would have to do. Once back across and on the road to Rjukan, they would not retreat the way they came, but head towards Rjukan. Closer to town they would climb up the Ryes Road, which zigzagged 1500 feet up the mountainside beneath a cable-car lift built before the war. The purpose of the lift was to enable the people of Rjukan to reach the Hardanger Plateau and enjoy sunshine in the long winter months when Rjukan itself lay in darkness.

All was quiet at Vemork on 27 February 1943. The weekly heavy-water production report, issued that day, indicates 4.6 kg for the previous 24 hours, up slightly from the average for the previous week [12-42].

The commandos struck that night, or during the night of 27–28 February 1943. Led by the covering party, weapons still unloaded to prevent accidental discharge, the men set off in single file on skis down the Fjøsbu Valley toward the Rjukan Road. After about 2000 yards, the woods thickened and they paused, leaving skis and rucksacks, as well as their white camouflage suits, hidden in the snow. In British battle

Figure 12.5. *The electrolysis building at Vemork (during its inauguration), as seen from the east. The Gunnerside party came down from the Hardanger plateau roughly in the hilly depression seen at the rear. They crossed the gorge at a location behind the photographer's vantage point, and approached the plant adjacent to the turbine hall on the upper left. Courtesy Norsk Industriarbeidermuseum, Vemork; negative No UF-237.*

dress, with guns and explosives pocketed, and a hefty pair of armorer's shears slung over Kjelstrup's shoulders, they proceeded on foot. The hum of Vemork grew ever louder, until the plant was in sight, bathed in moonlight across the gorge. At Rønneberg's silent command, they crossed the road and slid down the nearly vertical gully into the bottom of the gorge, using scraggy bushes and vegetation along the cracks as hand- and foot-holds. Fording the semifrozen Maane River, they began the difficult ascent of the south rockface, 600 feet up to the railroad shelf leading into the plant.

It was close to midnight by the time they reached the top. Once there, they advanced toward the barred gate, the throbbing hum of turbines masking any small noise they made. Hiding behind a small transformer station, they had a good view of the moonlit yard of the plant, and waited for the guards to change. Three minutes before midnight two German soldiers headed down the hill for the suspension bridge, followed some minutes later by two returning guards. All was quiet again. Kjelstrup severed the chain on the gate with his shears. They all stole into the yard, the covering party taking up preassigned posts, while Rønneberg led the demolition party to the electrolysis building, now looming above them. All doors were locked, but Jomar Brun had told them how to get into the heavy-water installation in the basement by way of a cable duct somewhere near ground level. Rønneberg and Kayser soon found the entrance, crawled in over a tangled maze of pipes and cables, and lowered themselves into the basement. An unlocked door opened upon the high-concentration apparatus, two parallel rows of nine cells each.

A speechless Norwegian night watchman gave them no trouble, just a blank stare. Rønneberg began pressing the sausage-shaped plastic charges around each of the water-jacketed electrolytic cells, just as they had practiced so often in Scotland, while Kayser covered both him and the watchman at gunpoint. About half the charges were placed when they were startled by the crash of broken glass. Strømsheim had smashed a basement window from the outside. Hurrying up, the remaining charges were laid, each was carefully checked, and fuses connected between them. The watchman was told to run for safety upstairs, while they scattered parachute badges as British calling cards in the little room. As they were preparing to light the fuses, the watchman pleaded that he could not find his glasses. A frantic search located them quickly, just as the night foreman appeared on the scene, startled to see the watchman with his hands up and three men in British uniform. Rønneberg ordered both Hydro men up the basement stairs at full speed, lit the final 30 second fuse, and the demolition party sprang for the door to the yard. Thirty paces from the building they felt the explosion, the sound muffled by the whine of the turbines and the basement walls.

The covering party heard the explosion as well, anticlimactic in its feebleness, they thought as they trained their tommy guns on the German barracks. The Germans heard something as well. The door to the hut was flung open, and a bareheaded soldier peered out. He listened intently, then disappeared back into the room. A moment later he reappeared with rifle and helmet, stepped outside, played a flashlight around, and tried a door to the electrolysis building. Finding it locked, he vanished around the building [12-43].

Swinging the gate shut behind them, the covering and demolition parties withdrew along the railroad track [12-44]. Just as they tumbled down the south face of the gorge, sirens broke into a piercing wail. Fortunately, no searchlights stabbed the night; the Germans must not have found the switch. Reaching the bottom, the saboteurs waded across the Maane River, now swollen by a thaw from a warm foehn wind blowing up the Vestfjord Valley. Up the north side they pulled themselves hand over hand. As they reached the rim, and sprinted across the road for their stashed equipment in the melting snow, German cars and trucks came roaring up the road from Rjukan. Making their way through the woods along the highway, the party found the abandoned zigzag trail and began the arduous climb up towards the Vidda, skis on shoulders. Struggling against a rising winter storm, they reached a vacant cabin where they rode out the weather.

Before the group broke up for all parts of the country, as planned, Rønneberg had one last, happy duty to perform as mission commander: dictating a report for transmission to London. The day after the attack, London was still in the dark about the outcome. In his journal for Monday 1 March, Tronstad noted simply. 'Gunnerside has presumably done its job Sunday night'. But the next day, Tuesday, there was no doubt. 'Enormous headlines today about G. Side, with fantastic accounts of attack on Vemork which undoubtedly must have succeeded from our point of view' [12-45]. Indeed, owing to some confusion, it took the better part of a week for Skinnarland and Haugland to get the coded message off to SOE-Norway in London:

HIGH-CONCENTRATION INSTALLATION AT VEMORK COM-PLETELY DESTROYED ON NIGHT OF 27TH–28TH STOP GUN-NERSIDE HAS GONE TO SWEDEN STOP GREETINGS. [12-46].

Alas, the mission would prove a qualified success. Vemork was down, but not permanently out of action as an electrolytic mainstay facility. Moreover, SOE-London had seriously underestimated the tenacity of German Army Ordnance. The Germans had a vested interest in ensuring the well-being of Norsk Hydro, their technological partner to the north.

Chapter 13

NEUTRONS DESPITE BOMBS

13.1. Aftermath at Vemork; neutrons in Berlin-Gottow

Shortly after midnight, 28 February 1943. The silence of the early Sunday hour was shattered by wailing sirens at Vemork. Chief Engineer Alf Larsen, who had succeeded Brun at the heavy-water plant, was about to make his exit from late evening bridge with a few friends in a residence adjoining the plant. On hearing the commotion, he got on the phone with Olav Ingebretsen, the night foreman still shaken up from his encounter with the saboteurs. Learning of the attack, placed at 1:15 a.m. in the laconic wording of Hydro's monthly report for February [13-1], Larsen at once called Bjarne Nilssen, the Norsk Hydro director for Rjukan, then headed for the main gate in his wood-burning automobile. Nilssen alerted the German garrison in Rjukan, and then he, too, took off for Vemork as fast as his own wood-burning motor car would take him. If he did pass the weary men in battle dress crossing the Rjukan Road from the gorge, he paid them no attention. Inside the plant, he joined Larsen, already surveying the wreckage.

High-echelon German officers and officials reached Rjukan from Oslo at the break of dawn, while *Gunnerside* and *Swallow* members were still struggling up the mountainside toward safety on the plateau above. The Germans were led by *General der Polizei* Rediess, and included Gestapo as well as German and Quisling police. On reaching Vemork, they herded Nilssen, Larsen, Ingebretsen, and Gustav Johansen, the hapless night watchman, into the wrecked high-concentration room for explanations.

The sappers had known what they were doing.   All 18 of the high-concentration electrolysers had their bottoms smashed, with the liquid contents flooding into the drains in the concrete floor underneath. Adding to the damage, flying shrapnel had punctured the copper tubing of the plant's cooling system; when Larsen first arrived on the scene, not long after the explosion, the room was one great shower, with the cooling water swirling what remained of the heavy water away [13-2].

Figure 13.1. *The morning after. The Vemork high-concentration cells after sabotage early on 28 February 1943. Photograph by Gunnar Syverstad a few hours later. Courtesy Norsk Industriarbeidermuseum, Vemork; negative No UF-227.*

The punctured tubing had been largely patched by the time Rediess and his party arrived, but the room was still a mess.

About 500 kg of D_2O in various stages of concentration (ranging between 10% and 99%, depending on the cell location in the cascade) had been lost, representing five months of production. Adding to that 400 kg in future production, the total loss came to nearly a ton of heavy water altogether [13-3]. Worse yet, the long-awaited expansion of the regular electrolytic heavy-water plant had only been operational since last summer, with monthly production finally set to double from 100 kg to 200 kg. (Stage 6, retrofitted as a catalytic exchange unit, was not damaged, as it was installed in an adjoining room.) Months of delay, perhaps a year, would be inflicted on Germany's uranium program. The timing was particularly unfortunate, in that it happened just as a new series of heavy-water-moderated pile experiments were under way at Berlin-Gottow.

Somewhat later the same morning, *Reichskommissar* Terboven arrived at Vemork from the High Command in Oslo, followed shortly by General von Falkenhorst, his military counterpart. Rediess explained what had happened, as it did not take long to piece together the train of events; obviously it was tied to Brun's sudden disappearance only months earlier. Enraged, Terboven had some 50 Vemork employees, including Johansen and Ingebretsen, and 10 Rjukan citizens, arrested and sent to the *Ortzkommandantur* at the Grand Hotel in Rjukan for interrogation. They might have fared worse, had not Falkenhorst intervened and affected their release, deciding it was a purely military operation staged from Britain [13-4]. While scathing in his public dressing-down of the German soldiers at Vemork, he expressed admiration for the operation as 'the best coup I have ever seen'.

> On the night of February 27–28, 1943 [reads Rediess's report to Berlin], about 1:15 a.m., an installation of importance to the war economy was destroyed at the Vemork factory near Rjukan by the detonation of explosive charges. The attack was carried out by three armed men wearing grey–green uniforms. They gained entrance to the factory by cutting a chain in the gate, and passed both German sentries and Norwegian watchmen undetected. From the effects left behind, it can be assumed that they came from Britain. [13-5]

Three armed men were, after all, the only commandos Johansen and Ingebretsen had seen. Falkenhorst interpreted the action differently, as the spearhead of Allied forces landing in strength on the Hardanger Plateau, and for the second time in only three months initiated a major sweep of the Vidda, only on a much larger scale. The plateau was sealed off, and upwards of 10 000 German *Wehrmacht* infantry, Alpine ski troops, German and Norwegian police, and Norwegian *Hird* detachments massed over the area, with low-flying aircraft crisscrossing and even some tanks thrown in. By then, however, Rønneberg's party was already on its way to Sweden. Haukelid and Kjelstrup were safely sequestered in a far corner of the Vidda. Haugland and Skinnarland, too, avoided capture in the mountains west of Lake Møs. That the Germans sealed the Vidda off permanently to Norwegian vacationers was to the advantage of the loosely reconfigured *Swallow* team, as it relieved them of attracting the attention of inquisitive countrymen in their travels around the snowy wastes.

Only Claus Helberg came nearly to grief in the sweep. On 25 March he ran into three Germans—they, like him, on skis. The Germans gave chase, but two of them soon gave up. One, however, gained slowly but surely on Helberg, who finally stopped in his tracks and the two of them shot it out, the German with his Luger and Helberg with his Colt. Helberg downed the German with a lucky shot, then pushed on toward a nearby lake. Failing to see a deep ravine in the gathering darkness, he plunged down a slope, breaking an arm as he fell. After a grueling trek for 36 hours he reached Rauland, where he went straight to the German Sergeant in command. Brazenly, he explained that he had been guiding a German party in search of Norwegian partisans, when he met with an accident. The German accepted his explanation, got him to a German first aid station and then to nearby Dalen where he could wait overnight in the hotel for a bus to Oslo the next day.

No sooner had Helberg settled down for the night, than he was awakened by loud trampling of boots in the hotel hallway. Terboven and Rediess had arrived with their entourage. The next morning all Norwegian guests in the hotel were arrested without further ado, Helberg included, and herded into a German bus bound for Grini Concentration Camp. At Drammen, after a short whispered conversation between Helberg and a loyal Norwegian girl on board the bus, she began flirting with the guard. With the guard distracted, Helberg jumped out of the bus and bounded into the woods, with Nazi bullets zinging past. He made it, found shelter in a hospital, and in due course escaped across to Sweden to resume his duties with Linge Company.

At Rjukan, Falkenhorst signed an order transferring the Rjukan *Wehrmacht* commander to the Russian front. Needless to say, the garrison was greatly strengthened and the minefields extended. The penstocks, still a weak link at Vemork against attack, especially from the air, were camouflaged with 800 artificial trees, and smokescreen pots installed on both sides of the Vestfjord Valley. Moreover, in view of RAF Bomber Command's subsequent attack on the Ruhr Dams, the Møs Dam was secured against aerial attack with a balloon barrage and torpedo nets. For his part, General Rediess shut down the Rjukan telephone exchange and closed the railroad station to passengers for the rest of the war. A state of emergency was declared in the town, with an 11:00 p.m. curfew in effect [13-6].

The ten members of *Grouse/Gunnerside* received their just rewards, on the initiative of Winston Churchill himself. On receiving a detailed report on the mission, including the fact that there was no loss of life on either side and that the heavy-water plant had been destroyed without harming the hydrogen plant needed for Norwegian agriculture, he scribbled across the bottom of the report: 'What rewards are due to be given to these heroic men?'. Mission commanders Rønneberg and Poulsson received the DSO, the other men the MM or MC. Tronstad, Wilson, and Brun each received a most secret appointment as Honorary Officers of the OBE [13-7].

Things were going better for German Army Ordnance in Berlin, just as the uproar was at its peak in Rjukan. One upshot of Heisenberg's falling out with Kurt Diebner, after taking over the KWI for Physics, was that Diebner—not as bad a physicist as his detractors would have him—returned to research. In so doing, German pile research received a shot in the arm. At the start of the war, Diebner had pulled together a group of young experimental nuclear physicists at the Army's laboratories in Berlin-Gottow, as may be recalled (see p 136). This small but enthusiastic team, grateful for doing their military service in the laboratory, began designing their own model uranium machine in the summer of 1942, using cast-off uranium oxide which was becoming more plentiful, *in lieu* of newly available, albeit still scarce uranium metal. G-I, the first experiment, used horizontal layers of compacted uranium oxide cubes cast in paraffin wax, alternating with layers of paraffin—25 tons of U_3O_8 and 4 tons of paraffin. The effort, casting paraffin-wood formers, and slowly filling them with toxic uranium oxide by the spoonful, proved worth the trouble; though the experiment gave no clear-cut sign of any neutron increase, it yielded a higher rate of neutron production than any of the earlier uranium oxide tests [13-8]. If nothing else, it demonstrated the importance of uranium cubes over flat plates as a first approximation to a lattice design.

The next attempt called for uranium *metal* and heavy water, which substances were widely known to be an important step towards ensuring a self-sustaining pile. At the same time, calculations by Karl-Heinz Höcker, a young colleague of Heisenberg now at German-occupied University of Strassburg, performed on Heisenberg's urging, suggested that a better lattice approximation would be one incorporating uranium *spheres*, providing the spherical radius equalled the neutron mean-free path; the Gottow group reached the same conclusion independently. However, certain practical problems attending a lattice of small spheres necessitated a compromise in the form of a spherical vessel, 75 cm in diameter, containing horizontal layers of metal uranium cubes embedded in frozen heavy water, alternating with layers of frozen heavy water alone, surrounded by a spherical mantle of light ice. (Frozen heavy water obviated the need for neutron-absorbing support material.)

The uranium metal for the new Gottow pile experiment came from Works No 1 of Degussa (Deutsche Gold und Silber Scheideanstalt) in Frankfurt am Main without much trouble, but where to obtain the heavy water? The bulk of the heavy-water stocks in Germany were then under Heisenberg's control, stored in the basement of the KWI for Physics in Berlin-Dahlem. According to Heisenberg, it was kept in a large tub, and he planned to pull the plug if a bomb ever seemed imminent [13-9]. However, soon after the Gottow group was transferred to the PTR under Esau, Diebner's boss since March and the official 'Plenipotentiary for Nuclear Physics', Esau phoned Heisenberg and ordered his 600 liters of heavy water moved from the KWI to the low-temperature laboratory of the Chemisch-Technische Reichsanstalt, ostensibly due to its vulnerability to air attack until the underground bunker was completed at the KWI. Behold, G-II, chilled to −11 °C, yielded the best neutron-production coefficient of any heavy-water pile to date; it even outperformed the famous L-IV by a factor of 1.5 in neutron counts [13-10].

Unfortunately, the 600 liters seized by Esau was well-nigh all of the Norwegian heavy water to be had in Germany, what with the Vemork disaster only a month earlier. Compounding the situation, any remaining heavy water in Germany would be urgently needed as catalyst to restore the equilibrium circulation and resume production at Vemork once physical repairs of the electrolyzers had been completed; such a step would avoid the very time-consuming process of starting from scratch. The only other potential source of heavy water in occupied Europe appeared to be two Montecatini electrolysis plants in Italy—one near Merano and one at Cotrone, with a combined power consumption of 68 000 kW, or half that of Vemork. Harteck thought it possible to use the Italian plants for an enrichment to around 1%, whereupon the material could be shipped to Germany for further processing at I G Farben's Leuna works [13-11]. Both Esau and Harteck had personally inspected the Merano plant during the winter 1942–1943, but Esau, for one, was not enthusiastic about the Italian option. Tests of Italian water samples were not very promising. Then again, he had good reason for believing that Vemork could be repaired rather quickly. Moreover, the plants at Såheim and Notodden were unaffected by the Allied attack, though not yet actually producing deuterium. (High-concentration apparatus was in place at both plants, and deuterium concentration had started; however, none had yet been drawn off.) At any rate, a Merano–Leuna arrangement was still a fall-back solution, in case of further troubles in Norway.

13.2. The Americans strike

To their dismay, Tronstad and Wilson learned from Rolf Sørlie via Einar Skinnarland, now a competent radio operator after much coaching by Knut Haugland on the Vidda, that Vemork would be back in full operation long before the one to two years anticipated by SOE. Norsk Hydro's problem was not so much rebuilding the physical plant. For one thing, only the high-concentration plant was knocked out, not the multitude of stages leading up to it—some hundreds of massive Pechkranz electrolyzers containing 43 000 electrolytic cells in all. No time was wasted. Within a week of the attack, Hydro engineers estimated approximately one month for completion of repairs, which were in fact well advanced by mid-March [13-12]. Working day and night under the watchful eyes of *Wehrwirtschaftstab Norwegen*, the repair crews had the high-concentration cells back in place by 17 April, in a cascade now expanded from 9 to 13 stages, scarcely six weeks after *Gunnerside* had struck. The problem was the precious liquid itself, every drop of which had gone down the drain. Starting from scratch with dry electrolyzers, filling the cells drop by drop from preceding stages in the cascade might take a year or more. Instead, the Germans resorted to the awkward but clever expediency of bringing back to Vemork 100 kg of 97.5% Norwegian heavy water from Germany, where its purity had been reduced somewhat through neutron-multiplication experiments; this heavy water was used to fill the empty electrolyzers so that the process could be restarted in a very efficient manner [13-13].

The first few kilos of heavy water were drawn off as early as May, and the production for June was no less than 199 kg, a record not to be repeated during the war [13-14]. The record output was in part due to the catalytic-exchange process, which by then was working well in one of the stages, augmented by deuterium regeneration by combustion in several other stages. Alas, the production for July fell to 141 kg, and by August was down to 100 kg, where the production would remain pegged for the next few months. The early setback was due to the US Eighth Air Force, by then open for daylight bombing business over Norway as elsewhere in occupied Europe. On 24 July, 120 American heavy bombers struck Norsk Hydro's light-metal and fertilizer production complex at Herøya, on Norway's southeast coast near Porsgrunn. It marked the Eighth Air Force's first attack on targets in Norway, and its longest mission undertaken to date [13-15]. Flying Fortresses of the 1st and 4th Heavy Bombardment Wings attacked in five waves and dropped 1500 bombs on the factories. The raid was carried out without prior knowledge of the Norwegian High Command in London, much to the annoyance of the Norwegian government in exile. However, it was scheduled for the lunchtime period to minimize loss of life; 55 workers were nevertheless killed, owing to a direct hit on an air raid shelter. The raid was a success from the point of putting *Nordisk Lettmetall*, the concern forced upon Hydro by the Germans, permanently out of business. Unfortunately, it was also partially successful in destroying the *Eidanger Salpeterfabrikker*, Hydro's synthetic nitrogen fertilizer plant. That plant, built in 1927 and which lay immediately adjacent to the light-metals works, was not a target in itself. At any rate, the raid, including the fact that it was planned without consultations with FO4 and SOE, disturbed Tronstad no end; for one thing, a similar strike on Vemork might be on the Air Ministry's 'Target List' as well [13-16].

The damage to the saltpeter works was, among other things, enough to indirectly affect heavy-water production at Vemork, owing to the fact that the Eidanger plant was the main outlet for Rjukan's ammonia production. Rjukan, in turn, drew its hydrogen by pipeline from nearby Vemork. As a result, Vemork was forced to reduce hydrogen production accordingly, and with it production of heavy water.

Reichskommissar Terboven, when informed, would not hear of any reduction in plant output at Vemork. Hydrogen production should continue at the normal rate; any excess hydrogen not needed for heavy-water production should be allowed simply to escape into the air. Upon hearing of this outrageous order, Bjarne Eriksen, who had succeeded

Axel Aubert as General Director for Norsk Hydro in late 1941, put down his own foot. Eriksen was at Herøya inspecting the damage when the *Reichskommissariat* order came through from Oslo. He immediately rescinded Terboven's order, a brave thing to do. He went as far as informing the *Reichskommissar* that he intended to recommend to Hydro's Board of Directors that Hydro cease heavy-water production altogether, since it entailed a constant threat of Allied air attack on Rjukan itself [13-17]. Eriksen even threatened to resign his position, if the board did not concur. Before the board had time to formally act on the matter, Eriksen was arrested by the Gestapo and sent to Germany for the rest of the war [13-18].

Eriksen's obstructionism nevertheless paid off. In due course, Berlin agreed that in the future Hydro would cease 'artificial' production of heavy water; that is, it would confine production to heavy water that was a natural byproduct of ammonia production. After the war, Jomar Brun learned from his German colleagues on the uranium project that they had cautioned against artificial production at Vemork (as at I G Farben); among other things, it gave the Allies a false impression of the importance attached to heavy water by Army Ordnance [13-19].

In his ceaseless campaign to establish closer cooperation between the Norwegian High Command, FO4, and the Allied military authorities in planning future attacks on industrial targets in Norway, Tronstad, too, seemingly met with some success. In particular, though he could not prevent Vemork from landing on the Air Ministry's Target List, he managed to have it listed as a special target, to be attacked by means *other than by bombing* [13-20]. However, the Allied supreme command remained gravely concerned over the German heavy-water situation, despite repeated advice to the contrary by Tronstad and his associates. Vannevar Bush, in a visit to Britain in June of 1943 had pressed General Ira C Eaker of the American Eighth Air Force to add Vemork to his own list of targets for daytime precision bombing. Eaker had demurred at the time, claiming that he would need higher-level authority before such an operation could be contemplated. Such authority was soon forthcoming, after Army Chief of Staff General George C Marshall, a member of President Roosevelt's Top Policy Group, was informed by the British representative in Washington, Sir John Dill, that the damaged Vemork plant would be back in operation far sooner than previously thought.

Marshall checked the matter with General Groves. Groves did not exactly trust the British where Vemork and Rjukan were concerned; he had been told nothing in advance of the disastrous *Freshman* operation, and only learned of *Gunnerside* somewhat casually on the eve of the raid itself. He persuaded Marshall that a high-priority bombing attack on Vemork was called for to end the heavy-water problem once and for all. Thence, the proposal for a strike slowly made its way down the

Allied chain of command, finally landing on General Eaker's desk at his Bushy Park headquarters, transmitted by Sir Charles Portal, Air Chief of Staff on behalf of the Combined Chiefs of Staff. As with the Herøya mission, the Norwegian authorities were kept in the dark about the final disposition of the raid. They were in good company; their ignorance was shared by the British Ministry of Economic Warfare, and even by the Supreme Headquarters of Allied Expeditionary Forces. In fact, in the end Vemork, all along a primary target, was upgraded to the status of additional backup target for two other simultaneous raids in southern Norway on 16 November: on the Knaben molybdenum mines 50 miles southeast of Stavanger, and Kjeller aerodrome north of Oslo. Vemork was now a target of opportunity as well, should either of those two primary targets be obscured by weather. The primary strike on Vemork was, as at Herøya, timed for a noon-time half-hour when most of the workers would be away at lunch. The striking force included the 100th Bombardment Group of the Third Bombardment Division, which had also struck Herøya on its historic mission four months earlier.

Four hundred and sixty B-17 Flying Fortresses and B-24 Liberators, representing three Bomb Divisions of VIII Bomber Command, 8th Air Force, rose into the cold predawn air over USAAF airfields in England on 16 November. Cumbersomely they stacked up in circulating layers at predetermined altitudes, while forming into groups and squadrons. Groups of the First and Second Bombardment Divisions were to attack the targets near Stavanger and Oslo, respectively. Groups of the Third Bombardment Division were bound for Vemork, as noted. The weather was bad over East Anglia: a minimum ceiling of 400 feet, turbulence, icing conditions. 'The success of our take-off and climb through the overcast was due to no skill of my own', wrote Major John M Bennett, in command of the 100th Bombardment Group operating out of Thorpe Abbotts, 'but to Lieutenant Owen D (Cowboy) Roane from Valley View, Texas' [13-21]. The forecast called for improvement in cloud conditions halfway to Norway over the North Sea, and target visibility over southern Norway until noon, with clouds closing in after then. Not the best of conditions, to be sure.

Of the total air armada droning out over the North Sea in divisional box formations, nearly two dozen bombers turned back with mechanical difficulties. For the rest, the trip across was uneventful, with no German fighters encountered. The formations approached the Norwegian coastline between Kristiansand and Stavanger, still without opposition, and altered course for their respective targets. Because of the smoothness with which the operation had proceeded, the 100th Bomb Group arrived over the target eighteen minutes early. Major Bennett accordingly ordered the formation to swing back out over the Skagerrak, where he led the aircraft in a large circle for about fifteen minutes, then headed in over the coast again. By then, enemy defenses were alerted, and antiaircraft shell bursts began blossoming among the aircraft. One bomber pulled out of the formation with a smoking engine; another plunged into the sea. Bennett counted ten parachutes, and was relieved to see German patrol boats headed in the direction of the descending chutes [13-22].

The 100th Group approached the target once more at 12 000 feet with the Vemork plant clearly visible in its Norden sights. It began its bombing run at 11:34, squarely in the prop wash of the 95th Group which had gone in ahead but turned away because of cloud cover. As the alarm sounded at Vemork, 9000 feet below, foreman Ingebretsen frantically phoned the power station to begin powering down machinery, then saw to it that the apparatus servicing the high-concentration plant and compressor house began shutting down as well. On hearing the powerful roar of numerous engines overhead he cracked the door to the third-story balcony. Just then bombs began bursting.

> I was tossed here and there by the air pressure, while glass panes, dust and everything else blew around the ears. I attempted to reach shelter, but groped for the door on account of atmospheric pressure from one explosion after another. [13-23]

During the next thirty minutes, 140 Fortresses released 711 500-pound bombs over Vemork. By mistaking the Rjukan nitrate plant for the Vemork electrolysis plant, an additional fifteen sorties dropped 118 250-pound bombs on Rjukan. Adding to the conflagration, about twenty Liberators of the Second Bombardment Division, finding Kjeller obscured by clouds, swung around for Vemork instead. Making the same mistake, they dropped additional 250-pounders on Rjukan, and plastered Vemork as well. (The drop on Knaben went according to plans, and caused extensive damage to the molybdenum mines.)

The material damage in Vestfjord Valley was widespread, albeit sparing of vital installations, in part due to the efficiency of the smoke generators emplaced after the *Gunnerside* raid. At Vemork, four bombs hit the power station, though the generators fared surprisingly well [13-24]. All 11 penstock pipelines were hit, and so were the sluices at the top of the mountain, but the automatic sluice gates slammed shut, preventing a major flood catastrophe in the valley below. The main hydrogen-electrolysis building was struck by two bombs at one end, and the four top stories were badly damaged. Providentially, the high-concentration plant in the basement escaped almost entirely, just as Tronstad had predicted, except for a single bomb splinter which flew in through a window and struck one cell with a loss of 120 pounds of heavy water. Five houses and four troop huts were destroyed. The suspension bridge over the gorge took a direct hit and collapsed, and another bomb tore up the railroad tracks leading into Rjukan. Gas pipes from Vemork to Rjukan were damaged in several places. At Rjukan, the packing plant, central workshop, and installation for acid concentration

Figure 13.2. *Rjukan after the American bombardment, 16 November 1943. Courtesy Anker Olsen Kr 1955 Norsk Hydro Gjennom 50 År (Oslo: Norsk Hydro) p 396.*

suffered varying degrees of damage. Water, gas, and steam pipes were broken, and electrical wires dangled in great profusion. Noxious fumes from burst ammonia pipes seriously delayed repair work.

Though the town of Rjukan, by something of a miracle, suffered no hits, the loss of life in the residential quarters at Vemork was not insubstantial. Twenty-one were killed in all, including one man on skis several miles away in the mountains, struck by splinters from a stray bomb dropped by a homebound aircraft. Among those killed at Vemork was the factory foreman, Thor Abrahamsen, and his wife; it was he who had assisted Brun in his escape from Vemork over a year earlier. As at Herøya, the greatest loss of life, sixteen—mostly women and children, resulted from a direct hit on a newly completed air raid shelter just outside the Vemork gates. None of the workers inside the plant itself were seriously hurt, probably due to the robust concrete structures. That so many fewer lives were lost this time compared to Herøya seems to be partly due to a greater state of preparedness of the civilian air-warning and fire-control systems in place.

Axel Aubert, who had resumed the duties of General Director of Norsk Hydro upon Eriksen's arrest, arrived at Vemork the day after the attack. What he saw made a deep impression on him. 'But this was not the intention!' he is said to have exclaimed [13-25]. Clearly his 'inside information' from Austrian sources about the military threat, or lack thereof, of the German uranium project (see p 155) had been wrong. In the chilly weather at Vemork he contracted pneumonia, and died shortly afterwards, at the age of 70.

To the Norwegian authorities in London, the bombing came as a bolt out of the blue, after British assurances that bombing was not on the cards for Vemork. Tronstad was in Scotland, attending a military course when the raid took place. 'U.S.A.F. has struck Rjukan despite promises to avoid these installations', begins his journal for 17 November, the day after the strike. 'Most disheartening, sure to have diplomatic consequences' [13-26]. Together with Hydro's Director Nicolai Stephansen he promptly saw General Hansteen in London, who professed to be equally surprised over the raid. On 1 December, the Norwegian government-in-exile lodged a formal protest with the British and American governments against the attacks, first at Herøya and now at Vemork and Rjukan, which 'were carried out without any advance notice being given to the Norwegian government... and seemingly out of all proportion to the desired effect'.

> The Norwegian military authorities have always done everything in their power to furnish the Allies with the most detailed and precise information on industrial installations and factories in Norway, as well as to what extent they were put to use by the enemy for waging war. On several occasions they have initiated action to put factories and other industrial installations in Norway out of working order if their production was essential for the enemy's war effort. ... If the reason for the attacks was the necessity of stopping other production than that of fertilizer, for example heavy water, better results could have been achieved by specialized methods of attack than by all-out bombing of the installations. [13-27]

The British reply came a month later. On 4 January 1944, they rejected both the Norwegian protest, and the Norwegian implication that they were in the best position to decide upon the relative merits of bombing versus sabotage of targets. Vemork had been bombed, the British note added, because accurate intelligence had shown that in the wake of the sabotage action the Germans had increased security precautions to the extent that sabotage could no longer possibly meet with success. The American reply was delivered in a note to the Norwegian embassy in Washington three weeks later. The Secretary of Defense had assured the State Department that thorough investigations had accompanied the planning for both the raid on Herøya and on Vemork. In both cases, the attacks had been judged necessary for the successful prosecution of the war against Germany; 'heavy water production at Vemork was the reason for the latter raid' [13-28].

Chapter 14

WAVERING OUTLOOK FOR HEAVY WATER

14.1. Penultimate pile experiments

The first casualty of the air raid on Vemork was Kurt Diebner's group at Gottow. With the success of G-II, the next experiment, G-III, was shaping up as the definitive test—perhaps even achieving a self-sustaining chain reaction. The bombardment of Vemork's heavy-water facilities, effectively shutting down its production line once more, dashed early hopes for such a test. Moreover, such stocks of heavy water that Diebner still possessed suddenly reverted back to Heisenberg's control, whence they had been expropriated half a year earlier by Abraham Esau. Worse yet, another administrative purge swept through the upper echelons of the tottering German nuclear power project just then.

Never easy to get along with, Esau, Hermann Goering's plenipotentiary for nuclear physics and head of the physics section in the Reich Research Council, had incurred the wrath of many, including Albert Vögler, president of the Kaiser Wilhelm Society; Vögler was disturbed by, among other things, the lack of adequate stocks of heavy water on hand. More seriously, Albert Speer, too, had lost patience with Esau, and persuaded Rudolf Mentzel, Esau's boss in the Reich Education Ministry, to find someone more to his liking to take over Esau's two posts. It was Mentzel, we may recall, who had fingered Lise Meitner as of Jewish extraction back in 1938. Mentzel, curiously, offered the job, not to a party hack but to Walther Gerlach, a soft-spoken academic physicist in Munich. Though widely respected, Gerlach was a complete newcomer to nuclear physics. An atomic physicist who had made his mark in a beam-splitting experiment with Otto Stern in 1922, his wartime efforts had centered on the development of magnetic proximity fuses for torpedoes; he currently held a position with the Institute of Physics at the University of Munich. Gerlach discussed the offer with Otto Hahn and Heisenberg, both of whom urged him to accept. Gerlach did, and on

2 December 1943 Goering signed the decree appointing him the new plenipotentiary for nuclear physics, effective as of the first of the year.

To placate Esau, Mentzel offered him the overall command of high-frequency research instead, which he accepted on condition that Speer had no objections. Esau had no wish 'to be shot at from that quarter again'. Adding to the irony of the reshuffle in command, Kurt Diebner was reappointed as administrative assistant under Gerlach at Berlin-Dahlem. His extensive background in the uranium project, both as an administrator and practicing nuclear physicist, made up for Gerlach's utter lack of experience in either capacity.

Gerlach's taking over the reins at this juncture has been considered somewhat of a turning point in the uranium project [14-1]. Though not a Nazi, he was committed to the German war effort simply as a citizen of the Reich; yet he also saw Germany's defeat as the handwriting on the wall. An idealist, he viewed his mission as one of saving German science (and scientists), and the shaky uranium project was an opportune vehicle for doing so. Though cool toward Heisenberg for his ambitious nature, Gerlach himself did not shrink from exercising absolute authority in allocating funds, or in having the last say with regard to military call-up papers issued to any of his staff—indeed, these were two of his stated conditions for accepting his new post. He was also on close terms with Paul Rosbaud. Though he did not share Rosbaud's uncompromising conviction that Germany had lost the war from the outset, the two agreed fully on the need for saving German science and its infrastructure.

Though he ruffled many feathers, Esau's dismissal drew a mixed reaction from colleagues within the uranium project. Among his positive contributions during his tenure had been a stable leadership, and several important technical judgments, as we shall see. Thus, late in his command he had presided over a rather fractious meeting at the PTR, called on Heisenberg's prodding while the latter was still smarting from Esau's expropriating his heavy water. Attended by the principal nuclear scientists from the far-flung factions of the Uranium Club, they thrashed out the next move in pile research. To Esau's credit, he had overridden Heisenberg's insistence that his layered pile design be emphasized at the expense of the new lattice design pursued at Gottow. The upshot was that *two* pile experiments would be pursued in parallel: a third lattice pile variant at Gottow based on uranium cubes cast by the Auer Gesellschaft in Berlin, and a larger pile model using uranium plates, also cast by the Auer Company, at the KWI for Physics.

Heisenberg, to be sure, knew full well that cubes, not plates, held the key to a working uranium engine. However, he argued, since the Auer Company was geared up for uranium-plate production, why not use them in a large-scale layer experiment, then cut the plates up into cubes for another test? [14-2]. In that way, much larger quantities of uranium could be brought to bear early on the technical problems at hand, especially the thermal stability of a uranium pile, without upsetting the tooling schedule at Auer. Besides, the amount of heavy water on German soil was at best sufficient for one definitive experiment at a time.

In the event, technical difficulties in manufacturing the large plates, exacerbated by the deteriorating war situation, held up their casting, rolling, and delivery. Since the uranium cubes cast by Auer were more straightforward to produce, the smaller experiment at Gottow went ahead as planned. G-III featured uranium cubes suspended in heavy water (this time at room temperature), housed in a hollow paraffin sphere. Once again Höcker's predictions were amply verified: G-III gave a higher rate of neutron production than any experiment in Berlin-Dahlem, Heidelberg, or Leipzig [14-3].

To repeat, Heisenberg was not unmindful of Höcker's calculations, or of the message conveyed by the Gottow experiments. That he nevertheless insisted on going through with the layer machine may have been partially due to personal disagreements with Esau and Diebner, partially to preoccupations with more fundamental physics problems [14-4], and perhaps as much a result of pure stubbornness [14-5]. In his interview with Ermenc long after the fact, he termed the decision simply a mistake [14-6]—albeit a costly one in the end, as we shall see. At any rate, the long-awaited delivery of the massive uranium plates from the Auer Company coincided with the completion of the underground bunker on the grounds of the KWI for Physics at the start of 1944. Authorized by Speer so long ago, the bunker featured 2 meter thick walls, floors, and ceilings of reinforced concrete. As such, it provided a fairly secure shelter in the midst of intensifying raids on the capital by the American Eighth Air Force by day and RAF Bomber Command by night. Heisenberg, Gerlach, and others who commuted from out of town closeted themselves in the damp chambers while they toiled away on the uranium pile.

The peak of the incessant British bomber campaign against Berlin, what Air Chief Marshal Harris termed the 'Battle of Berlin' and initiated with the longer nights in November of 1943, occurred on the night of 15 February 1944. Wave after wave of Lancasters, 806 bombers in all, led by Pathfinder Mosquitoes, hit the city with no less than 2643 tons of bombs in the space of only 40 minutes [14-7]. Max Planck's house in Grünewald was gutted, and with it all of his papers and possessions, though Planck himself was away in Rogätz near Magdeburg that night [14-8]. Bombs also rained down on the Kaiser Wilhelm Institutes in nearby Dahlem. Heisenberg's bunker was untouched, and his institute sustained only superficial damage. However, the KWI for Chemistry, where Hahn had teamed with Meitner, and Hahn and his assistants had continued their fission studies, took a direct hit and was severely damaged. 'I saw your former rooms', Max von Laue wrote to Meitner.

'Splinters! By the director's room . . . great heaps of ashes' [14-9]. Hahn, like Planck, lost all his papers, including his precious correspondence with Rutherford. Hahn was away in Tailfingen in southwestern Germany just then, but Heisenberg and many of the Dahlem staff organized a desperate effort to rescue the library in the basement of the smoldering ruins. Even Minister Mentzel pitched in, carrying sodden books to safety. The military fire brigade was of less help, wrote Laue to his son, as they 'fought the fire on top'.

> In my opinion they got things wrong and sought to save a part of the roof which could not be saved, while in other places the fire invaded sections which could have been held—hot water from a broken water pipe dripped into the cellar, which contained a large portion of the library. The water was already several centimeters high by the time I left, and it was a singular sight to behold how a pair of men from the fire police sat in a corner of the room and, calmly, as if everything around them were in order, telephoned someplace else. [14-10]

Elsewhere in Germany, members of the Uranium Club fared scarcely better under the mounting air offensive. In Hamburg, the infamous fire raids of the previous summer had forced Harteck and Suess to relocate their isotope-separation equipment to Freiburg. As late as December, 'Bomber' Harris's Lancasters had devastated the upper floors of the Institute for Theoretical Physics in Leipzig, where Heisenberg had taught since 1927, destroying most of his own papers. His home there, too, was destroyed, though Heisenberg was in Berlin at the time and his family in Urfeld in the Bavarian Alps. Nor did Walther Gerlach's home in Munich escape destruction later in 1944, at which point he moved into Harnack Haus, still standing in Berlin-Dahlem.

With the ever-worsening situation in Berlin, evacuation of the uranium project to some safer spot in the country became a pressing concern early in 1944. In the midst of the havoc, however, uranium for the outmoded pile experiment began arriving, and Karl Wirtz had no choice but to start assembling the first of several experiments collectively designated B-VI. The centerpiece of the bunker in the Institute grounds was a circular pit, about 2.1 meters deep and equally wide, down into which could be winched by an overhead crane a containment vessel of magnesium alloy. The vessel was filled with a total of 1.5 tons of heavy water, about 60% of the remaining stocks in Germany, and well below the critical 5 tons of D_2O considered sufficient by virtually everybody to sustain a self-supporting chain reaction. In the vessel were stacked horizontal layers of the 1 cm thick uranium metal plates, held apart by low-neutron-absorbing magnesium spacers. A Ra–Be neutron source, as always, was placed in the center of the pile assembly. The vessel and its contents were lowered into the pit, and the neutron propagation tests commenced with Wirtz in charge.

Four different uranium plate spacing geometries were tried, designated B-VIa through B-VId, and incorporating uranium metal weighing between 0.9 and 2.1 tons, respectively. Each arrangement gave a positive neutron mass-production coefficient, though none of them quite as high as the value for the latest Gottow lattice experiment. With regard to the neutron propagation factor, another figure of merit denoted Z in German reports, none of the latest values were significantly higher than the best Gottow value [14-11]. The easily perturbed Dr Vögler, for one, was not impressed and told Gerlach as much.

One who would have an important say in these experiments, conducted under the harrowing conditions of the Berlin blitz, was Walther Bothe, the man who had obtained such discouraging results with graphite, and whom we last met in Paris trying to coax a beam out of Joliot's recalcitrant cyclotron. As of late, Bothe had been dividing his time between his own cyclotron laboratory in Heidelberg and the pile research group at Berlin-Dahlem. Some time ago he had suggested that a reflecting mantle of carbon might do better than one of water, as used in B-VI [14-12]; more recently calculations by Fritz Bopp and Erich Fischer had supported his suggestion. Bothe proved right. B-VII, retaining the optimum ratio of uranium and heavy-water moderator utilized in B-VIb, but with a carbon reflector, did better yet, giving a substantially higher value for Z [14-13], though the overall performance was still not as good as the best lattice design.

Two conclusions were unavoidable by mid-spring of 1944. First, the layer design had to go in favor of one last, concerted push on a full-scale lattice design. Wirtz, Heisenberg's right-hand man in these experiments, concluded as much, but was in no position to argue with his boss. Esau had no such qualms; he had argued strenuously and vociferously to that effect before he was sacked. Now Gerlach, too, had to agree and so told the skeptical Albert Vögler. Delighted, for a change, Vögler informed Heisenberg that he was prepared to back a new effort to the best of his ability [14-14].

The second conclusion was the obvious need to evacuate Berlin. By mid-1944 the western Armies had breached the Atlantic Wall; in the east, the Soviet Army was poised to smash into German territory. In Berlin, saturation bombing by night continued to terrify the population as nothing else. English-born Christabel Bielenberg, married to a German, has left an account of phosphorus bombs 'that burst and glowed green and emptied themselves down the walls and along the streets in flaming rivers of unquenchable flame' [14-15]. Firestorm winds of hurricane strength were sucked in to replace superheated gases boiling above the rooftops, consuming people and everything around them in one great conflagration.

By the time the B-VI tests were under way, Diebner and his group had already left Gottow for Stadtilm, south of Berlin. There they began planning a pile experiment with a carbon–hydrogen compound as moderator—the brainchild of Harteck—*in lieu* of heavy water. Gerlach had turned the bulk of the remaining heavy water over to Wirtz, who would use it in one final, large-scale pile experiment in the bunker at Dahlem, while the rest of the Uranverein relocated piecemeal to the town of Hechingen in southern Germany. In the event, Diebner and Harteck's latest pile design never proceeded beyond the drawing board. Despite Gerlach's decree with regard to the heavy water, however, Harteck had remained optimistic about prospects for more. In the aftermath of the air raid on Vemork, he thought production at the plant could be restarted with modest effort; after all, the high-concentration apparatus was still intact. So was the newly completed plant at Såheim. What undermined his personal involvement in the effort was the shift in command of the uranium project from Esau to Gerlach. Shortly before, Esau had offered Harteck full responsibility for heavy-water production as his personal representative [14-16]. With Esau out, that fell through. All the same, Esau would have the last word in determining the direction of further heavy-water production developments—developments, moreover, in which Harteck would remain a central player for the remaining duration of the war.

14.2. The ferry

On 29 January 1944, Einar Skinnarland, still ensconced in a remote hut on the Hardanger Plateau, received a telegram from Leif Tronstad in London:

> IT IS REPORTED THAT THE HEAVY WATER APPARATUS AT
> VEMORK IS TO BE DISMANTLED AND TRANSPORTED TO
> GERMANY STOP CAN YOU GET THIS CONFIRMED? STOP
> CAN THIS TRANSPORT BE PREVENTED? [14-17]

Tronstad's information was correct. For some time discussions had been going on between German Army Ordnance, the Reich Research Council, and *Wehrwirtschaftstab Norwegen* about the wisdom of continued heavy-water production in Norway in the face of recurring threats of Allied intervention—whether by air attacks, commando operations, or simply internal plant sabotage. Not only would the heavy water remain in jeopardy, but so would electrolytic hydrogen and fertilizer production—of importance for the German war industry and Norway's economy alike. On 13 December, instructions were issued from the *Reichskommissariat* in Oslo's Parliament Building on Karl Johans Boulevard to Norsk Hydro's main offices at Solligaten 7 to the effect that heavy-water production would not be resumed at Vemork or elsewhere. Esau's last order while in office, conveyed to Goering's headquarters, called for shipping all electrolyte stocks remaining in the damaged Vemork plant, in *all* their different concentrations, to the relative safety of Germany for final processing [14-18].

Esau was acting on the advice of Friedrich Berkei of Army Ordnance laboratories at Gottow, as well as many others among the German scientists and officials who had some connection with Norsk Hydro; during the winter weeks of 1943–1944 Rjukan and Vemork 'fairly crawled with Herr Doktors, Herr Professors and Herr Generals' [14-19]. To be sure, Skinnarland and Knut Haukelid, the two remaining *Swallows* in the region, also had ready access to information at the idled Vemork plant. Two of their contacts, besides chief engineer Alf Larsen, were the engineer Rolf Sørlie, actually now in hiding in the hills, like them, and the Hydro technician Gunnar Syverstad, Skinnarland's brother-in-law. Another inside contact was Kjell Nielsen, the firm's transport engineer, whose advice would prove pivotal in the coming events. Thus confirmation of German intentions was not long in coming; the heavy water was indeed to be shipped, lock, stock, and barrel, along with much of the apparatus. The apparatus was of little concern to the Allies; the semi-processed material was, arguably, of great concern. Telegrams crackled back and forth over the airwaves between Skinnarland's Nilsbu hut and FO4, London.

> *Skinnarland to London, February 3:*
> FILLING OF DRUMS PROCEEDING... STOP REGARDING PREVENTION OF TRANSPORT, I DO NOT YET HAVE THE DETAILS ON THE TIMING, NOR IS THE MODE OF TRANSPORT DEFINITE. [14-20]

> *Skinnarland to London, February 6:*
> TRANSPORT IMMINENT, AND SABOTAGE OF SAME PROBABLY FEASIBLE. [14-21]

In his journal entries for 7 February, Tronstad wrote: 'The matter placed before Sir John Anderson (Minister of Nuclear Warfare) through Commander Welsh. Sir J. A. startled upon learning that so much juice remained, and expressed his desire that the matter must be carried through' [14-22].

> The possible repercussions of such an attack [added Welsh in his letter to Tronstad the next day] has been brought to the knowledge of the gentlemen in question. I have however to inform you that notwithstanding this the gentlemen in question are of the opinion that such an attack should be made. [14-23]

It did not take long for Kjell Nielsen to learn about the method of shipment, since it was up to him to arrange it! The heavy-water stocks were to be drawn off the various stages in the production line and piped into steel drums marked 'Potash Lye'. The drums would be loaded on railway flatcars at Vemork and taken by rail via Rjukan to the rail-ferry dock in the town of Mæl in a side branch of Lake Tinn (figure 10.2) near its northern end. There the railway cars would be put aboard the ferry and proceed to Tinnoset at the south end of the lake. From there the transport would continue by rail to Notodden and on to Herøya near the coast, and thence by ship to some port in Germany—probably Hamburg.

The shipment would comprise some 15 tons of potash lye distributed in 49 drums with deuterium concentrations per drum ranging from 0.9% to 99.24%. Thus the equivalent 100% deuterium contents in the entire consignment amounted to no more than about 607 kilos [14-24].

Haukelid and Sørlie stole down into Rjukan on skis, and with Syverstad and Nielsen they huddled in Nielsen's house, carefully going over the grim alternatives. Attacking the Vemork plant itself was out of the question after the new German security measures in place, at least with only a handful of men and with time running out; 20 February was now the scheduled day for the ferry passage. Blowing the train from Vemork off the ledge into the gorge had seemed at first the most effective way to destroy the shipment; Norsk Hydro's dynamite shed lay closely alongside the track. However, the security precautions for the transport were less certain; all they had to go on were vague rumors that a special Army unit was being mobilized to guard it all the way from Vemork to Germany [14-25]. Besides, with the destruction of the suspension bridge across the gorge in November's attack, the railway was used by Vemork workers coming to and from Rjukan. Attacking the train *en route* from Tinnoset to Herøya presented just as great risk of endangering Norwegian lives. Among other things, trains southbound out of Rjukan carried as a rule tanks of ammonia; setting them off was a sure recipe for disaster. Finally, the ship carrying the cargo to Germany might be sunk by submarine or aircraft. That possibility, however, opened up another one closer to home: the stretch of water by rail-ferry along Tinnsjø between Mæl and Tinnoset.

At its widest, the 18 mile long lake measures about one mile across and is virtually bottomless at 1300 feet, with no hope of salvaging the cargo if the ferry were sunk, nor of beaching it if the action were carried out swiftly. By the same token, the hope of saving lives from the ice-cold waters was equally slim. No warning of any kind could be given, on board or beforehand. Fortunately it would be Sunday morning, according to plans, with fewer than normal passengers. Nielsen had chosen the day and hour of crossing with just that in mind.

Skinnarland to London, February 9:
WE CONSIDER SAFEST SOLUTION THE SINKING OF FERRY BY CIVILIAN SABOTAGE STOP ARE WE PERMITTED STOP ONLY OTHER POSSIBILITY MAY BE DEMOLITION BY EXPLO-SIVES OF THE ENTIRE TRAIN, FOR EXAMPLE NEAR SVELG-

FOSS STOP WE MUST EXPECT REPRISALS IN CONNECTION WITH ACTION STOP ANSWER URGENTLY NEEDED. [14-26]

London to Skinnarland, February 10:
APPROVE SINKING OF THE FERRY WITH IMI [heavy water] CARGO IN DEEP WATER, PREFERABLY BETWEEN HESLEVIKEN AND DIGERRUD STOP IF BOTTOM VALVES ARE OPENED, THIS MUST BE COMBINED WITH AN EXPLOSION TO INDICATE LIMPET [plastic explosive] ATTACK FROM THE OUTSIDE STOP THE ENGINES MUST BE PUT OUT OF OPERATION SO THE FERRY CANNOT CONTINUE TO SHALLOW WATER STOP THE SINKING MUST NOT FAIL, BECAUSE THE METHODS OF ATTACK IS THEREBY REVEALED STOP LEAVE BEHIND BRITISH UNIFORM ITEMS IF POSSIBLE AT SUITABLE PLACE STOP GOOD LUCK. [14-27]

In his above reply on 10 February, Tronstad added that Lieutenant Haukelid, the only remaining commando on the Vidda skilled in handling explosives, was to take charge of the operation. The order was issued after due deliberations with Colonel Wilson, SOE, and with the concurrence of General Hansteen and Minister of Defense Oscar Torp of the Norwegian High Command.

On 14 February, Haukelid and Sørlie met again with Nielsen, going over details once more. (Besides Haukelid and Sørlie, and still unknown to Nielsen, a third man had been added to the demolition team, Knut Lier-Hansen, a former Sergeant in the Norwegian Army. He had fought in the battle for Rjukan and was one of those who doffed their uniforms and slipped away when the Germans marched into the town in May of 1940. Since then, he had been on the Vemork payroll, while moonlighting as a weapons instructor in the local resistance organization.) Nielsen reminded the two that since the operation involved removal of virtually all of the various concentrations, it was doubtful whether the Germans could extract much high-quality heavy water from such uneven stocks. Added to this was the strong likelihood of German reprisals at Rjukan, not to speak of loss of lives in the action itself.

Skinnarland to London, February 16:
OUR CONTACTS SAY THAT THE GERMAN METHOD IS INFERIOR TO ELECTROLYSIS AND THEY THEREFORE DOUBT WHETHER THE EFFECT OF THE OPERATION IS WORTH THE REPRISALS WHICH MUST BE RECKONED WITH STOP AS WE CANNOT DECIDE HOW IMPORTANT THE OPERATION IS, WE REQUEST A REPLY SOONER, IF POSSIBLE THIS EVENING STOP IMI SHIPS APPARENTLY GO DIRECTLY TO HAMBURG. [14-28]

London obliged them with an immediate reply. The Allied High Command remained adamant.

London to Skinnarland, February 16:
CASE CONSIDERED STOP VERY URGENT THAT IMI BE
DESTROYED STOP HOPE THIS CAN BE DONE WITHOUT
TOO SERIOUS CONSEQUENCES STOP OUR BEST WISHES FOR
SUCCESS IN THE WORK STOP GREETINGS. [14-29]

Chief engineer Alf Larsen, in charge of heavy-water production, had been appraised of the forthcoming action and agreed to lend a hand wherever he could. He had, in fact, been doing so all along recently. He had seen to it that Syverstad, whom he realized was in touch with the men on the Vidda, was well informed about developments in the plant, without suspecting that Larsen was onto his illicit role as contact man. It was also made clear to Larsen that he himself would have to flee Norway immediately after the operation; Haukelid would see to it that he was 'exported' to Sweden on the day of the event.

Larsen's predecessor, Jomar Brun, was, though in Tronstad's camp, less aware of all the preparations going on, engaged as he was in the planning of a British heavy-water production effort under the Department of Scientific and Industrial Research at ICI in Billingham, perhaps in Tasmania. (The project was never actually realized.) He happened to be in London, and dropped in on Tronstad, just as the decision was made to sink the ferry. Tronstad confided in him the forthcoming action, to Brun's consternation. Gravely concerned over the loss of lives to be expected, Brun argued for an eleventh-hour change in plans, but was persuaded that it was too late in any event, with the Allied authorities already having put their official stamp of approval on the operation.

Indeed, the British had gone to extraordinary lengths to prevent the heavy-water consignment from reaching Germany. Unbeknownst to Haukelid, the ferry operation was not the only effort afoot to assure success in the mission [14-30]. SOE headquarters had arranged for a second Norwegian Resistance party, *Chaffinch* in the district of Vestfold, to be prepared to sabotage the consignment near Herøya, should it get that far. The Tube Alloy Directorate also worked on both RAF Bomber Command and the Admiralty to be prepared to sink the cargo ship bound for Germany, if all else failed.

Meanwhile, Haukelid had gone openly to the ferry dock at Mæl to size up the situation for himself. Dressed as a workman, but with a Sten gun wrapped in a sleeping bag and hand grenades in his rucksack, he booked a ticket and took a round-trip ride on *Hydro*, the old ferry plying Lake Tinn. It required half an hour to reach really deep water, and 20 more minutes to cover the deep-water stretch between Hesleviken and Digerrud; thus a detonation 45 minutes out of Mæl would be about right, with a margin of approximately ten minutes before and after.

As for explosives, Haukelid had cached away plenty of Compound 808, the putty-like, nitro-cellulose plastic left over from the air drop the year before, and used so effectively by Rønneberg and Strømsheim on the high-concentration plant. However, fuses could not be used this time; the time delay was too long. Instead, he and his companions improvised a timing mechanism out of two alarm clocks with the help of a retired Hydro inspector and friend of Sørlie, John Dieseth, who had a little workshop in town. Instead of striking a bell, the hammer would strike a metal contact, closing a battery-powered circuit. Dieseth worked on the clocks, while Haukelid and Sørlie worked alongside him in the shop preparing the explosive package. They kneaded 19 pounds of the plastic substance into a 12 foot long sausage—enough to blow a hole four feet across in the metal skin of the ferry. The plan was to place the charge in the bow of the vessel. A hole blown there would cause the stern to rise as the forward compartments flooded, and raise the propeller out of the water. The freight cars with their drums would roll forward and tumble into the lake as the bow went under. D_2O being heavier than ordinary water, the drums should sink about as fast as the railway cars.

After retiring to their mountain hideout above town with all their equipment, Haukelid and Sørlie wired an electric detonator to their timing device, and set the clocks to go off some hours later. Dead tired, they went to sleep, but were violently awakened by rifle-like cracks. They sprang to their feet, Sten guns cocked, then sheepishly nodded to each other. The timing mechanism worked just fine!

On Saturday afternoon, 19 February, all 49 drums of potash lye were loaded onto two railroad flatcars at Vemork and prepared for the short run to Rjukan, with armed guards actually sitting astride some of the drums. At Rjukan the freight cars would be sidetracked in the railroad yard, floodlit and still under heavy guard, until early Sunday morning, when they were to be coupled to several other railroad cars bound for Mæl. In Mæl, they would be pushed by a small switching-engine aboard the ferry for the trip to Tinnoset. Sailing time was ten in the morning.

About an hour past midnight that same evening, a car stopped in the shadows a little way back from the ferry dock at Mæl, carrying Haukelid, Sørlie, Lier-Hansen, and Larsen. Larsen was in sports clothes, and had with him a bag with a few essentials. A bachelor, he was not returning to Vemork. Getting out, Haukelid handed Larsen a pistol and told him to wait for them. If the rest were not back in two hours, he was to drive off to Kongsberg, board the train for Oslo, and then find his way to Sweden. Haukelid, Sørlie, and Lier-Hansen cautiously advanced toward the station at the dock, heavily loaded down with Sten guns and hand grenades, explosives, tools and equipment underneath their coats. The night was clear and bitterly cold; a thin layer of crackly ice on the road snapped irritatingly as they trod gingerly along. Amazingly, there were no floodlights on the dock; only a pale lamp marked the outboard end of the gangway onto the darkened hulk of the silent ferry. More amazingly, no guards were apparent on the dock.

Figure 14.1. *The ferry D/F Hydro at the railhead in Mæl on Lake Tinn. Courtesy Norsk Industriarbeidermuseum, Vemork; negative No UF-354.*

A few *Wehrmacht* soldiers could be discerned playing cards inside the station house. Had they stumbled across the weak link in the elaborately orchestrated transportation scenario?

Casually, the three sauntered up the gangway. On board, nothing stirred on deck. Silent reconnaissance revealed most of the crew playing poker below, with the engineer and stoker absorbed in the engine room. Knowing the ferry well from his earlier passage, Haukelid led his companions to the third-class accommodations. While Lier-Hansen stood watch, Haukelid and Sørlie began opening a hatch leading down to the bilges. Before they got very far, a Norwegian night watchman stepped in as they scrambled for cover. 'The situation was awkward, but not dangerous', recalls Haukelid laconically. 'We hurriedly explained to the watchman that we had to hide and were looking for a place' [14-31]. The watchman nodded to the half-open hatch, remarking that this was not the first time they carried passengers on the run. Lier-Hansen, who in fact knew the watchman from the athletic club back in town, remained chatting with him, while Haukelid and Sørlie wriggled down through the hatch. Closing it above them, they crept forward along the keel up into the bows. About a foot of water was standing above the bilge plates. The sausage-shaped plastic charge was pressed against the corrugated iron hull well under water. The alarm clocks and some batteries were taped to some fore and aft girders, detonators tied to each end of the coiled-up charge, and all wiring tied together. The alarm clocks were set to go off 45 minutes past 10 a.m.

> Making the last connections was a dangerous job; for an alarm clock is an uncertain instrument, and contact between the hammer and the alarm was avoided by not more than a third of an inch. Thus there was one-third of an inch between us and disaster. [14-32]

With everything set and the clocks ticking away, the two crawled back and returned to the passenger saloon.

The watchman, still with Lier-Hansen chatting away, was relieved to see them. They explained that they had to return ashore to fetch some things, but would be back. Obviously, the good man could be told nothing. Haukelid was sorely tempted to bring him along to safety, but his absence was sure to arouse suspicion a few hours later. They simply shook hands. 'Don't be late', the watchman called softly after them as they stole down the gangway.

Hurrying off the dock, they could hear the noises of the arriving train in the night. Larsen was waiting for them by the car; in just three more minutes he would have left according to instructions. They drove off, with Lier-Hansen behind the wheel. After a short ride, Sørlie got off. He was to join Skinnarland up on the Hardanger Plateau. (Gunnar Syverstad, back in Vemork, would head for the hills as well, joining the resistance.) Outside Kongsberg, Haukelid and Larsen got off to catch the morning train for Oslo. Kjell Nielsen was already in Oslo, waiting in a hospital for the removal of a perfectly healthy appendix. When the Gestapo came calling on him, as they assuredly would, he would be under anesthetic with a solid alibi. He, too, could have escaped to Sweden, but agreed to stay behind as a future contact for Skinnarland at Vemork.

Lier-Hansen turned around and drove back to Mæl. He, too, had elected to take his chances at Vemork. His present job was not quite finished, however. Parking in the same spot as earlier, he climbed into the woods above the hamlet to watch events and make sure of no last-minute hitch. The train stood there puffing away, and one by one the cars were being gingerly nudged onto the ferry, with German guards everywhere. A little after ten o'clock the venerable old ferry with twin stacks, dating from 1914, churned away from the dock and headed south across the icy waters. On board were 53 people and nine railroad cars lashed to the deck—two the flatcars with the drums containing roughly 15 000 liters of potash lye, some of them tank cars filled with ammonia, and some ordinary boxcars.

'At 10:30 I was on the bridge when I heard an explosion down in the belly of the ship', Captain Erling Sørensen reported later. 'I understood right away what it was—a bomb' [14-33]. Right on schedule and quite as intended, the vessel went down bow first in under five minutes. As she did, the railroad cars broke loose *en masse* and rolled crashing into the water. Time permitted the launching of only one lifeboat; those who could jumped and clung to whatever was handy. Eighteen lives were lost, including twelve passengers, two crew members, and four Germans [14-34]. More would have been lost, had it not been for the heroic efforts of several farmers setting out from the shore. Of the 49 drums, four partially filled ones bobbed to the surface and were saved, rendering the effective loss of heavy water perhaps 500 kilos, in Brun's estimate [14-35].

The question of the actual loss of heavy water in the *Hydro* sinking does not end there, however. After the war, Brun was informed by Karl Wirtz that a consignment of drums, containing heavy water in various degrees of concentration, arrived in Berlin [14-36]. Wirtz also claimed, now citing Kurt Diebner according to David Irving, that the Germans had been alerted to a forthcoming sabotage attempt on the shipment of heavy water from the drained electrolyzers, and switched ordinary water for heavy water in the drums [14-37]. The ferry was, accordingly, sunk with a 'dummy' consignment containing no heavy water at all. The consignment reaching Germany with heavy water, albeit in a highly diluted state, came by road. Brun dismisses Diebner's claim of a dummy consignment out of hand; too many sharp-eyed Norwegian colleagues bore testimony to the filling of the drums with *bona fide* electrolyte and the circumstances of their shipment.

On the other hand, all accounts agree that four drums did survive the ferry sinking, as noted. In addition, a small quantity of heavy water was shipped from Såheim to Germany by road in March of 1944, along with some further small amounts from Vemork. Hydro records show a consignment of 19 drums shipped, containing a total of 6200 kilos, presumably of greatly varying concentration [14-38]. Combined with the four drums saved from the ferry, some two dozen drums appear to have reached Germany and may account for what Wirtz observed in Berlin, and which Brun doesn't question. Amounting to well over 200 liters of heavy water, at least half of this material was, however, so diluted with potash lye and water to be of no use to the Germans without further concentration.

The ferry went down 300 meters from land near the community of Rudsgrenda, and lies at a depth of 430 meters in the icy waters of Lake Tinn. In 1994 two heavy-water drums were recovered from the wreck by divers. One of the two drums, pitted and streaked with rust, is on display at Norsk Industriarbeidermuseum in Vemork, along with a dramatic rendering of the sinking and subsequent recovery operation. The ferry *Ammonia*, a sister ship of the *Hydro*, is no longer operational. It is permanently berthed at the rail head at Mæl, while discussions continue over whether to preserve it as a museum piece. The same

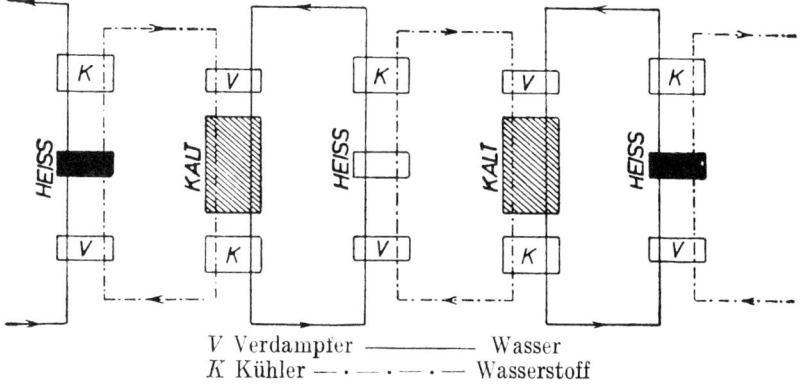

V Verdampfer ——————— Wasser
K Kühler —·—·——·— Wasserstoff

Abb. 57. Versuchsanordnung des Heiß-Kalt-Verfahrens

Figure 14.2. *The 'hot–cold' exchange process between hydrogen and water, utilizing the disparate equilibrium constants at two different temperatures. Bothe W and Flügge S (ed) 1953 Kernphysik und kosmische Strahlen, Naturforschung und Medizin in Deutschland 1939–1945 vol 14 (Weinheim: Chemie) part 2, p 184.*

discussion concerns cars of the the long disused Rjukan–Mæl railway, rusting away alongside the railway depot in Rjukan.

What, then, were Germany's remaining options for heavy water by mid-1944? Not inconsequential, it turns out. Though the war may have, at long last, put Norsk Hydro out of business as far as the German uranium project was concerned, the same was not necessarily so for Hydro's German partner, I G Farben, as we shall see.

14.3.   Prospects for heavy-water production in Germany

The sinking of the *Hydro* dramatically brought home to Paul Harteck the prudence of relying neither on a sole source of heavy water, nor on a single scheme for its production. With Norsk Hydro largely out of contention as a viable source of future stocks, the spotlight shifted to I G Farben's Ammonia Werk ('Leuna Works') in Merseburg near Leipzig, though Linde Eismachinen Fabrik in Höllriegelskreuth might well be in the running as well. Fortunately, an alternative to Hydro's energy-intensive, electrolytic-production scheme was seemingly also in hand. By late September 1943, a small pilot plant, code-named *Stalin-Organ* for its many pipes and columns, was ready to go into operation at Farben's Leuna Works. Operating on the so-called 'hot–cold' principle, it depended on a dual-temperature version of the catalytic water–hydrogen exchange process of Harteck and Suess.

In the dual-temperature, water–hydrogen exchange process, depicted in figure 14.2, water and hydrogen gas flow in closed, counter-circulating loops between a high- and a low-temperature region in the presence of a catalyst; the two streams are brought into contact at points in both the cold and warm sections [14-39]. An exchange equilibrium is established, with deuterium transferred from a water molecule to a hydrogen molecule and vice versa, according to $H_2O + HD \rightleftarrows HDO + H_2$. The chemical exchange is quite like the monothermal process described on page 152, but more efficient, at the cost of greater technical complexity. By a sequence of circuits, the equilibrium balance is progressively adjusted; as a result, the water is correspondingly enriched in deuterium in the cold section and depleted in the warm section, but with a net gain in concentration.

Stalin-Organ had been proposed by the physical chemist Paul Herold of I G Farben in February of 1943, after the major conference at the KWI for Physics that month, and after discussions with Karl-Friedrich Bonhoeffer, Harteck, and Suess. Herold had been sufficiently enthusiastic to offer the plant free of charge, an offer vetoed by Heinrich Bütefisch, director of the Leuna Works, as well as a physical chemist and a Major in the *Waffen-SS*. Bütefisch agreed to go ahead with a pilot installation aimed at an enrichment of one per cent and modest yearly output of 50 liter, for the nominal sum of 150 000 Reichsmark, provided Army Ordnance kept him fully appraised of all aspects of the nuclear power project, and granted Farben a share in the patent rights. Army Ordnance concurred. This agreement, in effect, marked the start of I G Farben's stranglehold on the uranium project in Germany—an agreement with serious consequences for the project and Farben alike before a year had passed [14-40].

By September, when the plant was ready to go on line, Harteck was sufficiently satisfied with the design to push for a full-scale effort—that is, a plant 100 times as large, capable of the five or six tons of heavy water deemed adequate to sustain a viable uranium-power project. Harteck suggested the cost should scale as the two-thirds power of the number of liters of heavy water produced, implying a cost for the full-size plant of around $2\frac{1}{2}$ million Reichsmark [14-41]. Herr Direktor Bütefisch and his Farben colleagues disagreed, arguing that the cost went linearly with the output, and quoted a figure around 20 million Reichsmark. Even so, warned the gentlemen of Farben, the cost represented such a small fraction of their annual revenue to hardly be worth their bother. Esau, too, then still in charge of the power project, vacillated on recommending such a costly undertaking, with the nuclear physics budget for all of 1943 amounting to 3 million Reichsmark.

Then again, at least two other heavy-water production processes appeared viable. One was an alternative chemical-exchange process devised by Karl-Hermann Geib, a student of Harteck on the staff at I G Farben. (Geib, and other heavy-water colleagues at the Leuna Works, was captured by the Russians in 1945; he was later liquidated while trying to reach the Canadian Embassy in Moscow [14-42].) Geib had gone over his idea with Harteck, and Hans Jensen had subsequently performed supporting calculations that lent credence to his scheme [14-43]. A dual-temperature process using hydrogen sulfide instead of water in the exchange with hydrogen, this time without a catalyst, Geib's scheme became eventually known as the Girdler-Sulfide or GS process, and the basis for all large-scale, heavy-water plants after World War II [14-44]. On paper, the process looked even more promising than the dual-temperature Harteck–Suess process, since it offered a high separation factor with lower energy requirements [14-45]. In practice, the corrosive (and toxic) nature of hydrogen sulfide and other problems ruled it out for implementation anytime soon.

Another contending heavy-water process was the hydrogen distillation scheme developed by Klaus Clusius in collaboration with Linde's Ice Machinery Factory; it, too, figured prominently on the agenda of Army Ordnance's research conference in February of 1943 [12-7]. It was also carried to the pilot-plant stage, in Munich, by Clusius and the young radiochemist Kurt Starke, for whom Clusius had obtained a reprieve from military service. (Starke later wound up extending his deferral with the group of German scientists at the Paris cyclotron.) The chief problems with the Clusius–Linde process were that it required scarcely obtainable pure hydrogen gas as input material, and costly low-temperature apparatus. I G Farben scientists disclaimed any interest in following up on the process, ostensibly due to the electrical power requirements, which were substantial though less than for the GS process. Their disinterest may, in fact, have been largely because the patent rights were held jointly by Clusius and Linde.

Harteck, meanwhile, had not written off Norsk Hydro entirely, even after the ferry sinking. Early in the spring of 1944 he set off for Norway once more. Increasingly miffed at the hard-line attitude of I G Farben regarding annual revenues and patent issues, he felt, at the same time, a certain responsibility for Farben's Norwegian partner in just such matters. Norsk Hydro, he learned, had received no payment for the heavy water they had produced under ceaseless German pressure, much less any compensation for the destruction from bombers and saboteurs, and had pretty much written the whole matter off as a loss. Informed of this, Harteck urged his superiors in Berlin to come up with at least $2\frac{1}{2}$ million Norwegian crowns—a token amount already agreed to by Army Ordnance and the Reich Research Council, but not acted on. His concern was, to be sure, not without ulterior motives; he was himself a driving force behind Hydro's exploitation. The main stock in trade of the Norwegian giant was, after all, nitrogen products on a large scale, and thus Hydro was important for Germany in more ways than just as a source of heavy water. Any future cooperation would demand a certain demonstration of good will on Germany's behalf.

Another Norwegian matter was brought to Harteck's attention by Paul Rosbaud that January. Odd Hassel, who in some sense had helped launch the heavy-water venture in Norway, had been arrested at the end of 1943, when the German authorities closed the University of Oslo and rounded up faculty and students [14-46]. His arrest came at a most inopportune time in his personal research, shortly after publication of what is generally considered his seminal paper on molecular structure [14-47]. His arrest may also have been for more than simply being a professor, however, as he appears to have acted as a 'consultant' on German affairs in Norway to the military and naval attachés in the US embassy in Stockholm [14-48]. (Another Norwegian scientist sought by the Americans, in vain as it turned out, was Hassel's colleague Victor Goldschmidt. He too had been arrested, a year earlier as a Jew, released, had escaped in a hay wagon into Sweden and on to England, where Eric Welsh took him in hand without informing his American counterpart, the OSS. Harald Wergeland, also of the University of Oslo, and one who most definitely *was* in contact with Allied sources in Stockholm, had been arrested too in an earlier teacher roundup, but was released after some unpleasant months with a German labor gang in northern Norway.)

At any rate, Rosbaud asked Harteck to do what he could for Hassel when he visited Norway again [14-49]. Harteck took the matter up with German authorities in Oslo, and with van der Bey, the German I G Farben official at Norsk Hydro, whom he had other reasons for seeing as well. Bey suggested perhaps Hassel could be put to work at Norsk Hydro under SS supervision. In the event, the matter went no further, and Hassel was released toward the end of 1944, though unable to resume his work until after the war.

However, the main purpose of Harteck's latest visit to Norway was his recurring interest in any residual heavy water at Vemork, or perhaps Såheim. Even 1% material, which might readily be drawn off a restored hydrogen production line, was of interest. And Norsk Hydro had not remained idle after the bombing of Vemork and Rjukan the previous November. The very day after the raid, Hydro management met to plan the reconstruction. Already in July of 1944, the Vemork power station was humming once more, and by August ammonia production was back to normal as well. To be sure, the heavy-water apparatus at Vemork was not returned to operation; the same month of August, the high-concentration apparatus was dismantled for shipment to Germany. Nevertheless, Harteck had been hopeful all along that a compromise could be worked out for the tapping and storage of one per cent heavy water for eventual German use. Any heavy-water production process contemplated for startup on German soil would benefit from this level of concentration as starting material, and there was no lack of potential processes to choose from when Harteck reviewed them for a prospective heavy-water program in mid-April of 1944.

There was, first of all, his own simple process for distillation of water at low pressure—the process dismissed by Urey as too expensive for the

236

fledgling Briggs Committee to contemplate, but finally adopted by du Pont as a simpler alternative to the catalytic exchange plus electrolysis approach pursued at Trail in Canada. Harteck had himself developed a low-pressure column for the purpose, and the method had subsequently been taken up by I G Farben at Bitterfeld [14-50]. There was also the Clusius–Linde process for distilling liquid hydrogen, as well as the two dual-temperature, chemical-exchange processes involving water and hydrogen sulfide, respectively. Harteck was inclined to recommend either method two or three as being most readily adopted for a medium-size installation producing two tons a year for a power-producing pile. The decision was, however, up to Gerlach, who authorized the expenditure of 1.3 million Reichsmark on the Clusius–Linde process in Munich, and nearly as much on a low-pressure column at Farben's Leuna Works.

Trying, however, to distance the nuclear power project from the quarrelsome people at Farben, and acting on Harteck's advice, Gerlach cancelled plans initiated by Esau for a full-size electrolysis plant, to be erected by Farben *in lieu* of Vemork; without ready access to even low concentrations of SH.200 from either Norway or Italy, the plan no longer made sense [14-51]. Moreover, Farben had insisted on erecting the plant in Upper Silesia, much too close to the eastern front, in Harteck's opinion. Instead, Gerlach was also persuaded to support a much smaller electrolysis device at the KWI for Physics in Berlin-Dahlem—one just large enough to re-concentrate heavy water that had been diluted during experiments. The firm of Lüde received that particular contract. To further lessen the dependence on Farben, Gerlach and Harteck, finally, initiated a contract with Linde for a large-scale plant using a combination of the hydrogen distillation and Harteck–Suess catalytic-exchange processes [14-52].

The agreement with Linde was signed without the knowledge of Farben. When Bütefisch and company got wind of the deal through Gerlach, they went out of their way to placate the plenipotentiary for nuclear physics, offering to build a small high-concentration stage for the Linde plant 'wherever Gerlach wanted it' [14-53]. However, by that time Harteck was no longer satisfied with Farben's progress on their low-pressure column, which the company hinted would take an inordinate length of time to reach full-scale production, and transferred further work on that approach to the Bamag-Meguin Company. Needless to say, Bamag, already involved in prototype isotope separation, was more than happy to take on that effort as well and broaden its involvement in the uranium project.

I G Farben's troubles were mounting, and climaxed on 28 July. On that day, 569 Flying Fortresses of the Eighth Air Force hit Merseburg-Leuna with a massive daylight raid. The hydrogenation works were totally destroyed, as was the *Stalin-Organ* pilot plant [14-54]. Harteck and Gerlach, along with Diebner, arrived in Leuna on 11 August, and encountered Herr Bütefisch who ranted and raved about a conspiracy between heavy industry in Germany and abroad, and about the Allies having been unfairly alerted through murky channels of a heavy-water plant on the premises [14-55]. Henceforth, declared Herr Bütefisch, Farben was no longer interested in heavy water [14-56]. In fact, Farben was not alone in the disaster inflicted upon it. The Bamag-Meguin Works had been similarly wrecked in the final days of the Battle of Berlin just months earlier, but the firm had elected to carry on in relative safety at Butzbach near Frankfurt.

The same day Harteck and Gerlach had their encounter with Bütefisch in the smoking Leuna ruins, Consul Schoepke and a party of German mechanics in uniform arrived at Rjukan without giving Hydro any prior notice, and began dismantling the 18 high-concentration cells at Vemork and 9 more at Såheim [14-57]. In just one day, the apparatus was packed in 18 crates, each weighing between 600 and 700 kg, under the watchful eyes of German military police. From Rjukan the cargo went south by military trucks, bound for Germany. Seven tons of nickel catalyst dispersed in pumice stone followed four days later. Nine of the high-concentration cells went to Berlin-Dahlem for incorporation in the heavy-water re-concentration plant, while the rest wound up in Hechingen in the Bavarian Alps.

In retrospect, German plans for heavy-water production at home visualized up to three stages of concentration: chemical exchange (probably at Leuna) for the lower, most costly stage from 0 to 10%, followed by electrolysis using some or all 18 last-stage cells relocated from Vemork, and supplemented if need be with an exchange unit or even distillation of hydrogen gas [14-58]. In Harteck's own opinion, Geib's H_2O–H_2S dual-temperature process was the most promising of them all. However, 'in 1943 (sic) nobody in Germany was willing to start developing a new method. As time went on the goals became more and more modest' [14-59].

Chapter 15

CANADA ENTERS THE RACE

15.1. ZEEP

As B-VII, the penultimate German heavy-water pile, was coaxed through its sub-critical paces in the bunker at Berlin-Dahlem in early 1944, two other disparate heavy-water reactor projects occupied scientists in North America. That March, the world's first operating heavy-water reactor, P-9 or CP-3, was only two months from going critical at Argonne outside Chicago. In Montreal, Canada, another heavy-water reactor project was going nowhere.

The central figure in the beleaguered Montreal effort was Hans von Halban, whom we last encountered on an ill-fated British mission to the United States two years earlier. The purpose of that mission had been to convince leaders of the American fission project that he should take charge of British heavy-water research in America. The Americans squelched that idea without further ado. Halban was welcome to come, but only as a team member; if the British insisted on an initiative under their own control on North American shores, let them take it up with the Canadians instead. That admonition, in fact, was fine with Halban. In Canada he stood a good chance of creating a project purely under British and Canadian direction, yet seemingly with the formidable resources of the United States to draw on.

The Directorate of Tube Alloys, the successor to Britain's MAUD Committee, also thought an Anglo-Canadian project under Halban a good idea. First, however, the Canadians had to be convinced as well. In August of 1942 Sir John Anderson, who oversaw Tube Alloys on the cabinet level, opened discussions with Mackenzie King, the Canadian Prime Minister, Clarence Decator ('C D') Howe, Canadian Minister of Munitions and Supply, and the physicist C J Mackenzie, acting President of the Canadian National Research Council. The upshot was an agreement that Canada would administer a substantial reactor project, on the understanding that the scientific personnel would be drawn equally from the United Kingdom and Canada, and that a scientific director be chosen jointly by the two parties. That Halban was the man of choice was a foregone conclusion, however, and so it was that on 1 December 1942 he set up shop in a partly finished medical wing at the University of Montreal. Outdoing the British in their innocuous coining of Tube Alloys, the project was initially known as 'High Velocity Corrosion'.

The agreement on jointly contributed personnel notwithstanding, the core of the initial staff would of necessity have to come from Cambridge [15-1]. Alas, Lew Kowarski would not be one of them. Halban had offered his old partner a position of distinctly junior rank in the project, instead of granting him charge of the physics division, as Kowarski had been led to believe, and he declined the invitation. In Kowarski's opinion, for that matter, the only sensible way for the British and French to maintain a viable presence in a realistic Allied reactor project was to join Fermi at Chicago on Fermi's terms. Halban himself had been offered just such a chance, along with one assistant of his own choosing, tacitly understood to be Kowarski. Halban had flatly refused, and in so doing denied Kowarski what he looked upon as an opportunity of a lifetime [15-2].

The strain between Halban and Kowarski had, according to Kowarski, been building ever since their disagreement over the choice of a moderator in late 1939 [15-3]. (Halban had argued for carbon, Kowarski for heavy water.) More likely, the tension was rooted in their divergent personalities—Halban the aristocratic physicist, Kowarski the uncultured latecomer to physics with rash scientific insight. Skeptical, in any case, of what could be accomplished by an upstart reactor team in Canada, Kowarski elected to remain at the Cavendish Laboratory, pursuing slow-neutron physics mostly on his own, but under loose supervision by James Chadwick at Liverpool.

Other, younger physicists at the Cavendish sided with Kowarski, fermenting a minor rebellion against the call to Canada; the Montreal project was good and well, but did not seem right without Kowarski. Only when Sir Edward Appleton, head of the Department of Scientific and Industrial Research under which Tube Alloys functioned, persuaded them that they had a wartime obligation to rally behind the Montreal effort, and that a suitable role for Kowarski would be worked out as well, did they acquiesce and departed—most of them as a group in January 1943. Their three-week Atlantic crossing by convoy through submarine-infested sea lanes was not without risk; that month alone, Hitler's U-boats accounted for 200 thousand tons of Allied shipping losses in the storm-tossed North Atlantic waters. Nor was their voyage without irritations. Manhandling cans of the much-traveled heavy water of Halban and Kowarski, originally spirited out of Norway [15-4], as well as crates of uranium metal lumps, and worrying about wives and families, passport formalities, and payments in Canada, all took its toll. In the event, Kowarski was left 'puttering about with some small experiments; it was not very interesting work' [15-5].

A prominent piece of equipment accompanying the final shipment out of British ports on 14 April was a 60 cm spherical aluminum vessel—the centerpiece of Halban and Kowarski's somewhat controversial experiment at the Cavendish over two years earlier. It bore the code name 'Pinnochio' [15-6].

Unable to recruit enough senior British or Canadian scientists, Halban came up with a distinctly cosmopolitan team. For a starter, he turned to the many French refugee scientists in the United States. Among them was the cosmic-ray physicist Pierre Auger, well known for the effect which bears his name. A former student of Jean Perrin and Francis Perrin's brother-in-law, he had been involved with the CNRS and continued research and teaching at the École Normale after France fell. Learning that the police were after him for his connection with the underground publication *L'Université Libre*, he fled via Spain to the United States where he joined the Free French and found a position at the University of Chicago. There Halban looked him up, and persuaded him to come to Montreal and head his experimental physics division.

Cleared for secrecy by Compton shortly before leaving for Montreal, Auger had inspected Fermi's pile in the west stands of Stagg Field just a few days after it reached criticality of 2 December, and thus cheated one or more of Joliot's former team-members of the distinction of becoming 'the first Frenchman to see a functioning chain reaction' [15-7].

As for Francis Perrin, Halban's former colleague, he wound up teaching Urey's chemistry classes at Columbia, before becoming heavily involved with the Free French in Algiers. Montreal was not in the cards for him; Halban bore him a perpetual grudge for taking out an independent patent on reactor cooling systems. (Perrin would, however, play a major role in French nuclear energy circles after the war.) Perrin's father, Jean Perrin, followed his son to New York where, despite his advancing years, he campaigned for the French cause. He died there in 1942.

Another French scientist in Chicago when Halban came calling was the radiochemist Bertrand Goldschmidt. Like Joliot ten years before him, he had gone through the École de Physique et de Chimie Industriele, and then been steered by Paul Langevin into radioactivity under Marie Curie [15-8]. With the fall of France, and being Jewish, he narrowly escaped on a liner jammed with refugees bound for Martinique. From there he hopped to Puerto Rico aboard a Pan American flying boat, and finally reached New York on the regular mail packet. In New York, determined to avoid the coterie of French refugees, he applied unsuccessfully for a position with Fermi at Columbia. In due course he ran into Halban in a hotel, who was nursing the early stages of pneumonia contracted during his Atlantic crossing. In one of his more thoughtful gestures, Halban secured Goldschmidt an invitation to spend a few weeks as a trainee with Glenn Seaborg's radiochemistry group under Compton in Chicago, learning the basics of plutonium chemistry. When Halban later passed through Chicago sounding out Auger for a position at Montreal, he also recruited Goldschmidt; he would be responsible for a small section devoted to plutonium in a chemistry division set up under Friedrich ('Fritz') Paneth, the respected Austrian radiochemist.

Among other non-French recruits to Halban's project was Bruno Pontecorvo of Fermi's old team in Rome. When that team broke up in 1938, he joined the Curie Laboratory in Paris; he had left France in July 1940 after the German invasion, and as of late was employed by an oil company in Oklahoma engaged in, of all things, well logging, the detection of oil-bearing strata by neutrons. The Canadians and British objected to Halban taking on yet another non-British national, but Pontecorvo's reputation and the shortage of physicists carried the day. Another physicist of first rank joining the team, and definitely a feather in Halban's cap, was George Placzek, whom Halban had befriended in Niels Bohr's laboratory in 1937 and who had been of inestimable value in supplying Halban, and everybody else for that matter, with theoretical nuggets crucial for understanding chain reactions. Placzek took charge of the theoretical physics division. He also became Halban's closest friend at Montreal, even though he married Halban's first wife.

Next to assembling a team of crack scientists, the most important order of priority in launching a heavy-water reactor project was obviously securing the necessary heavy water. By fortuitous timing, construction of the heavy-water plant at Trail began in September of 1942—to be sure, as an initiative between the United States Government and Cominco, a private Canadian corporation, and thus had no official link with the Canadian authorities at Ottawa. At the beginning of November, Wallace A Akers, Research Director of Imperial Chemical Industries and also director of Tube Alloys, arrived in Canada to check on Halban's growing organization. Stopping in Washington, he approached Bush and Conant on reserving heavy water for the fledgling project. A formal request for six tons of heavy water, the entire anticipated first year's production at Trail, followed by letter from Mackenzie to Conant. No formal reply was immediately forthcoming, though the Anglo-Canadian leadership remained cautiously optimistic in light of sympathetic utterances by Bush and Conant [15-9].

Conant's reply to Mackenzie on 2 January 1943, came as a shock, if not quite as a bolt out of the blue. It confirmed what the British had been privately fearing for some time [15-10]. A new policy of severely curtailed exchange of information, applying to the whole American uranium program, would apply to Halban's heavy-water project as well. Strictly speaking, the policy was a consequence of turning the problem of producing an atomic weapon in the United States over to the US Army, and concerns over military secrecy thereby engendered. In reality, the policy reflected the widening gulf between the American and British nuclear programs, and concomitant American reluctance to needlessly share scientific data with its allies. The Americans were growing increasingly wary of the British slipping information to Germany or the Soviet Union. They were also mistrustful of the strong presence in the Tube Alloys hierarchy of individuals on loan from ICI, among them Akers, and even Sir John Anderson himself, and correspondingly reluctant to reveal information that could prove important in postwar industrial competition [15-11].

The new policy, Conant explained to Mackenzie, affected the Montreal project specifically on account of two recent American decisions. First, the ongoing Chicago graphite-reactor program was to be supplemented with a strong effort to produce plutonium in a heavy-water program based on a construction project entrusted to du Pont. Second, interchange on design and construction of new weapons was predicated on the recipient of the information being able to take advantage of it in the current conflict. Since clearly neither the Canadians, nor the English, were in position to produce elements '49' (plutonium) or '25' (uranium-235) on a timely schedule, exchange of information with Halban's team would be limited to basic scientific questions. It would not cover heavy water, uranium, or plutonium production, nor fast-neutron physics. The Montreal team *could* have heavy water from Trail for a period, but only enough for purely scientific research, and only to the extent that the results were useful for du Pont's own reactor engineers.

Outraged, Sir John Anderson protested all the way to the Prime Minister, imploring him to take the matter up with Roosevelt, unaware that Roosevelt himself had approved the new American policy. Churchill, for that matter, was preoccupied with the North African campaign and other weighty issues, all of which crowded the agenda of his meeting with Roosevelt in Casablanca just then. Mackenzie, on the other hand, was less inclined to argue with Washington's conditions. His own concern was to avoid Canada becoming embroiled in a nasty Anglo-American confrontation, or, worse, becoming stuck with a white elephant in Montreal. He was also in closer touch with Bush, Conant, and Groves. He suspected that there were other factors underlying the American decision, not disclosed, and that the British were exaggerating the importance of their contributions relative to those of the Americans. An intimate conversation with Bush and Conant over dinner in Washington convinced him that the American restrictive measures were motivated by high-level, military concerns, not by peacetime industrial interests; they were, in any case, not open to negotiations. Akers, too, was mollified after meeting with Conant and Groves, or at least saw that it was fruitless to protest.

In the end, the Montreal team plodded along as best it could during the spring and summer of 1943. The laboratory was put in working order, and gradually a modicum of experimental work picked up speed in neutron physics, nuclear chemistry, and reactor engineering, accompanied by a lingering unease that it was a wasted effort without meaningful American cooperation [15-12].

As time went by, another concern increasingly worried the demoralized Montreal team: Halban. His dictatorial and secretive style put most everybody off [15-13]. The Canadian authorities in Ottawa, too, found him a burden for his excessive demands and vacillating decisions, and the Americans distrusted him intensely as a security risk, because of his Austrian background and especially because his ties to ICI. Halban himself sensed the hostility and felt less and less at ease with the administrative burden; in addition, he suffered from recurrent bouts of chronic heart problems.

The resolution of the quagmire in Montreal had its origins shortly after the conference in Quebec City between Roosevelt and Churchill. There, an agreement on better Anglo-American cooperation was reached in principle, but rendered moot in practice, owing to Groves's steadfast insistence on strong security measures. More important for the Montreal laboratory proved a visit to it by the two senior British scientists in North America, James Chadwick and Mark Oliphant, in advance of the first meeting of the Combined Policy Committee (CPC), an outgrowth of the Quebec agreement [15-14]. Chadwick and Oliphant, the latter on his way to join Lawrence at Berkeley, had several long sessions with Mackenzie, first in Montreal and then in Ottawa, where they reviewed the state of the laboratory in considerable detail [15-15]. The mood of discontent and woes plaguing Halban's team were subsequently further spelled out by his senior Canadian staff member, George Laurence. Halban was acting without consulting or even informing an advisory committee set up the previous fall. Canadians on the staff, especially native French-speaking Canadians, were being subordinated. The strong ICI connection, in particular, often kept British circles informed about Canadian technical developments while the Canadians were left in the dark. Chadwick, especially, was disturbed over all he heard, and returned to the problem during repeated visits with Mackenzie. The two agreed that a larger Anglo-Saxon and United Kingdom presence on the laboratory staff would do a lot of good. Indeed, wondered Mackenzie to himself, might not the solution be to invite someone of Chadwick's or Oliphant's caliber and background to take charge? [15-16]

John Cockcroft was another possibility. His name first surfaced in connection with the Montreal project in late 1943, when Chadwick broached his name to Groves as a possible replacement for Halban [15-17]. Nothing came of it until April 1944, when Halban himself agreed under pressure to step down, retaining control of the physics division instead. (Auger would be returning to France in the wake of liberating Allied forces.) Cockcroft, released from his radar work then coming to a head in preparation for the Normandy invasion, agreed to take over the moribund laboratory, largely as the result of a dogged campaign on his behalf by Chadwick.

The reasons for Chadwick's successful role in these events were several. As both Technical Advisor to the British representative on the CPC, and as head of the British Mission in Washington on atomic energy, he carried exceptional weight among the British contingent of scientists in America. Another reason was the curiously close friendship between himself, a retiring, soft-spoken man, and the tough General Groves—a relationship born of a sense of mutual respect, rather like Oppenheimer and Groves. It was, to be sure, a relationship misunderstood by a few back in Britain who suggested Chadwick 'had sold himself to the Americans' [15-18].

An outcome of the CPC discussions between Groves, Chadwick, and Mackenzie was a decision to scale down the projected Canadian reactor. With Hanford capable of producing enough plutonium for the current war requirements, the need for a production facility in Canada was no longer there. Instead, it was agreed to build a small pilot plant operating at nominally zero power—small enough to be built on a reasonable time scale, yet one that could keep heavy-water pile research going. Ostensibly, it would be the pilot plant for a much larger National Research Experimental (NRX) power reactor dimly foreseen in the more distant future.

Once in charge, Cockcroft had little trouble persuading Kowarski and those scientists still languishing in Cambridge to join him in Montreal. Kowarski crossed the Atlantic on one of the Empress boats, arriving in Montreal in time for a Canadian feast of broiled steaks, served up by Cockcroft himself in chef's apron and cap. Cockcroft had more in mind for Kowarski, however. Taking him aside, he asked him to lead the design of the pilot plant—undoubtedly to the annoyance of Halban in his lessened role. A suitably remote site for the plant, and other reactors that might follow, had already been selected by Cockcroft, with Groves and Mackenzie in tow, and approved by the Canadian authorities: a place called Chalk River on a secluded and wooded bank of the Ottawa River, 130 miles west of Ottawa. A construction firm for actually building the nuclear plant was also chosen: Defense Industries Limited, a crown corporation. Design work would continue in Montreal for the time being, pending completion of the first facilities at Chalk River.

To bring himself up to date, Kowarski paid a visit to CP-3 at Argonne Laboratory, and consulted with its builder, Walter Zinn, and his colleagues. On returning to Montreal, he threw himself into the task of directing the design of what he dubbed the Zero Energy Experimental Pile, or ZEEP for short. In fact, it was designed to operate at all of one watt of output power, smaller by a factor of 250 than Zinn's CP-3. As a result, ZEEP was simplicity itself compared to, say, the complexities of a power-producing, heterogeneous graphite reactor with its complex lattice, or even a heavy-water reactor on the scale of NRX: rods of aluminum-clad uranium metal suspended in a tank of heavy water [15-19]. The tank was an aluminum cylinder, 206 cm in diameter and 260 cm in height, or of necessity about the size of Karl Wirtz's tanks at Berlin-Dahlem, still unknown to the Allies. The volume may have been the same; the liquid housed within was not. ZEEP's tank was slated to contain the full five tons of 99.7% heavy water calculated in Canada (and Berlin) as just adequate for a divergent chain reaction, or about three times the heavy water available in Germany. Neither a cooling system, nor heavy radiation shield was needed. What little heat was generated in ZEEP was dissipated in the heavy water; the graphite reflector for neutron economy, some three feet thick, sufficed as a radiation shield.

Simple, perhaps, ZEEP was nevertheless the first nuclear reactor to go into operation outside the United States, thirteen months after authorization by the Combined Policy Committee in Washington. It went critical on 5 September 1945, a week after the end of World War II [15-20].

Chapter 16

FEARS AND FACTS ON THE CONTINENT

16.1. Alsos

If the Americans were growing weary of their wartime partners, they grew downright frenzied over the strong possibility of a German atomic bomb in the making. (Curiously, the Germans, believing themselves ahead in uranium research, simply discounted the possibility of significant Allied progress in this field.) With the American Fifth Army under General Mark Clark advancing up the Italian boot in the fall of 1943, and the capture of Rome seemingly within sight, General Groves saw the opportunity of learning something about the German uranium program from Italian scientists and military officials. Thus was born the *Alsos* mission, named after the Greek for *grove* without Groves's prior knowledge and to his 'horror' [16-1]. It was, indeed, a name poorly disposed to ward off undue inquisitiveness.

Led by the flamboyant intelligence officer Colonel Boris T Pash [16-2], the mission left for Naples in mid-December, 1943. In Taranto, Pash and his staff met with Italian naval officers; they also questioned a few Neapolitan and Genoan university professors. They learned a fair amount about German weapons developments generally. However, as to atomic research in Germany, the Italians could tell them little more than they already knew. For instance, Major Mario Gasperi, who had served in Berlin for six years as the Italian Air Attaché, remembered a casual conversation with a German industrialist about German heavy-water activity in Norway, and that I G Farben was somehow involved. 'Seems to remember D_2O not used for biological purposes' [16-3]. As to any heavy-water plants in Germany, Major Gasperi was of no help.

Not long after Pash's group reached Naples, it became clear that Rome was not about to be liberated anytime soon. Incessant rains, the onset of winter, fierce German resistance, all combined to slow the Allied armies to a crawl, as they painfully battled a series of natural obstacles one by one: the Volturno River north of Naples, then the Garigliano; the natural fortress of Cassino; the wild hill-country reaching north through Frosinone. How then to make contact with the two senior physicists in Rome, Edorado Amaldi and Gian Carlo Wick? [16-4]. An elaborate scheme was concocted to smuggle the two out of the city by submarine, involving all of Fifth Army G-2 (intelligence), General Eisenhower's Allied Force Headquarters (AFHQ) in Algiers, SOE and its American counterpart, the OSS (Office of Strategic Services), and even the British Navy (which had control over naval operations in the Italian theater). In the end, these machinations came to nothing [16-5], and on 22 February Pash left for the United States in a huff. Thus ended the first *Alsos* mission without having learned very much about German nuclear research.

At this point, it may be recalled that the British had been onto the German uranium project ever since Cockcroft was alerted to the Uranium Club by Paul Rosbaud in 1939. With the warning of stepped-up Norwegian heavy-water production in 1941, British Intelligence shifted into high gear. Michael Perrin, deputy chief of Tube Alloys, assigned Rudolf Peierls the task of looking into 'German plans for the heavy water they were obtaining from Norway', soon after Peierls and Otto Frisch came up with their sober estimate of the critical mass of uranium. Peierls prepared a series of reports on physicists and their activities to be watched within the German scientific community. He did so by scrutinizing German journals obtained through neutral countries, including the *Physikalische Zeitschrift*, which each semester published a list of lecture courses in physics in all German universities, and papers abstracted in *Physikalische Berichte* [16-6]. At the same time, the Norwegian section of the SIS under Eric Welsh was forming a picture, albeit an ambiguous one, of German activities, especially in Norway, from a frail network of agents: Leif Tronstad, Jomar Brun, Brunulf Ottar, and Harald Wergeland in Norway, Njål Hole in Sweden, Paul Scherrer in Switzerland. Much of the information was channeled through Rosbaud, and evaluated in greater depth by Tronstad after he himself took over Section IV of the Norwegian High Command in London, in charge of industrial intelligence, espionage, and sabotage.

The information flowing from Milorg and XU painted a variously reassuring and disturbing picture of the importance of heavy water for German military purposes. The picture emerging proved sufficiently alarming to Michael Perrin and his superiors to activate the upper echelon steps that led to the heroic campaign against Norsk Hydro with widely differing effectiveness, but with the end effect of denying Germany all but enough heavy water for small-scale exponential experiments. Consequently, by early spring of 1944 the British grew more relaxed about a German nuclear threat. For one thing, the code breakers at Bletchley Park failed to come up with a single item suggesting

German atomic bombs in the making. On 21 March, Sir John Anderson reassured Churchill that, while it seemed certain that the Americans would have an atomic bomb in time for use in the war, 'fortunately all the evidence we have goes to show that the Germans are not working seriously on the project' [16-7].

With *Alsos*, the British monopoly on nuclear intelligence had come to an end. In fact, the British welcomed the marshalling of American resources in this area, as well they might. In typical American fashion, more would be spent on the *Alsos* mission than the entire German outlay on their uranium project. For a starter, Groves opened a Manhattan Project liaison office in London. In its wake a joint Anglo-American Intelligence Committee was established, whose formal members were Michael Perrin, R V Jones, Eric Welsh, and two American Majors, Horace K Calvert and Robert R Furman. Before long, Major Calvert, a tough no-nonsense intelligence officer with a broad knowledge of the Manhattan Project, had a staff of officers and agents manning desks in the Tube Alloys building in London's Old Queen Street.

The failure of *Alsos I* can be variously blamed on the slow pace of the Allied advances up the Italian peninsula, bureaucratic friction and imprecise divisions of responsibility, and a lack of qualified scientists on the mission. When *Alsos* was revived in the spring of 1944, for the purpose of following the Allies into France, the Low Countries, and ultimately into Germany, a team of scientists was assigned to the mission as well. It would be headed by Samuel A Goudsmit, a respected physicist, though, like Gerlach, not a nuclear physicist. Indeed, he was only vaguely aware of America's atomic bomb project. Born in the Hague in 1902, Goudsmit was the son of a prosperous dealer in bathroom fixtures [16-8]. He studied physics under Paul Ehrenfest at the University of Leiden, and published his first paper on atomic spectra at the age of nineteen. Ehrenfest also arranged that he work part-time in Pieter Zeeman's laboratory in Amsterdam, where he honed his spectroscopic skills. In 1925, on the suggestion of Ehrenfest, he and George Uhlenbeck began their regular collaboration on the theory of atomic spectra, and within the year published their first joint paper, introducing the concept of electron spin for explaining complex spectra. Two years later they both received their doctorates in Leiden, and sailed for the United States where, still together, they began their academic careers at the University of Michigan in Ann Arbor. Goudsmit would remain on the Michigan faculty until after World War II.

When the United States entered the war in late 1941, Goudsmit joined the Radiation Laboratory at MIT in Cambridge. Late 1943 found him in Britain, on loan from the 'Rad Lab' while sorting out radar problems for the American Eighth Air Force. Thus, when Vannevar Bush called on him in connection with *Alsos*, Goudsmit assumed the job called for looking into German radar developments, and physics in general. Indeed, the revived mission would cover a much broader range of scientific topics than *Alsos I* in Italy, but nuclear physics was still the main target of investigation. It was Major Furman, Groves's personal representative on *Alsos II*, who took Goudsmit aside and briefed him on the real purpose of the mission: German atomic bomb research. Goudsmit only learned some months later that his dossier credited him with 'some valuable assets, some liabilities'. Among the assets, 'I was expendable and if I fell into the hands of the Germans they could not hope to get any major bomb secrets out of me' [16-9].

About 40 key scientists were associated with *Alsos II*, though only a small number of them were permanently attached to the mission. The rest were sent to Europe as the need arose. Their task was to ferret out important German scientists, their laboratories and their location; the task of the military contingent, under Colonel Pash as before, was to get to the individuals and their institutions 'before any one else [i.e., the Russians] got there' [16-10]. Only Goudsmit was officially briefed about the Manhattan Project, though many if not most of the physicists and chemists involved had more than an inkling of the project—certainly more than the security officials attached to the mission suspected.

An advance party of the *Alsos* mission entered Paris on 25 August, with the 2nd Armored Division of the Free French Forces under General Jacques Leclerc. They lost no time looking for Frédéric Joliot, as had the Germans five years earlier. We last heard of Joliot in 1941, with his brief arrest by the German authorities who questioned him about his relationship with left-wing groups at the Collège de France. By that time, German repressions were on the increase in occupied France, with frequent arrests, torture, and executions. Joliot, however, was left pretty much alone; the Germans wished to make use of the distinguished Professor. Putting up with minor inconveniences, he occupied himself with the day-to-day needs of his research groups in Paris and Ivry, and keeping French science going generally by teaching, lecturing, and putting laboratories back in working order. His research program now concentrated exclusively on topics compatible with the strained circumstances of enemy occupation and of non-military significance—mainly biomedical applications of artificial radioactivity. At the same time, he became increasingly involved in resistance activities, working closely with militant Communists. In May 1941 he became president of the *Front National*, a broad Communist movement called to oppose the Germans, and the next year he joined the Communist Party. Since it was an illegal party, his membership could not be made public until after the liberation of Paris [16-11].

By 1944, Joliot's clandestine activities reached a point where he had to go into hiding as an electrical engineer under various pseudonyms, while resistance groups escorted the ailing Irène and the two children into Switzerland. Holing up at night in an apartment in the Belleville section of Paris, he now spent his time preparing for the uprising in Paris ahead of the Allied liberation. He and his assistants manufactured Molotov cocktails at the École de Physique et Chimie and Collège de France: collecting flasks of sulfuric acid and potassium chlorate, emptying champagne bottles, refilling them with gas and acid, recorking and wrapping them in rags soaked in potassium chlorate, running the deadly bottles onto the rooftops of the Préfecture of Police in Paris [16-12]. On the day of the liberation, Colonel Pash and Major Calvert found Joliot waiting for them on the front steps of the University, wearing a Forces Françaises brassard; his laboratory was cluttered with cotton powder drying on mattresses.

Goudsmit, having stayed behind in London, reached Paris on 27 August and saw Joliot immediately. Within days, Joliot was flown to London and interrogated by British Intelligence; however, they learned disappointingly few hard facts from him about uranium research in Germany. More of substance was learned when Goudsmit and Calvert entered liberated Brussels on 9 September. In the files of Union Minière they traced various shipments of uranium compounds to Auer and *Degussa*, an Auer subsidiary. More material had subsequently passed to an agency directly connected with the German Ministry of Trade and Finance. Moving on into Holland, they stopped at the huge Philips works in Eindhoven, learning of high-tension equipment ordered by the Reich Research Council for the University of Strasbourg (Strassburg in German). When Strasbourg finally fell in late November, they headed straight for the University. Professor Carl-Friedrich von Weizsäcker was not to be found, nor was his assistant Karl-Heinz Höcker, the young reactor theorist. However, von Weizsäcker's office, left in hasty disarray, yielded a treasure trove of papers. Letters to and from Bothe, Wilhelm Groth, Rudolf Fleischmann, von Weizsäcker, and Heisenberg testified to the slow progress of the German nuclear power project. Scanning them by candlelight as the Germans began shelling the city from afar, and analyzing them in greater detail back in Paris, they came to an incontrovertible conclusion.

> Germany had no atom bomb and was not likely to have one in any reasonable time. There was no need to fear any attack either from an atomic explosive or from radioactive poisons. [16-13]

The Strasbourg raid, and the denouement of Germany's uranium research program, marked a turning point in Goudsmit's *Alsos* affiliation. He had satisfied himself that Germany had no atomic bomb on the drawing board, much less in the field. General Groves was less convinced. Might not the bountiful paperwork fished out of von Weizsäcker's files and wastebasket be a plant? Several other pieces of evidence suggested that, at the very least, the Strasbourg papers did not tell the whole story. In particular, files seized in Paris before the fall of

Strasbourg pointed to two centers of suspicious activity: in Oranienburg, a suburb northeast of Berlin, and in the sleepy village of Hechingen in the Swabian Jura, not far from Hohenzollern Castle. Oranienburg was the site of the Auer Company's uranium processing and rare-earth plants. Not only had 60 tons of uranium compounds been shipped there early in the war from Union Minière, but, as the *Alsos* team learned in the offices of the Société des Terres Rares in Paris seized by Auer, the entire French stock of thorium had gone there as well. The commercial uses of thorium were limited, but the element was a possible alternative to uranium in the construction of atomic bombs.

Oranienburg was seemingly not the only place of importance for the Auer Company. Senior Auer personnel were in the habit of passing in and out of Hechingen, labeled a 'restricted area' in captured documents. Hechingen had caught the attention of R V Jones of SIS Air Intelligence some months earlier, and now photo interpreters pored anew over the photographs of the Hechingen–Bissingen area, obtained under the direction of RAF Wing Commander Douglas Kendall. New photographs were taken, and carefully examined for tell-tale signs of some kind of nuclear production facility. A small building, closer to Stuttgart and known from other intelligence to be connected with the KWI in Berlin-Dahlem, seemed harmless enough. But wait! Not far from Hechingen fresh ground had been broken, and a number of small factory-like buildings were springing up with amazing speed, along with storage tanks and a network of pipes on the ground. There was a new railroad spur, and power lines—all adjacent to a forced-labor camp.

An apparent key to the Hechingen activities turned up among the Strasbourg papers: fragments of a letter from Heisenberg, obviously the brain behind the German uranium power project, to von Weizsäcker, confirming that Heisenberg had relocated from Berlin-Dahlem to Hechingen. That Heisenberg had left the KWI for Physics in Berlin-Dahlem for somewhere in southern Germany they already knew, from Wick in Rome and Rosbaud via the Norwegian underground, but exactly where to had remained a mystery until Strasbourg. Now they even had his address and telephone number: Weiherstrasse 1, telephone 405, Hechingen.

Even if there was no German atomic bomb, all evidence pointed to a German atomic power project. If at Hechingen, however, where was the necessarily conspicuous central reactor hall or structure? The only thing clear was that the new construction was unique, commanding top German priority. That being the case, it was of paramount concern to Groves that the United States seize control of both Hechingen and Oranienburg. At the Yalta meeting of Roosevelt, Churchill, and Stalin, a general agreement was reached on the subject of German zones of occupation for each Allied power. As a result, much of Berlin, including Oranienburg, would fall within the Russian zone. To deny the Auer plants to the Soviet Union, Groves set wheels in motion for the plants to be destroyed by air attack. Hechingen, however, came safely within the Western Allied zones, albeit in the French zone, subsequently carved out of the British and American zones. To forestall the French, not a nuclear partner with the United States and Britain, Groves and Pash began plans for a parachute drop into Hechingen, and, as if that were not enough, a bomb strike on Hechingen as well!

In the event, Goudsmit was able to talk Groves out of this absurd operation, considering the German project 'not worth even the sprained ankle of a single Allied soldier' [16-14]. Instead, the Haigerloch–Hechingen operation would develop into a close overland race between Anglo-American forces, spearheaded by a US Engineering Battalion under the command of Colonel Pash himself and perhaps appropriately code named operation 'Humbug', and a contingent of French and Moroccan forces. Word of the impending race for the principal German atomic research laboratory, between ostensibly friendly forces, quickly made the rounds; all the leading lights of the American and British intelligence communities converged on the scene by all possible means: RAF Dakota, small observation plane, Jeep. Who would get there first, and what would they find?

16.2. In the Haigerloch cave

On 8 January 1945, Hitler authorized the withdrawal of the vanguard of his troops in the Argonne Forest northwest of Bastogne—a clear sign that he was giving up his desperate counteroffensive launched on 16 December. On the eastern front, Russian forces were poised for a powerful drive into Poland. In the air, Allied squadrons roamed Germany at will, destroying communications, plants, and installations.

In Berlin, all but a handful of the *Uranverein* scientists had deserted the ruins for safer spots to the south of the city. Already in July of 1943, Albert Speer had ordered all war research industries to identify places to which they could relocate, should the mounting air raids render further work impossible in the capital and its environs. Heeding the call, Heisenberg had looked for a suitable location well to the south and west—somewhere closer to his family in Urfeld in the Bavarian mountains, and a place one could assume would be safely out of reach of the Russians when the war ended. It should not be too close to western industrial centers, due to the constant threat of air attacks, and a good agricultural region when foodstuffs became scarce [16-15]. Gerlach had earlier taught in Tübingen south of Stuttgart in Baden-Württemberg, and from him Heisenberg learned of several little villages south of Tübingen, among them Hechingen, nestled in the backwater Balingen–Hechingen depression in the Hohenzollern region of the Swabian Jura. As a result, the KWI for Physics had moved piecemeal, relocating gradually in a large and nearly vacated textile factory at Weiherstrasse 1 in Hechingen [16-16]. By the end of 1943, about one third of Heisenberg's staff were installed in Hechingen, including assistant director Max von Laue. Following the disastrous RAF raid of early 1944, Otto Hahn's KWI for Chemistry had evacuated too, moving lock, stock, and barrel to Tailfingen, just south of Hechingen.

Heisenberg himself settled into two rooms in Hechingen, while his wife Elisabeth and their seven children remained in their 'eagle's nest', high on a forested slope in Urfeld. Joining him in Hechingen were von Weizsäcker and Karl-Heinz Höcker, who escaped Strasbourg just ahead of the Allied bombing and subsequent capture of the city. About the only ones left in Berlin-Dahlem were Karl Wirtz and Kurt Diebner, along with a small group of technicians. In January 1945 they started assembling B-VIII, Germany's last heavy-water pile experiment, in the still-intact bunker on KWI grounds. Harteck, meanwhile, set out for one more visit to Norsk Hydro on 9 January, in a last attempt to secure some low-concentration heavy water without upsetting the firm's nitrogen production, vital for the explosives industry as well as Norwegian agriculture [16-17].

Late January found Gerlach, who lent what support he could extract from Goering's Reich Council office in support of the eleventh hour reactor effort, back home in Munich. He surveyed the bomb devastation of the city, his home, and the Institute. Then and there he decided the time had come to evacuate the rest of the KWI for Physics to safety in the Baden-Württemberg hills west of Munich. Wirtz's B-VIII assembly, though agonizingly close to startup, would have to be disassembled as well, and all vital parts laboriously transported south. Time was indeed running out. On 17 January, Marshal Zhukov's central striking force was nearing the main rail line and highway between Warsaw and Berlin. The next day, General Vasily Chuikov, the hero of Stalingrad, had Lodz, Poland's second-largest city, in his binoculars. The German frontier was only 140 miles away.

The loss of Upper Silesia on 27 January was the last straw for Albert Speer, Hitler's economic czar, who prepared a note for his boss ending 'The war is lost!'. The same day Gerlach, still in Munich, telephoned his reluctant decision to Hechingen, then to his office in Harnack Haus in Dahlem, saying he would be traveling north all night. Arriving in Berlin the next day, he talked things over with Diebner, his administrative assistant, even as the KWI building shook and windows burst from an ongoing air raid. On the 29th he called his close friend Rosbaud at Springer Verlag, telling him he was leaving Berlin, and taking 'the heavy stuff' with him. 'Where to?' asked Rosbaud. 'To Werner', replied Gerlach simply [16-18]. The next day, 30 January, he ordered the pile dismantled, and everybody ready to leave Berlin the next morning. He also assured Rosbaud that he had impressed on Heisenberg the importance of safeguarding the precious heavy water, obtained in Norway 'under false pretext' in Rosbaud's opinion [16-19].

On 31 January, a small convoy of cars sped out of the city, carrying Gerlach, Wirtz, and Diebner, the last in Army uniform, along with Fräulein Guderian, Gerlach's secretary. Close behind them rumbled several overloaded military trucks carrying the heavy water, uranium, and pieces of heavy equipment. Rosbaud, remaining in Berlin for the time being, had been unable to learn from Gerlach the convoy's ultimate destination. He sent word to this effect via the Norwegian underground to the British physicist P M S Blackett, another close friend. Apparently the message did not reach Blackett, an outspoken and influential critic of many of Britain's wartime scientific policies, though it apparently reached General Groves.

> We had learned from both aerial reconnaissance and a Berlin scientist, who got word to us through the Norwegian underground, that research on uranium had been moved, presumably to a safer location, but where he did not know. Until that time, our intelligence had come in on a fairly regular basis, but then it virtually ceased. We were confronted with the problem of finding out where the Kaiser Wilhelm group had moved and what they were up to. [16-20]

The destination of the crated pile was, in fact, Haigerloch, a picturesque village above the Eyach River, about two miles west of Hechingen. The village had been chosen by Gerlach fully six months earlier as the eventual location of the pile laboratory, because of its remote location in a mountain valley, seemingly safe from aerial attack. Gerlach was prepared to have a new bunker blasted out of the cliff-like promontory rearing up behind the lower village, topped by the baroque Schlosskirche where Heisenberg, an excellent pianist, would sometimes play the organ. Conveniently, however, there was already an old wine cellar hewn out of the base of the cliff face, which could be enlarged for just that purpose with relatively modest effort. The cave, along with the adjacent Swan Inn, its owner, was requisitioned on 29 July; by the end of 1944, civil construction was far enough advanced for the facility to start receiving equipment.

Barreling south on the ice-slicked autobahn, Gerlach's convoy halted the next day in Stadtilm outside Thuringia, about 140 miles southwest of Berlin. Stadtilm was where Diebner's former group had relocated late the previous summer, installed in an old schoolhouse. The group had intended to carry on their Gottow exponential experiments there in relative safety from bombers and Soviet troops, this time with uranium oxide blocks produced under contract by *Degussa* over a year earlier. Solid carbon dioxide would have to do, *in lieu* of heavy water. Believing that Diebner's laboratory was further advanced than the Haigerloch facility, Gerlach decided on the spur of the moment to off-load the pile equipment. Too late, he learned that he was wrong; the 'laboratory' was no more than a small pit dug in the schoolhouse cellar, and a few scientists holed up in an inn. While Gerlach hurried off to nearby Weimar to obtain assistance from the local authorities in putting some of the laboratory infrastructure in working order, Wirtz called Heisenberg in Hechingen, highly miffed at the sudden change in plans. Equally upset, Heisenberg called Gerlach, insisting the pile experiment be brought to Hechingen forthwith, not be subsumed in a last-minute push under Diebner.

To underscore his displeasure, Heisenberg set off for the nearest operational railway station by bicycle, accompanied by von Weizsäcker. They reached Stadtilm after a three-day harrowing journey by train and car, dodging bomber and fighter–bomber attacks all the way. Heisenberg and Gerlach had a long talk, and then left together by car for the south. After dropping Heisenberg off in Munich, Gerlach went on to Haigerloch to organize the transport from Stadtilm; Heisenberg would spend a little time with his family in Urfeld. Later that month, Erich Bagge accompanied a column of trucks from Haigerloch to Stadtilm to fetch the heavy water and uranium. Assisted by Party officials, the trucks were loaded and headed back for Haigerloch. Wirtz rode along, making quite sure the material actually left Stadtilm. By the end of February, $1\frac{1}{2}$ tons each of heavy water and uranium cubes were safely stowed in the Haigerloch cave, along with 10 tons of graphite and a small quantity of cadmium metal. Some additional heavy water, uranium, and uranium oxide briquettes remained in Stadtilm, where Gerlach had set up office as the plenipotentiary for nuclear physics. Completing the Haigerloch inventory were 500 mg of radium–beryllium preparation, flown down from Berlin shortly before by Fritz Bopp.

The amount of heavy water and uranium remaining at Stadtilm is of passing, if academic interest in light of the subsequent experimental results at Haigerloch. Fully ten tons of uranium oxide blocks, produced by *Degussa* in Frankfurt, were still on hand at Stadtilm when the Haigerloch experiment took place, as well as rather small amounts of uranium metal and heavy water [16-21]. Bringing this material to safety would preoccupy Gerlach's as well as Goudsmit's team in the closing days of the war.

Reconstruction of B-VIII began at once in the Haigerloch cave. The ill lit, damp and gloomy interior of the working space featured a circular water-filled pit in the floor of the cave, as in the bunker at Berlin-Dahlem, and an overhead crane; a diesel generator had been installed in the vacated bier stube across the street. The same cylindrical aluminum tank that earlier housed B-VII, 2.1 m high and equally wide, was lowered into the pit, where it rested on a wooden grating [16-22]. Lining the bottom and inner walls of the aluminum vessel was a neutron-reflecting layer

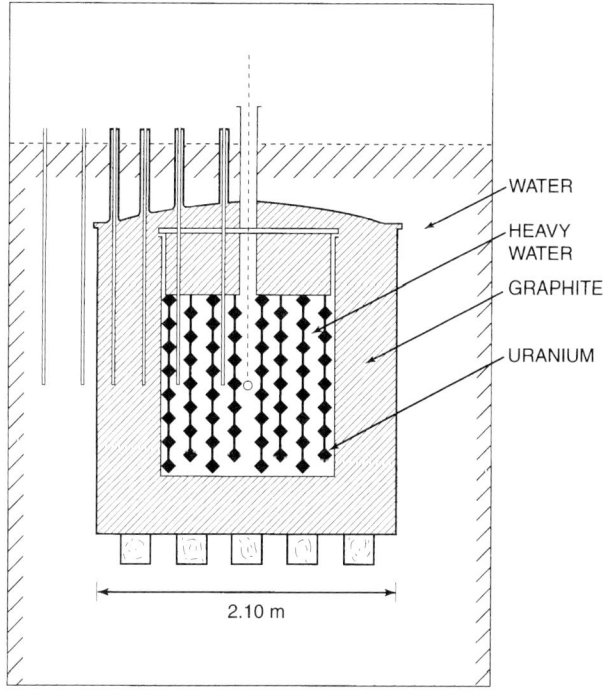

WATER

HEAVY
WATER

GRAPHITE

URANIUM

2.10 m

Figure 16.1. *Experiment B-VIII, the Haigerloch pile. Adapted from Bothe W and Flügge S (ed) 1953 Kernphysik und kosmische Strahlen, Naturforschung und Medizin in Deutschland 1939–1945 vol 14 (Weinheim: Chemie) part 2, p 159.*

of carbon, 40 cm thick, forming a cylindrical cavity open at the top. Into this cavity was snugly inserted a second, smaller magnesium alloy vessel—the reactor vessel proper. Measuring 1.24 m in diameter and 1.64 m in height, this was the magnesium vessel first used in B-VI. In the present arrangement, it was capped off with a 40 cm thick cover of graphite sandwiched between two magnesium plates. Hanging from this graphite–magnesium cover were 78 alloy-wire chains; each chain supported 8 or 9 uranium-metal cubes, as shown in figure 16.1. (The scheme of suspending uranium cubes by thin wires in heavy water was first used by Diebner and company in their G-III pile at Gottow.) Ideally, they should have been 7×7 cm^2 uranium cubes supplied by Auer. In the event, the Haigerloch team had to make do with 680 smaller, 5×5 cm^2 cubes, mostly resurrected from Diebner's late Gottow experiments. Bagge's Ra–Be neutron source could be lowered into the center of the pile through a chimney extending through the composite lid; through it, heavy water could also be poured into the magnesium vessel. Neutron probes hung inside and outside the reactor vessel, as also depicted in figure 16.1.

A startling feature of B-VIII, and all the German pile experiments for that matter, was the woeful inadequacy of safety precautions. The neglect was not predicated on lack of knowledge, but on necessity. It was in part due to scarcity of resources in the wartime economy and, in the case of Haigerloch, lack of time before the Allied victors were upon them. There were no cadmium-sheathed control rods, nor was there even a provision in the bottom of the reactor vessel for rapidly draining the heavy water, should the reaction run away. After all, ran the casual thinking, the pile was still viewed as a sub-critical experiment. A single lump of neutron-absorbing cadmium metal would have to do. If, against all expectations, the reaction threatened to get out of hand, the lump could be dropped down the chimney into the pile.

Final-assembly operations commenced during the first week of March. The bottom layer of carbon was placed in the aluminum vessel, and the magnesium vessel centered on top of the carbon; next, the remaining blocks of the carbon mantle were packed around the smaller vessel. The composite lid with its uranium-hugging chains was winched into place, and a top cover plate, shaped like a shallow inverted dish, was bolted into position. Ordinary water with an anti-corrosion additive was pumped into the concrete pit. Only when all joints were pronounced watertight was the neutron source lowered down the chimney, and the irreplaceable heavy water was slowly pumped in. At regular intervals, the heavy-water pumping was stopped, and the neutron multiplication count measured with silver and dysprosium probes as a function of radial distance from the central neutron source. Tension mounted among the onlookers in the little cavern.

Plotting the reciprocal of the neutron intensity against the quantity of heavy water admitted, Heisenberg and Wirtz could anticipate the liquid level at which the pile must go critical—that is, at which the chain reaction would become self-sustaining, independent of the neutron source [16-23]. Alas, when all the heavy water had been pumped in, the neutron multiplication factor read 6.7. In other words, for every 10 neutrons introduced at the center by the Ra–Be source, 67 neutrons reached the surface of the pile. This was over twice as good as B-VII, or six times that achieved with L-IV, but still short of self-sustaining performance. A simple calculation indicated a mere 50% more heavy water, or about 750 kg, and about the same amount of uranium, would suffice to reach criticality. Was there enough at Stadtilm, just 200 miles to the north?

Chapter 17

SWABIAN JURA AND UPPER TELEMARK: FINAL EVENTS

17.1. The rush for Haigerloch

The Anglo-American armies under General Eisenhower crossed the Rhine in force on 22 March, two weeks after the US First Army captured the Ludendorff Bridge at Remagen. Immediately before crossing, I G Farben's plants in Ludwigshafen were occupied by elements of the US Sixth Army, and the next day, 23 March, Colonel Pash led a team of *Alsos* scientists into the plants as well. Once across the Rhine, Farben's shattered 'Leuna Werk' in Mannheim, for all intents and purposes out of the heavy-water business due to bombs and administrative wrangling, was also taken [17-1]. Pash's T-Force paused only long enough there for a cursory inspection, then pushed on eastward. Heidelberg, a bigger prize, beckoned.

Heidelberg was only 10 miles away. American troops entered the ancient university city in the last days of March, and *Alsos* occupied Walther Bothe's laboratory in the KWI for Medical Research on the thirtieth. Goudsmit had a friendly but delicate discussion with Bothe, his colleague from years past. Bothe proudly showed him his now functioning cyclotron, but respectfully declined to speak about his military research. Wolfgang Gentner, who had coaxed the cyclotron into operation just before the Americans marched in, and new to Goudsmit, was more forthcoming in their first conversation on 1 April. In particular, he informed Goudsmit about the Haigerloch pile, adding that it was not yet self-sustaining.

Gentner was evidently better informed than Gerlach, who happened to be in Berlin when premature news arrived from Hechingen that B-VIII was on the brink of going critical. 'Die Machine geht!' he exclaimed to Rosbaud on 24 March. At first stunned by the news, Rosbaud had listened patiently to his enthusiastic friend and soon saw the experiment for what it was: a last, valiant try, but too late for a meaningful test—certainly to affect the outcome of events.

In a follow-up conversation on 9 April, Gentner also drew Goudsmit's attention to Diebner's rival pile laboratory in Stadtilm, and offered his opinion that Diebner's group was not on a technical par with Heisenberg's. Goudsmit could not afford to relax on that reassuring note; any German laboratory in the wrong hands was bad news, and it was not clear which of the two laboratories would be captured first by the Allied troops. But not for long. Back in Paris, he learned that General Patton's Third Army was sweeping through Thuringia with such speed that Stadtilm was sure to fall any day. Colonel Pash and Frederic Wardenburg, a du Pont chemist with *Alsos*, set off for Stadtilm from Heidelberg by Jeep, and Goudsmit followed from Paris in a small plane piloted by the physicist David Griggs, another *Alsos* regular and a civilian adviser to General Carl Spaatz, the US Eighth Air Force Commander.

The *Alsos* team reached Stadtilm on the heels of the retreating German forces, and too late. Informed by Reich Research Council headquarters in Berlin that the Stadtilm group was about to be overrun by the enemy, the SS had dispatched an armed truck unit to Thuringia. It reached Stadtilm on 8 April, with orders to evacuate everybody to safety in the 'Bavarian Redoubt' south of Munich. Diebner differed on the desirability of an SS escort out of town. While the exhausted SS soldiers caught up on some sleep, machine pistols across their laps, he organized a mini-convoy of his own. He and his colleagues hastily loaded all remaining uranium, radium, and heavy-water stores, and headed in the general direction of Munich. When Pash and his party pulled into the courtyard of the old schoolhouse building in Stadtilm a few days later, the only scientist of note to greet them was Friederich Berkei, who had elected to stay behind.

The reactor pit in the schoolhouse cellar was empty, as was a smaller pit with a lead cover, used to store radium. A few cubes of pressed uranium oxide, and a block or two of paraffin, was all there was left [17-2]. Some standard laboratory instruments upstairs, plus considerable documentation, mostly reports to the Reich Research Council in Gerlach's hand, bore out what had been learned at Strasbourg—mute testimony to the modest scale of the German uranium effort. The whereabouts of Diebner and his convoy was unknown. Berkei did confirm what Gentner had told them about Hechingen and Haigerloch. The Haigerloch reactor was subcritical for all he knew, but Diebner's heavy water and uranium could alter that. In particular, Diebner's uranium-oxide briquettes, all of ten tons worth, introduced into the graphite reflector, might raise the multiplication count by just the right amount, judging by recent neutron intensity measurements by Wirtz [17-3].

Now a new difficulty arose: the French might beat the Americans to Heisenberg's laboratory. On 18 April, French-Moroccan troops were already in Horb on the Neckar River, only 12 miles from Hechingen.

Pash lost no time organizing an extemporized party of Sixth Army combat engineers; Operation 'Big', the hunt for the last-ditch German atomic effort, was under way. His team, too, reached Horb three days later and headed straight for Haigerloch, while the French troops stalled, looking for survivors of the Vichy Government in the ancient castle city of Sigmaringen. As Pash and his Task Force took off on their final dash, General Devers, Commander of 6th Army Group, gave him a friendly send-off. 'Well, Pash', he shouted. 'I hope you hit the jackpot!' [17-4]. The British were not far behind. Once they, too received news of the capture of Stadtilm, R V Jones and Eric Welsh commandeered an RAF Dakota. They picked up Michael Perrin and Sir Charles Hambro, a former SOE commander, in Paris and flew on to Rheims, the forward base of Supreme Headquarters, Allied Expeditionary Forces. There they had a brief conference with General Bedell Smith, Eisenhower's Chief of Staff, before proceeding to Horb, alias '*Alsos* Forward'.

Pash and his combat engineers, accompanied by General Eugene Harrison, Chief of Intelligence in the Sixth Army Group and General Devers' observer, overran Haigerloch on 23 April. The *Alsos* scientists and their English colleagues had remained behind, under Pash's orders not to move out until he sent the word. His advance team quickly located the padlocked steel door to the cave laboratory off Bahnhofstrasse near the bend in the Eyach River [17-5]. Rounding up the supervisor, whose identity was blazoned on a scrap of paper stuck to the door, they had it open in no time. Lighting candles to illuminate the damp interior, they entered cautiously. No radiation shield barred their way, except for a heavy cover over the 3.4 m wide concrete pit dominating the otherwise empty floor. With some trepidation, they removed the cover, with radiation monitors at the ready. Within the pit hung a thick-walled cylindrical shell of graphite blocks, topped by a heavy-metal shield, and enclosing a pot-like metal vessel with a metal frame fitted on top. The vessel was empty, with no trace of uranium or heavy water. A few drums of heavy water were subsequently found stashed nearby in the main chamber of the laboratory.

Making an effort to conceal his emotions over having hit General Devers' jackpot, the objective of Hitler's Uranium Club for six years almost to the day, Pash sent word for the impatient scientists and the rest of the party in Horb to join them. Fascinated, they all picked their way through the Haigerloch facility. Next to the main chamber of the cave was a room outfitted as a laboratory, a workshop, and an auxiliary room. Cluttering the latter were three metal containers, each about 0.8 m in diameter and a little higher, connected by a system of pipes to the main chamber. But still no sign of the main stock of uranium and heavy water.

Dismantling and crating the meager booty—graphite blocks, a little heavy water, even a pair of Rowland optical gratings inexplicably lying around—they loaded it onto Army trucks. To deny the French access to the useless pit in the floor, Pash had suggested blowing up the chapel astride the overhanging cliff, thereby blocking the entrance to the cave. Once again, Goudsmit, looking for the German scientists in nearby Hechingen, objected. To him in retrospect, blowing up the Rococo Schlosskirche was about as irrational as the US Army's destruction of the Japanese cyclotrons at the end of 1945. (Bothe's cyclotron in Heidelberg was spared.) Instead, after carefully mapping the interior, the cave itself was blown up by the engineers, and the party joined Goudsmit in Hechingen. Pash was still elated. The German atomic pile was secure; now to round up the German atomic scientists.

In fact, the French had beat them to Hechingen by one day, though they showed little interest in the makeshift KWI for Physics in the textile factory at Weiherstrasse 1, or the scientists themselves. As in Stadtilm, the chief scientist eluded Pash and his little band. Heisenberg was gone, having set off by bicycle for Urfeld to the southeast three days earlier. Nor was Gerlach to be seen. Among those still lolling around at the Institute were Erich Bagge, Karl Wirtz, and Carl Friedrich von Weizsäcker. Two days later, Otto Hahn, emaciated from recent strain, along with Max von Laue, was found waiting for them in Hahn's own Chemistry Institute in one of the ubiquitous schoolhouses in nearby Tailfingen.

After close questioning, Wirtz was coaxed into revealing the whereabouts of the missing uranium and heavy water, hidden on Heisenberg's instructions. A small but high-level group of British and Americans, including Perrin, Welsh, and Colonel Lansdale, Groves' Chief of Security, sped back to Haigerloch. The $1\frac{1}{2}$ tons of heavy water was found in gasoline cans in the cellar of an old water-driven grist mill. The 700-odd uranium cubes were found buried in a plowed field near the Haigerloch cave. It took a few days longer to track down certain scientific reports still missing. On the eve of their departure from Hechingen, von Weizsäcker relented, and admitted they were sealed in a large metal canister submerged in the cesspool of one of the scientist's homes in town. Goudsmit assigned a Lieutenant and an enlisted man the smelly task of retrieving the documents. More to their liking, the soldiers also uncovered a supply of fine wines belonging to the Weizsäcker family; those supplies were appropriately disposed of in a V-E day celebration back in Heidelberg [17-6]. The rest of the haul—documents, uranium, heavy water, prototype isotope-separation apparatus—was shipped to Paris for transhipment to the United States, over the vigorous objections of the British, who felt entitled to a preliminary scrutiny of the documents in London.

As for the German scientists, Goudsmit made some decisions on the spot. Weizsäcker, Wirtz, Bagge, Horst Korsching, Laue, and Hahn were detained and escorted to Heidelberg before the French colonial troops had time to react. Laue was a staunch anti-Nazi and had been involved only peripherally in the fission project; however, Goudsmit decided that he, like Hahn, would prove valuable in future discussions about postwar German science with American authorities. Bagge and Korsching, too, were on the edge of the nuclear-power project, but Goudsmit chose them for their work on isotope separation. Bothe, on the other hand, was left alone, in part since Heidelberg would fall within the American zone of occupation, and, besides, Goudsmit felt he had nothing more of substance to reveal [17-7].

But where were Heisenberg, Gerlach, and Diebner? Gerlach, hunted by the SS, had himself been hunting for Diebner's truck convoy out of Stadtilm. He finally caught up with Diebner on 25 April, and split the convoy in two. Diebner continued towards the southeast still in German hands, while Gerlach returned to Munich with some of the supplies. There he was picked up in his office at the University on 1 May by Pash and his men. Diebner was detained the next day in a small town near Munich. That left Heisenberg, who had bicycled for three days straight through the war-torn countryside to Urfeld. On the night of 30 April, he and Elisabeth celebrated the radio news of Hitler's death with a bottle of wine long saved for some occasion, and three days later he received Pash and his party, to his considerable relief and with suitcase packed [17-8]. Heisenberg, slightly arrogant but cooperative, Gerlach, fully cooperative, and Diebner, silent, were escorted to Heidelberg; the rest of the scientists were already in Rheims.

Paul Harteck was picked up in Hamburg a few days later by Major Russell Fisher of *Alsos* and brought to Paris. Goudsmit, who had not met him before, recalled their first encounter.

> At first he was rather reluctant to talk. He was not aware of how much we already knew about him and his work.
>
> 'Tell me something about your efforts to make volatile uranium compounds', I said, knowing that he had undertaken this research in the hope of finding some substance suitable for isotope separation by gas diffusion methods.
>
> Harteck admitted having directed such work. 'It was not successful', he said. 'The nearest we came was with some complicated organic compounds containing a uranium atom surrounded by atoms of hydrogen and carbon. Very complex molecules indeed, hard to describe', he said. 'I'll try to show you', he continued. Pointing at the paperweight cube on my desk he said: 'Now let us assume that that represents uranium'. He moved to pick it up and its weight took him by surprise. 'But this *is* uranium!' he cried. [17-9]

Thus ended Harteck's wartime career on a positive note; indeed, he was taken into custody 'in high spirits' [17-10]. A scientist of first rank among the *Uranverein* members, he not only got the nuclear-power project started, but was, among other things, arguably the principal German expert on heavy water and its mobilization for Germany's wartime purposes. His Norwegian counterpart, if we may so refer to Leif Tronstad, did not fare so well.

17.2. Interrupted Sunshine

By the summer of 1944, the eventual defeat of Germany could no longer be doubted. Among the Allies, it was mostly a question of time; many Germans saw the handwriting on the wall as well. A major, lingering concern of the Allied High Command in the West, and most definitely of the Norwegian Government in exile, was the eventual outcome of the German occupation of Norway. 'Festung Norwegen' was literally just that. As late as May of 1945, there were no fewer than 340 000 German troops stationed there, or more than in the hotbed of Yugoslavia with five times the population [17-11]. They included General Lothar Rendulic's 20 *Gebirgsarmee*, numbering 200 000 seasoned troops—veterans of Finland and of the recent fighting in northern Norway. Nor were they adverse to the scorched earth policy, as they amply demonstrated during their westward evacuation in Finnmark. As German Naval bases fell in France, the *Kriegsmarine* withdrew much of its remaining strength to Norwegian waters. Augmenting the regular land, sea, and air forces under overall command of General Franz Böhme, were thousands of German SS and police troops, as well as German and Norwegian (Quisling) paramilitary forces of various kinds—all armed and under command of fanatically disposed *Reichskommissar* Josef Terboven. With all these forces at the ready, what would happen in Norway? [17-12].

The specific concern of Leif Tronstad from his vantage at the Norwegian High Command in London was safeguarding Norwegian industry from wholesale or sporadic destruction in the final stages of the war [17-13]. In particular, he was concerned over Norsk Hydro installations in Upper Telemark, including dams, power plants, factories and facilities, and the safety of personnel. In May of 1944 he placed before the Anglo–Norwegian authorities a plan for an armed incursion into Telemark, to strengthen and augment existing Milorg forces in the region with instructors, specialists, weapons, and equipment [17-14]. The expedition would consist primarily of personnel who had seen previous action in the territory of concern. Tronstad proposed that he himself, by now a Major in the Linge Company, should be in charge of the team. For second-in-command, he suggested Lieutenant Jens Anton Poulsson, who had led *Grouse*, the SOE advance party in support of the ill-fated Operation *Freshman* in the fall of 1942.

On the recommendation of Colonel Wilson of SOE, the Norwegian and British authorities concurred in the plan, code-named Operation *Sunshine* [17-15], with a single exception: they objected to Tronstad's personal participation, in charge of the operation or not. He was, after all, far too valuable for the Allied cause in his capacity as chief of FO4 in the Norwegian High Command. However, Tronstad insisted, and that was that [17-16]. In the event, details of the operation, such as its territorial extent, were worked out by Tronstad and his colleagues, in consultation with the overall commander of Milorg since late 1943, Jens Christian Hauge. Hauge was in London just then, on his first of several visits with the High Command during the last year of the war [17-17].

In a last letter to his wife back in Norway, dated London, 27 August, Tronstad explains that he is off on a vital mission to Norway, but for security reasons will be unable to meet with 'Bassa' in the near future. His last entry in his wartime journal, dated 30 August, ends as follows:

> Farewell to London for now. I bring the diary to a close here, hoping to continue it at a later time. If not, it will hopefully be of pleasure for my family and of use in throwing light on conditions in London during 1941–1944. Till we meet again! [17-18]

On 5 October 1944, Tronstad and a party of eight parachuted onto the Hardanger Plateau near Ugleflåt, about 30 km west of Lake Møs (the drop zone too frequently used by past commando operations to be safe any longer). Also dropped at Ugleflåt from the two Short Stirling bombers assigned to the mission were 24 'containers', or parachute-equipped supply canisters, and six additional packages of equipment. The *Sunshine* team quickly established contact with the local commanders of the Home Forces, as Milorg was now called, among them chief of operations at Rjukan, Thor Viten, known to most simply as Viggo. None were more enthusiastic in welcoming them back than Einar Skinnarland, the redoubtable radio operator still at large on the plateau.

Using Skinnarland's home-built hut at Skriubotn near Lake Møs as their base of operations, Tronstad and Poulsson traveled far and wide, meeting with, among others, Hydro transport engineer Kjell Nielsen, still on the job after the dramatic action on Lake Tinn, and engineer Rolf Sørlie. Sørlie, it may be recalled, had fled into the hills after the ferry operation, joining Skinnarland. Subsequently, he had gone over to Sweden, but was called back by the Home Forces to take charge of internal security at Rjukan (B-org [17-13]). While Poulsson met with Rjukan Director Arne Enger, Tronstad traveled as far as Kongsberg, where he again met with Home Force Chief Jens Chr Hauge, and with Director Kjell Meinich-Olsen from Hydro headquarters in Oslo.

Tronstad's trip was not without a bit of drama [17-19]. He traveled brazenly by Hydro car with a company chauffeur. On the road they were stopped by a suspicious looking, armed fellow. As Tronstad groped for his pistol, the fellow explained that he was the local bailiff or district sheriff (*Lensmann*), on the lookout for a 'Rottenikken', and asked if they had seen anything of him. 'No we haven't', answered the chauffeur,

Figure 17.1. *Leif Tronstad. Courtesy Leif Tronstad, Jr.*

adding on a lark, 'we thought you might be him'. The sheriff waved them on, not amused. Rottenikken was a sought-after, dangerous criminal.

The rest of the *Sunshine* team proceeded by ones or twos to their assigned areas of operation, keeping in touch among themselves and with Home Station in England by wireless. Gradually the local resistance network was evolving into a force to be reckoned with, as skepticism in the Allied High Command over the wisdom of anything more than a passive Home Force gave way to active support of units capable of putting up a stiff fight *in lieu* of regular Allied forces desperately needed on the Continent. *Sunshine* was but one tiny element in the burgeoning outside aid. Platoon commanders, often non-commissioned officers from 1940 (since the Norwegian officer corps had been rounded up in German camps), were assigned, sections assembled and began training in secluded regions. New landing grounds for parachute drops sprang up and equipment and food supplies began streaming from the moonlit skies in hitherto undreamed quantities, quickly to be stashed in well hidden depots. All the while, the airwaves crackled with cryptic messages in Morse or openly over the regular Norwegian BBC news transmissions, signaling yet more incoming operations from across the North Sea. 'Fresh cakes stink the most!' 'The football players strum the guitar!' 'Skiing in the cottage!' [17-20]. To be sure, open conflict with the *Wehrmacht*, other than in a link-up with an Allied invasion, was sure to have a disastrous outcome for the lightly armed Home Forces, and for local civilians. Equally likely, and a better match for 'the boys in the hills', was isolated action with rearguard, fanatical SS troops, if recent experience in liberated France was a guide.

The main mission of *Sunshine*, however, was to assist in the prevention of eleventh hour Nazi sabotage of industrial facilities essential in Norway's post-war economic recovery. The tables had turned from two years ago; now the High Command in London had good reasons to suspect that the Germans had something up their sleeves. Stepped-up German photography and measurements of key pieces of machinery suggested that they were prepared to place explosive charges here and there when the time was ripe [17-21]. Ironically, the person in charge of this ominous activity, a man by the name of Loritz assigned to Wilhelm Rediess's police HQ in Oslo, went under the improbable title *Werkschutzführer bei SS und Polizeiführer Nord*, or 'protector' of vital plant equipment—much like B-org, one might say [17-22].

Unknown to Loritz, those Norwegians assigned to assist him and his men on their rounds kept careful records of their inspection schedules. At the same time, preparations were made for active internal plant countermeasures, to go into effect if and when explosive charges were actually laid. In this they drew on the wisdom garnered by Special Training School No 17 (the School of Industrial Sabotage) in England. The preparations ranged from sealing off areas under lock and key, extinguishing lights and de-energizing machinery, to release of ammonia gas, a plentiful commodity at Rjukan, or collapsing ceilings in machine halls. Inactive machinery is considerably harder to damage than equipment under power. Ammonia, though lethal in large doses, has a very limited active range. A collapsing ceiling would do less permanent damage to machinery than carefully placed explosive charges [17-23]. In all of this, *Sunshine* could mainly offer advice; the actual countermeasures would have to be carried out by Hydro plant engineers.

Tragedy struck soon after Tronstad had his encounter with the suspicious sheriff [17-24]. By then, early March 1945, the German authorities had ample evidence for mounting underground activity in Upper Telemark. A parachute supply drop for *Sunshine* fell in German hands, as did a truckload of ammunition. Intensified German search paid off: a *Sunshine* operative, surprised at night, had to shoot his way out of a residence heavily surrounded by *Wehrmacht* soldiers. Equally disturbing, the sheriff in the Rauland district south of Lake Møs, a known quisling, suspected something afoot, and gave signs of notifying the Gestapo. Instead of 'liquidating' him then and there, the Home Forces decided to give him a stern warning to keep quiet, or else. Sergeant Gunnar Syverstad, then a weapons instructor in the area, paid the sheriff a visit, impressing on him the consequences of reporting anything amiss to the Germans. Syverstad assured the sheriff as well that his cooperation would be taken into account in the inevitable postwar settlement with Nazi sympathizers.

Before long, however, underground intelligence received word that Torgeir Lognvik, the Rauland sheriff, had squealed after all, reporting Syverstad's visit and who knows what else, to Gestapo in Rjukan. On 10 March, Syverstad, Skinnarland, and Jon Landsverk, another Milorg man, decided to do something about it. Landsverk dropped in on Lognvik's office, explaining he had come across some stolen goods stashed nearby. Would the sheriff care to come and have a look? The two set off on skis. A way down the road, they ran into Skinnarland and Syverstad. Smelling a rat, Lognvik attempted to flee, but Syverstad drew his pistol and fired a warning shot. They brought him to a cheese farm hut near Lake Møs, where Syverstad and Landsverk kept an eye on him, while Skinnarland went looking for Tronstad, occupied just then with some business in the neighborhood.

Tronstad, in full Major's uniform, decided to interrogate the sheriff himself, on the chance of learning something about the intentions of Quisling's Party of 'National Unity' (Nasjonal Samling or NS) in the event of an Allied invasion. He would then decide whether they should dispose of him, or hold him for the remaining duration of the war. Accompanied by Sveinung Landsverk, Jon's brother, he proceeded to the two-room hut where the sheriff was being detained. The details of what happened next were only pieced together in an investigation well after the war ended [17-25].

Essentially, the sheriff's somewhat unstable brother, Johans Lognvik, got wind of Torgeir's having gone off with Landsverk, and set off in search of them. Armed with his brother's revolver, he had no trouble following the fresh ski-tracks, straight to the Syrebekkstøylen hut, as the place was called. Spotting his brother's skis among those stacked against the wall of the little hut, he quietly slipped open the unlocked door to the small outer room. Empty. Cocking the revolver, he threw open the inner door, shouting 'Hands up!'. The room was dimly lit by a fire in the hearth, with window shutters partially drawn. 'Who is there?' demanded Tronstad, as the intruder rapidly got off two shots. One struck Syverstad in the forehead, the other missed. In the ensuing melee spilling into the outer room, more shots rang out. As Tronstad wrestled Johans Lognvik to the floor, the sheriff grabbed a carbine off the wall and struck him with the butt. Dazed, Tronstad went down, giving Johans Lognvik time for a last shot, striking Tronstad squarely in the head.

It had all happened very fast. The sheriff and his brother fled the scene. Jon Landsverk had been knocked down in the scuffle, also dazed but otherwise unhurt. Coming to his senses, he seized up the situation. Tronstad was dead, Syverstad barely alive, still breathing. By the time

Landsverk got hold of Skinnarland in a hut some distance away, and they returned, Syverstad, too, was dead. Landsverk covered Skinnarland while he was inside rounding up incriminating papers and effects. They found a sledge and pulled the bodies of Tronstad and Syverstad to a small nearby lake, where they sank them in a hole in the ice. The day after, Skinnarland tapped the following wireless message to London:

Skinnarland to London, March 12:
DEEPLY REGRET TO INFORM YOU THAT JULIUS [TRONSTAD] AND KÅRE [SYVERSTAD] WERE SHOT YESTERDAY BY UNKNOWN PERSON... SHERRIF IN RAULAND ARRESTED, AND INTERROGATED... WITH A VIEW TO LIQUIDATION STOP SURPRISED BY UNKNOWN PERSON STOP FULL REPORT FOLLOWS, THOUGH PAPERS SECURE AND BODIES UNLIKELY TO BE FOUND STOP AWAIT INSTRUCTIONS. [17-26]

Skinnarland also passed a warning to the local underground contacts, and went off to find Jens Poulsson, the latter now in command of *Sunshine*. With the Germans on full alert, thanks to the sheriff, the situation was precarious. However, because of quick action by, among others, veteran Knut Haukelid (alias 'Bonzo'), who was commander of Home Forces in the district bordering the *Sunshine* area [17-27], all local forces retired without incident to safety at Songevatn to the west, under strict orders to avoid provoking the Germans if at all possible. When things did quiet down, the *Sunshine* mission resumed operations under Poulsson, promoted at once to Captain by the High Command in London.

In the event, the Germans surrendered without offering any resistance whatever, and with exemplary discipline at that. On 8 May 1945, 20 000 men of the Home Forces proudly stepped forth, flouting armbands *in lieu* of uniforms and Sten guns slung across chests, and assumed temporary control of the country—among them the *Sunshine* groups. Neither Major Tronstad not Syverstad lived to see that glorious day. Though the *Sunshine* operation brought Tronstad more satisfaction than his desk job in London, the latter should not be overshadowed by the drama surrounding his untimely demise [17-28]. He is perhaps best remembered by his countrymen for his role in the heavy-water saga; of equal importance ranks his role in the unveiling of the Peenemünde flying-bomb complex. Then again, suggests N A Sørensen in his memorial address soon after the war, there is another point of view. When Tronstad fell in action on the Hardanger Plateau he left behind something of more lasting value for his countrymen than the fruits of the well executed task entrusted his *Sunshine* operation—namely an inspired and lasting example of devotion to one's native land [17-29].

Sledge tracks led the Germans to the little lake on the plateau. The bodies of Tronstad and Syverstad were recovered, photographed, and cremated. After the war, the Home Forces raised a large stone monument of traditional Nordic type at Syrebekkstøylen. The chiseled inscription [17-30] reads simply:

Leif Tronstad   *Gunnar Syverstad*

11 March 1945

Chapter 18

HIROSHIMA REVEALED; FURTHER CONTESTANTS FOR NUCLEAR ENERGY

18.1. Farm Hall: Operation Epsilon

Heisenberg, Gerlach, and Diebner were in *Alsos* hands by 3 May; Harteck was the last of the group nominally under Gerlach to be taken into custody, a few days later. By 8 May, when Supreme Allied Commander Dwight Eisenhower publicly declared Victory in Europe a *fait accompli*, the following ten were detained in relative comfort at the Chateau du Chesnay near Versailles—a center organized for interrogating German scientists and industrialists known as the 'Dustbin': Heisenberg, Gerlach, Diebner, Weizsäcker, Hahn, Laue, Harteck, Wirtz, Bagge, and Korsching.

8 May proved a better day for Norsk Hydro General Director Bjarne Eriksen, incarcerated in Germany ever since he raised a fuss over German demands in the wake of the Allied air raid on Herøya in the summer of 1943 [18-1]. Arrested in a roundup of former fellow Norwegian Army officers, he was held with the rest in a military internment camp for Allied officers in Germany—Eriksen in reality as a political hostage. As it happened, an adjoining section of the camp housed American officers. As a result, it did not prove difficult for Eriksen to convince liberating American troops of his civilian status, and of the urgency of his making contact with Norwegian and Allied authorities in London. No time must be lost in planning start-up strategies for Norsk Hydro in the immediate postwar period. Thanks to quick American action, Eriksen was a free man and on his way much sooner than could normally be expected [18-2].

Others among the German nuclear scientists met different fates, including Robert Döpel (now a widower), Karl-Hermann Geib, Gustav Ludwig Hertz, and Manfred von Ardenne. Caught in Berlin, or electing not to flee the Red Army for various reasons, they were rounded up by

271

a Soviet mission in the vanguard of the Soviet troops slogging their way into the city—a mission much like *Alsos*. Consisting of some 20 or 30 scientists, the leading ones in the ill fitting uniforms of NKVD Lieutenant-Colonels, the mission was, like its Western counterpart, charged with determining precisely how much the Germans knew about the atomic bomb, and exploiting German science and technology generally [18-3]. In the event, they, too, learned preciously little from German nuclear science. For one thing, the leading *Uranverein* personalities, among them Hahn and Heisenberg, were in Western custody. What German scientists they could lay their hands upon were escorted to the Soviet Union during May and June, along with much laboratory equipment from the Kaiser Wilhelm Institutes; the laboratories were largely stripped by the Russians before reverting to American control in accordance with the agreed-upon partitioning of Berlin. When *Alsos* pulled into Dahlem on 30 July, Goudsmit found that what little was left at the KWI for Physics had been dumped by American GIs in the back garden. The reactor pit in the bomb-proof sub-basement was empty [18-4].

The ten *Alsos* detainees, meanwhile, remained in custody under the loose British wartime decree 'at his Majesty's pleasure' [18-5], while the British and Americans pondered what to do with them. Eric Welsh argued for bringing them to England, and R V Jones, his superior, agreed. While the unhappy (except perhaps Harteck) scientists were shuttled back and forth between several locations in France and Belgium, Jones convinced his own boss in SIS, Sir Stewart Menzies, to have a British country manor in the village of Godmanchester near Cambridge prepared for their extended internment. We have encountered the estate, Farm Hall, before in the guise of STS No 61, an SOE staging area for agents who were to be air-dropped behind enemy lines in occupied Europe. Ironically, it was the very spot where *Gunnerside* members had whiled away time waiting to be parachuted onto the Hardanger Plateau. Among other things, Jones had the rooms outfitted with hidden microphones, and arranged for bilingual British military personnel to monitor any conversation judged of intelligence value. Anything of interest would be recorded on reusable shellacked metal disks—the state-of-the-art recording medium of the day.

On 3 July, the party of ten scientists was flown by a British Dakota to England, and thence went by auto to Farm Hall. There they were held incommunicado for exactly six months under conditions which could really not be faulted. Each man had his own room, and was assigned a prisoner-of-war manservant to look after his every need. The food was excellent, there was a rose garden for walks and contemplation, and a common room for mutual company and discussions; even a piano for Heisenberg.

The secretly recorded conversations, transcribed, translated and excerpted in weekly or biweekly installments, were forwarded over the signature of Major T H Rittner to Michael Perrin and Eric Welsh, with Copy No 1 dispatched to General Groves in Washington DC via the US Embassy in London. Portions of the published reports contain transcriptions in the original German as appendixes. The existence of these Farm Hall Transcripts (as they became known) remained a secret until 1962, when General Groves referred to them in his own aptly named memoir *Now It Can Be Told* [18-6]. A newly published version, edited and annotated by Jeremy Bernstein and with an introduction by David Cassidy, is based on the copy of the report in American hands, which differs slightly from the British version [18-7]. The original recordings are no longer extant; they were reshellacked at Farm Hall, and the disks reused. From the transcriptions, however, according to Cassidy,

> we gain unequaled insight into the personalities of these scientists, into their knowledge of fission physics, into how they viewed themselves and their work, into how they came to terms with their past in Nazi Germany, both for themselves individually and for public consumption, and into how they prepared to influence the course of science in a future Germany. At the same time, these transcriptions and commentaries provide us with new information and perspective regarding many of the heated controversies that have long surrounded the motives and aims, rationales and failures of German wartime research on the utilization of the awesome power unleashed by nuclear fission. [18-8]

Two weeks minus one day after the German scientists settled in at Farm Hall, and unbeknownst to them, their American and British counterparts huddled in shelters on a barren site on the Alamogordo Bombing Range, 210 miles south of Los Alamos, New Mexico. Only days before, a 'gadget' crafted at Project Y, as the central laboratory of S-1 (the Manhattan Project) was known, was delivered in stages to a commandeered ranch house in the middle of Trinity, a code word for the desert site appropriately known as Jornada del Muerto, or Journey of Death. Among the last components to arrive was the plutonium core of the implosion device to be tested at Trinity. Final assembly of the gadget took place in the McDonald ranch house on 13 July, and by 5:00 p.m. two days later the test device had been hoisted into a small corrugated-iron shack atop a 100 foot steel tower. The Arming Party completed its work as lightning bolts stabbed the black night, then retreated to S-10 000, an earth-sheltered control bunker 10 000 yards due south of Ground Zero. The main sequence timing switches had been locked hours before. Vannevar Bush, James Conant, E O Lawrence, Enrico Fermi, and James Chadwick (head of the British contingent at Los Alamos) all waited nervously at scattered observation points as the rain drizzled down.

The shot was scheduled for 4:00 a.m. J Robert Oppenheimer, chief scientist, discussed the weather situation with General Groves at S-10 000. Thirty minutes before firing time, they pushed zero hour back sixty minutes. At 4:00 a.m. the rain stopped. Groves immediately left Oppenheimer by Jeep and joined Bush and Conant at Base Camp, another five miles south of S-10 000, prudently separating himself from Oppenheimer in case of common disaster. Once the special arming switch was thrown, the final countdown began, enunciated over a public address system and broadcast over FM radio. A one-minute warning rocket burst atop the command bunker at 5:29 a.m. At 5:30 minus 15 seconds on 16 July 1945 the firing circuit closed. The gadget exploded, heralding the dawn of the Atomic Age. A flash of light, brighter than a dozen suns, bathed the New Mexico desert as far as the scenic overlook crowded with VIPs on Compañia Hill, twenty miles northwest of Zero. The yield of the bomb was subsequently estimated to equal 20 000 tons of TNT, equivalent to the bomb load of 2000 B-29 Superfortresses then pounding Japan.

The Farm Hall scientists learned of the birth of the new age as they sat down for dinner, three weeks later. Shortly before 7 p.m. on 6 August, Major T H Rittner, the British officer in charge of the manor, took Otto Hahn aside after the six o'clock news, and informed him that an announcement had been made by the BBC that an atomic bomb, based on Hahn's discovery of fission, had been dropped on Japan. Shattered by the news beyond words, it took a stiff portion of gin to calm Hahn before he could venture downstairs and inform his assembled colleagues. The announcement was greeted with considerable incredulity, if not outright disbelief—especially on Heisenberg's part. As the news sank in, dinner was all but forgotten. The ensuing conversation groped for technical answers. Separated uranium, element '93' (neptunium) or '94' (plutonium)? If the latter were used, one or more 'engines' (reactors) must have operated for a very long time.

About 9:00 p.m., they reassembled to hear the official BBC announcement of the Hiroshima bombing and the Manhattan Project. So it was true! Reported Rittner, 'they were completely stunned when they realized that the news was genuine' [18-9], and as the scale of the undertaking dawned on them. The first declaration by von Weizsäcker, when the free-wheeling discussion resumed, to the effect that the German physicists did not want to build a bomb on principle [18-10], is generally seen as the germ of the much taunted German argument, led all along by Weizsäcker, that they did not succeed because they did not wish to succeed. 'The ineluctable fact, borne out by the documentation . . . ', counters Bernstein, 'is that, at least in the beginning, von Weizsäcker and the others did want the project to succeed' [18-11]. Later the same evening, Weizsäcker offered a more interesting observation from our preoccupation in these pages. For them to have made a bomb, they should 'probably have concentrated more on the separation of isotopes and less on heavy water' [18-12]. Curiously, graphite is conspicuously absent in most of the Farm Hall discussions.

Still later that evening, Hahn and Heisenberg continued the discussion among themselves alone.

> One is struck, on the one hand, by Heisenberg's admission that he had never calculated the critical mass of U-235 needed to make a bomb and, on the other hand, by his clear understanding of the difference between a reactor and a bomb.

> This *tête-à-tête*, recorded unbeknownst to the two, succinctly captures what Heisenberg did and did not understand about bomb physics at the time of Hiroshima. [18-13]

18.2. Belated entries: Russia and Japan

Concern with heavy water, in the context of World War II, was seemingly limited to the Allied powers of the United States, Canada, Britain and France, plus Germany. In point of fact, the importance of heavy water did not escape the leaders of lesser-known wartime efforts to build an atomic bomb in the Soviet Union and Japan, respectively. In the event, lack of heavy water for all but basic experimentation ruled it out as an important ingredient in the Soviet and Japanese programs. Nevertheless, a cursory look at two additional, stillborn heavy-water efforts seems appropriate in rounding out our chronicle.

An excellent treatment of the Soviet nuclear-energy program is that of David Holloway, and most of the following is based on his account [18-14]. Serious nuclear weapons work in the Soviet Union only began in 1943; prior to the battle of Stalingrad, during the winter of 1942–1943, national attention was mainly focused on survival [18-15], much as wartime priorities had curtailed German scientific research two years earlier. However, Soviet interest in nuclear physics dates, as elsewhere, from the 1930s; indeed, the cyclotron at the Radium Institute in Leningrad was the first one in Europe, becoming operational in 1937. The news of the discovery of fission in Berlin the following year had similar impact in the Soviet Union as in the West. It caused great excitement and launched nuclear programs in several institutions, among them Igor Kurchatov's laboratory at the Leningrad Physico-Technical Institute. The first program to be tackled experimentally under Kurchatov was determining the number of secondary neutrons released during fission, if slightly after the same study by Joliot, Halban, and Kowarski, and with essentially the same conclusion [18-16]. Another important result reported by Kurchatov's group early in the summer of 1939, was confirmation of Bohr's conclusion that it is the rare isotope U^{235} that is responsible for fission, not the common isotope U^{238}.

The most important Soviet theoretical work in nuclear physics in this period was a series of chain reaction studies by Iakov Zel'dovich and Iulii Khariton at the Institute of Chemical Physics next door to the Physico-Technical Institute. In one of their papers on slow-neutron chain reactions in natural uranium, they reinterpreted the results of Joliot and colleagues in Paris, and of Fermi, Szilard, and Anderson at Columbia. For a chain reaction to occur, they concluded that

> it is necessary to use heavy hydrogen, or, perhaps, heavy water, or some other substance which will ensure a small enough capture cross-section, in order to slow down the neutrons.... The other possibility consists in enriching uranium with the isotope 235. [18-17]

In an article written during the early summer of 1940, Zel'dovich and Khariton calculated that about 2.6 tons of uranium oxide and 15 tons of heavy water would be needed for a slow-neutron chain reaction [18-18]. Later the same year, in a discussion before the *5th All-Union Conference on Nuclear Physics*, Kurchatov himself argued that a chain reaction could be achieved in a mixture of water and uranium enriched in the isotope 235, or in a mixture of natural uranium and heavy water [18-19].

The problem was the lack of uranium as well as heavy water. In July of 1940, a Commission on the Uranium Problem was established, attached to the Presidium of the Academy of Sciences and headed by the radiochemist Vitalii Khlopin, co-founder of the Radium Institute. For the time being, however, the Commission could do no more than study the problem on paper, and allocate a few paltry sums here and there for others to do likewise. Kurchatov, who got along poorly with the Commission in any case, was forced to send his younger colleagues out buying up all the uranium nitrate in the photographic shops in Leningrad [18-20]. As for heavy water, that problem was taken up at the first *All-Union Conference on Isotopes* in April 1940. There, lofty plans were discussed for producing heavy water by electrolysis in a plant of the electrochemical works of the Chirchik Nitrogen Combine near Tashkent in Central Asia, at the rate of 15 kg per year—a modest rate, to be sure, but sufficient for an experimental subcritical pile program well short of an actual reactor project [18-21]. The only actual research on the production of heavy water in this period seems to have been by S Levina of the Karpov Institute in Moscow. A report of his on 'the exchange reaction between deuterium and hydrogen on nickel' [18-22] dates from early 1941, or about the same time Harold Urey and Hugh Taylor were looking into the same process at Columbia and Princeton.

The German military drive on the Soviet Union in June of 1941 effectively halted all basic research not considered of immediate relevance for the war effort. Only in mid-1942 did revival of the moribund Soviet nuclear project come up for renewed discussion at the highest political levels in Moscow. By that time the Soviet government, alias the NKVD, knew, via the 'Fifth Man' of the 'Cambridge Five' recruited by Guy Burgess, John Cairncross, and perhaps from Klaus Fuchs, then working with Rudolf Peierls in Birmingham, that Britain had decided to build an atomic bomb. The revival of Soviet nuclear research may also have been provoked by the capture of a German officer's notebook dealing with the German uranium project [18-23].

The high-level discussions of renewed research, in consultation with leading scientists, dragged on during the winter of 1942–1943 as the Red Army fought desperately to hold its ground against the mighty *Wehrmacht* Army Groups B and A, driving, respectively, east toward Stalingrad and south into the Caucasus. In March 1943, with Russian forces riding high after their first strategic offensive of the winter war, Kurchatov was placed in charge of a central nuclear-energy project. His appointment, over more prominent physicists like Peter Kapitsa or Abram Ioffe, caught him by surprise, as he considered himself 'no match for either of them' as a physicist [18-24]. In fact, his appointment was made on the urging of Ioffe, the father of Soviet physics, who considered himself, at 63, too old to play a leading role in the new initiative.

In looking for guidance by reviewing the extensive Soviet intelligence files on Britain's research on the uranium problem that spring, Kurchatov saw confirmation that a divergent chain reaction was possible in a mixture of uranium and heavy water—something his fellow scientists had concluded was impossible. The problem of arriving at an informed decision had not been the theoretical calculations of Zel'dovich and Khariton, but the paucity of cross-section data at their disposal. However, the early subcritical experiment by Halban and Kowarski at Cambridge had given tantalizing evidence that such a chain reaction was indeed possible. Unfortunately, Soviet physicists could not repeat the Cambridge heavy-water exponential pile experiment, with only two or three kilos of heavy water on hand in the country as a whole [18-25].

In pondering these matters, Kurchatov noted that all published work abroad on the design of a nuclear 'boiler' (*kotel* in Russian), until Allied publications on the subject of nuclear physics generally abruptly ceased in mid-1940, presupposed a homogeneous mixture of uranium and moderator. Might not a *heterogeneous* system work better? Unbeknown to Kurchatov and his close associates who looked into the problem theoretically and even experimented with fragments of uranium oxide in paraffin, this was by then commonly assumed by all three competing parties abroad—the team under Fermi, Joliot's former colleagues, and key members of Heisenberg's Uranium Club.

Before long, Kurchatov met with his closest colleagues to decide upon the main lines of research for tackling the specific problem of producing an atomic bomb, including what type of pile to build for producing 'element 94'. (In choosing between the alternative routes to a bomb, separated U^{235} versus plutonium bred in a reactor, it appears that the Soviets initially concentrated on the latter as the technically simpler approach.) In his April report to Mikhail Pervukhin, Deputy Premier and People's Commissar of the Chemical Industry, as well as the government official with whom he dealt directly, Kurchatov repeated the estimate of Zel'dovich and Kharton that a heavy-water pile would require 15 tons of heavy water and 2 tons of natural uranium. By contrast, a graphite pile might consume as much as 1000 tons of graphite and possibly 100 tons of uranium—twenty times Szilard's estimates for a carbon pile in 1939. Despite its much higher uranium estimate, Kurchatov sided with the Americans in opting for a graphite pile, since the electrolytic plant in Chirchik showed no sign of producing heavy water any time soon. Moreover, again like the Americans, having no preconceived hangup about graphite as a moderator, as did the Germans, he too decided on a parallel program, with heavy water relegated as a more distant backup to graphite. Kurchatov himself took charge of the graphite-pile project. The task of building a heavy-water pile was entrusted to Abram I Alikhanov, an experimental physics colleague and cosmic-ray expert from the Leningrad Institute. In the event, and perhaps as well for the good of the Soviet nuclear project, serious work on that system did not get started until after the war [18-26].

Even obtaining graphite of the requisite purity turned out to be a formidable task. An initial batch of 3.5 metric tonnes acquired from the Moscow Electrode Factory in August of 1943 exhibited neutron capture cross sections, from ash and boron impurities, too high by orders of magnitude. When pressed for better material, the plant managers refused to entertain what they considered an 'impossible' task. Not until late summer of 1945 was the factory able to start turning out graphite of adequate purity with the help of physicists from Laboratory No 2 (later the Kurchatov Institute) on the outskirts of Moscow.

With respect to heavy water, the Soviets had requested small batches via Lend-Lease from the United States 'for research purposes', perhaps viewed as 'delicate feelers concerning American willingness to share officially the atomic secrets that were being shared unwittingly via Fuchs and others' [18-27]. The United States obliged them with all of 1 kg of heavy water in late 1943, and a further 100 grams in February 1945 [18-28].

Production of heavy water at Chirchik began at long last in 1944, and about the same time research on the physical chemistry of electrolytic cells was started at the Institute of Physical Chemistry in Moscow. In October of 1945, the NKVD-led mission in Germany (see pp 271–2) assembled a group of German heavy-water experts at I G Farben's Leuna works in Merseberg, for work on the design of a heavy-water plant. A year later, the group was transferred to Moscow, where it continued the work at the Institute of Physical Chemistry. In 1948, when the Chirchik plant began delivering heavy water in respectable quantities, the group of German scientists turned to other tasks [18-29].

Kurchatov's graphite reactor, F-1 [18-30], quite similar in design to Fermi's first pile CP-1, went critical at Laboratory No 2 on Christmas Day, 1946—the first self-sustaining chain reaction in Europe. Alikhanov's heavy-water reactor, built at Laboratory No 3, went critical in April 1949. The first Soviet atomic bomb, RDS-1 (or 'Joe-1' in American accounting)—a plutonium-implosion device closely modeled after the Trinity 'gadget' under Khariton—was detonated on 29 August 1949, on the steppes of Kazakhstan. The blast occurred 32 months after F-1 went critical, or virtually the same length of time (31 months) between the Trinity test and the first chain reaction in Chicago's Stagg Field.

In March 1943, as Sergei Kaftanov, plenipotentiary for Soviet science and Chairman for the Scientific-Technical Council of Academicians charged with organizing military research, recommended an all-out atomic bomb project under Kurchatov, a corresponding committee of scientific experts in Japan reached a contrary conclusion: that construction of an atomic bomb would not be feasible in time to affect the outcome of the present war. Despite the latter conclusion, a desultory Japanese nuclear weapons project, to the extent that it can be considered a single project, or a project at all, limped along to the end of the war. It was the smallest of all the wartime nuclear projects, minuscule even compared to the German effort.

The Japanese fission project had its formal start in April of 1941, when Lieutenant General Yasuda Takeo, chief of the Army Aviation Technology Research Institute, acted on Prime Minister Hideki Tojo's order to investigate possibilities of a nuclear weapons program in Japan. Yasuda passed the order on down the chain of command, until it landed on the desk of Okochi Masatoshi, director of the Riken Institute. 'Riken', the Institute of Physical and Chemical Research (*Rikagaku Kenkuūjo*), had been established in Tokyo in 1917 to promote basic research in response to the economic boom and economic challenges of World War I [18-31]. In 1931, the respected nuclear physicist Yoshio Nishina, who had studied both under Rutherford and Bohr, founded the Nuclear Research Laboratory within Riken. His laboratory grew at a steady pace, acquiring a Wilson cloud chamber, then a Van de Graaff generator followed by a Cockcroft–Walton high-voltage installation, and in 1937 the first cyclotron in Japan. Measuring 27 inches in pole–tip diameter and weighing 23 tons, this accelerator was also the first working cyclotron outside the United States. The same year, Nishina's laboratory began assembling a 220 ton cyclotron with a 60 inch magnet, with the assistance of E O Lawrence, whose own laboratory was just then erecting a 60 inch cyclotron as well [18-32].

By 1941, when the Army inquiry came through, Nishina's laboratory had a staff numbering well over 100 researchers. The director of Riken turned the problem over to Nishina, who took it up in a very leisurely manner [18-33]. The Navy Technology Research Institute, meanwhile, had independently engaged professors at Tokyo Imperial University for advice on nuclear weaponry. The upshot was the formation of the aforesaid Committee on Research in the Application of Nuclear Physics, under the chairmanship of Nishina, that met in similarly unhurried fashion, convening ten times between July 1942 and March 1943. Its conclusion, in the words of the report submitted to the Navy, was that, though an atomic bomb was realizable in principle, 'it would probably be difficult even for the United States to realize the application of atomic power during the war' [18-34].

This pessimistic conclusion caused the Navy to lose interest in the problem, concentrating instead on radar research. The Army was less easily discouraged. Soon after the Committee rendered its judgment, an experimental task, designated the NI Project, was initiated at Riken. NI was aimed solely at separating uranium-235 by thermal diffusion, the process Otto Frisch had studied while marking time in Birmingham at about the same time. It quite ignored alternative methods based on electromagnetic separation, gaseous diffusion, or the ultracentrifuge [18-35]. By February 1945, a small group of young scientists numbering less than 15 succeeded in producing a small amount of some kind of material in a rudimentary separator constructed in Building No 49 in the Riken complex—material which Riken's small cyclotron indicated was *not* U^{235}. Two months later, Building 49 caught fire following a bombing raid on Tokyo, and the separator project was no more. For all practical purposes, the NI Project was dissolved in May 1945 [18-36].

While these desultory experiments were still in progress, the Army and Navy undertook equally futile searches for uranium-bearing ore, in locations ranging from Fukushima Prefecture in the homeland to Korea, China, and far-flung Burma. The Japanese also requested pitchblende from their German partners. Reputedly, a U-boat bound for Japan, carrying 560 kg of uranium oxide and appropriately numbered U-234, surrendered to the United States in the Atlantic upon learning of Germany's surrender. This abortive mission appears to have been the only formal contact between the Axis Powers involving potential nuclear weaponry [18-37].

Late in the war, a different Japanese Naval command took up nuclear weapons research under Arakatsu Bunsaku at Kyoto Imperial University, designated the F-Project [18-38]. Its sponsorship dated from mid-1943, though the actual work, development of an ultracentrifuge for separating U^{235} (relying, as did the NI-Project, on uranium hexafluoride as the input material) was barely under way at the time of Japan's surrender. Only the design of the centrifuge was completed; no machine was actually built.

That, then, was the sole extent of Japan's wartime work on an uranium bomb: one failed and one uncompleted isotope separation experiment, a few vials of uranium hexafluoride, a sliver of metallic uranium, and a few theoretical papers. No attempt was made to assemble a uranium pile. Lacking heavy water for a moderator, Takeuchi Masa, in charge of Nishina's isotope separator, calculated (as did the Soviets) that light water would do if the uranium were enriched to 5 to 10% in U^{235} [18-39]. While uranium separation got nowhere, a belated heavy-water effort had, it appears, been launched years earlier by the Japanese industrialist Jun Noguchi. In 1926, he founded the Korean Hydro Electric Company at Konan (Japanese for Hungnam), subsequently fed by the massive Chosin and Fusen Reservoirs on the Yalu River. It became the site of an industrial complex producing ammonia, among other things [18-40]. In 1943, anticipating a shortage of electric power in Japan once the B-29 raids began in earnest, Noguchi stepped up ongoing hydrogen production for ammonia by the Korean Nitrogen Fertilizer Company at Konan, augmenting a smaller-scale effort on Kyushu, Japan's southernmost island [18-41].

Shortly after the surrender of Japan, the Manhattan Project's *Atomic Bomb Mission* deployed there in September reported that Arakatsu Bunsaku of the F-Project had been obtaining 20 grams per month of heavy water from electrolytic ammonia plants, based on the Haber process, in Korea and Kyushu [18-42]. Despite the availability of a heavy-water production program potentially rivalling Norsk Hydro's at Vemork, neutron-multiplication studies with heavy water or any other form of moderator seems not to have been undertaken at Kyoto. The industrial facilities on the Korean peninsula fell into communist North Korean hands before they could be fully explored and assessed by the US occupation authorities [18-43].

Chapter 19

EPILOGUE

19.1. Whither heavy water; what if?

Though the bulk of Heisenberg's heavy-water stores, roughly 2 tons, were uncovered by the Allies in 1945 in the Haigerloch grist mill cellar, the story of the hunt for the heavy water had a minor sequel in 1948. That year a Norwegian, Major Anton Schweigaard, arrived in the American zone of occupation in Germany. His job was to look into Norwegian property confiscated by the Germans, ranging from works of art to private radio sets, and see to it that it was returned to Norway [19-1]. In that capacity, he was assigned to the Restoration Branch, US Military Government, in Karlsruhe. He not only learned of *Alsos* having beaten the French to the heavy-water booty, and its subsequent shipment to the US, but from the American military press picked up the following item. It seems that a certain lady had been arrested in Munich, attempting to peddle a container of heavy water for an outrageous price. Closer investigation disclosed that the lady was the widow of a colleague of Walther Gerlach, who had brought the half-liter bottle of heavy water home and buried it in his garden, but was killed during artillery bombardment shortly before VE-Day. The trial in 1948 uncovered a protracted black-market affair involving a dozen questionable characters, including a Yugoslavian intermediary and a Spanish chemist purportedly adept in deuterium analysis [19-2].

Already in 1944 plans were discussed for the resumption of Norwegian heavy-water production after the war [19-3], and Norsk Hydro lost no time in picking up the pieces once the war was over. During the next decade production of heavy water at Vemork was restored and grew rapidly, along with the ravishing postwar demands for synthetic fertilizers at home and abroad, and the concurrent, newfound enthusiasm for atomic energy. In 1953, heavy-water production began as well at Glomfjord, north of the Arctic Circle, in Hydro's brand new ammonia plant, supplementing the expanding ammonia plant in Rjukan. An important incentive for Hydro's post-war heavy-water industry was

Norway's own entry into the atomic age. In 1951, Norway became the first country, after the United States, Canada, Russia, England and France, to operate a nuclear reactor, JEEP [19-4]. The reactor, at Kjeller north of Oslo, was actually a Norwegian–Dutch joint venture, with the Dutch supplying the uranium [19-5]. In 1948, Norsk Hydro contributed 7.7 tons of heavy water for the project, delivered in the winter 1949–1950, and making Hydro, too, a partner in the venture. Soon orders poured in for heavy water from home and abroad, and by the mid-1950s Hydro was producing 12.5 tons per year—8.2 tons coming from Rjukan and 4.3 tons from Glomfjord [19-6].

As an aside, the early 1950s marked the end of Hydro's prewar and wartime partner, I G Farben, if partner is the right term. Between 1945 and 1951, the German chemical giant, so closely linked to the Nazi regime, was effectively broken up, by the four Allied powers, into disparate units differing in make-up from one occupying zone to the next. In time, Farben's replacements, revitalized by new forces set in motion by postwar Allied and German policies, and a new international socio-political climate generally, emerged as a new multifaceted German chemical industry experiencing spectacular growth; however, that is another story [19-7].

Growth and diversity characterized postwar Norsk Hydro as well, as indicated. Alas, in the long run, heavy water played only a minor role in Hydro's new endeavors. Within a decade its heavy-water production was in rapid decline, and would eventually cease altogether. The new situation was largely in consequence of a somewhat painful transition by Hydro to an entirely new method of producing ammonia from petrochemicals, forsaking electrolytic production of hydrogen. In 1963 Hydro inaugurated its first oil-based ammonia factory at Herøya, and more soon followed. In the new approach, hydrogen was extracted by combustion of hydrocarbons in oil from the plentiful North Sea fields, and combined with nitrogen to form ammonia in more or less the usual way. Before long, variations in the process were introduced, but the burgeoning oil industry was the underlying element. Norsk Hydro was entering an industrial watershed. Just as in 1905, when the fledgling concern set out to exploit Norway's boundless hydroelectric resources, in the 1960s Hydro turned to the vast potentials of oil power, inaugurating a new epoch in the petrochemical industry as well [19-8].

As hydropower gave way to oil power as the basis for Hydro's fertilizer industry, heavy-water production rapidly fell off, and Norway found itself no longer the No 1 supplier. By 1977, the cumulative world production of heavy water (excluding the USSR and China) had been redistributed by country, with the USA in the lead, producing 59%, followed by Canada, 35%, Norway, 4%, and India and France producing 1% each [19-9]. That very year, the venerable electrolysis building at Vemork, the scene of much drama and rebuilt after the American bombardment in 1943, was demolished, having sat idle since 1971, when electrolytic production ended at Vemork. Consequently, the same year, 1971, the old Vemork power station went off the air for good. When inaugurated in 1911 after four years of construction, Vemork was the largest power station in the world, with an output of 108 MW or 145 000 horsepower from 10 giant turbines. In 1971 it was replaced by a new power station, still at Vemork but built in a cavern deep within the mountain, producing half again as much energy as the old station from the same mass of water and only two turbines.

Today, no trace of the electrolysis building remains at Vemork. Visitors find it hard to believe a towering eight-story glass-steel-and-concrete building, plus a score of other houses and structures, once hugged the sloping, narrow ledge on the south side of the Vestfjord valley. By a happy circumstance, however, the old 110 meter long, granite-faced power station building, the architectural handiwork of John Olav Nordhagen [19-10], is no longer obscured from view by the less inspiring electrolysis building in front of it. The 700 meter long penstocks leading down the mountain side are still there as well, albeit no longer carrying water; the new power station is fed by internal conduits within the rock face.

The old power-station building, still accessible by car and foot via the 75 foot suspension bridge across the 600 foot deep gorge carved by the Maane River, today houses the Norwegian Industrial Workers Museum, founded in 1988. The museum is dedicated to telling the story of the rise of industry in the Vestfjord valley and the hydropower that fed it, the history of the labor movement in Norway, questions of energy and environment, and the heroic yet difficult wartime history of Vemork and environs; it also houses a valuable archive documenting these variegated themes. Machinery of the former power station, restored to its 1930 status, still occupies the 110×20 m^2 machine hall, with the majestic row of old turbines glowing in the subdued light.

As of December 1988, Norsk Hydro ended hydrogen production altogether, and with it heavy water. At the time of writing, Hydro still *sells* heavy water, albeit heavy water made in Canada [19-11]. Canada is, in fact, currently the largest producer of heavy water in the western world, a legacy of its venerable power-reactor program. NRX, Canada's true pilot reactor, went critical in 1947, and ran reliably for an unprecedented 45 years [19-12]. It was followed by a series of demonstration power reactors, culminating in the CANDU (Canadian-Deuterium–Uranium) commercial reactor program. To sustain the latter program, a large-scale Canadian heavy-water production capability was imperative, and was met by three major production centers brought on line in the 1970s, at Glace Bay and Port Hawkesbury in Nova Scotia and Bruce, Ontario. Between 1970 and 1985 these plants produced more than 80% of the world's heavy water [19-13], though since then only one unit

Figure 19.1. *Vemork today. The building housing the Norwegian Industrial Workers Museum, formerly the Vemork power station, is no longer obscured by Norsk Hydro's giant electrolysis plant. Photo by the author.*

of the Bruce facility has remained operational, to better match supply and demand [19-14]. All of these plants utilized dual-temperature exchange between water and hydrogen sulfide, followed by fractional distillation [19-15]. (Trail, by contrast, was based on steam–hydrogen exchange and electrolysis.)

Price per unit volume being an essential figure of merit for heavy water, another word on the subject may be appropriate here, particularly since it has been paid relatively scant attention in these pages. Assuming with H K Rae that the price of heavy water per kilo was $500 at Rjukan in 1935, a kilo at Trail went for $130 in 1945, and at Port Hawkesbury retailed for $48 in 1970 [19-16]. In 1985 dollars, these costs per kg translate to roughly $4300, $900, and $150, respectively.

Today's principal suppliers of heavy water, in addition to Canada, are the United States, Russia, India, and Argentina. Still considered a strategic material, its sale and use is subject to a degree of international control, under the aegis of the International Atomic Energy Agency in Vienna [19-17]. Like weapons-grade fissile materials, ensuring that its use is limited to peaceful applications has never been an easy task

[19-18]. One such peaceful application is on a heroic scale, even by atomic standards: the use of heavy water as a medium for detecting neutrinos—elusive massless, uncharged members of the subatomic zoo of fundamental particles in nature. The loosely bound neutron in the deuterium nucleus results in a relatively large cross section for interactions between neutrinos and deuterons. The interactions provide a telltale optical signal for spotting neutrinos originating in supernovae in the far reaches of space, in an ongoing international high-energy physics experiment at Canada's Sudbury Neutrino Observatory (SNO). Sited at the 6800 foot level of an active nickel mine in Sudbury, Ontario, SNO houses 1000 tons of 99.92%-enriched heavy water in an acrylic sphere surrounded by 7000 tons of ultra-pure light water as a radioactive shield. The 300 million dollars worth of heavy water is on loan for $1 from the Canadian government agency AECL [19-19].

Returning, finally, to the German heavy-water effort in World War II, it appears that the German-controlled production in Norway, until the bombing of Vemork in November 1943, amounted to approximately 2.5 tons of high-grade material. Adding roughly 0.5 tons of heavy water lost in the *Hydro* sinking on Lake Tinn, we have for the total German wartime production, in round numbers, some 3 tons of deuterium [19-20]. (The American heavy-water effort, by contrast, yielded 55 tons by late 1945: 32 tons by the Trail process, and 23 tons by water distillation [19-21].) The 24-hour production at Vemork, at the time of the American strike, was 5.6 kg, and, had not the strike effectively halted all production, it seems likely, according to Jomar Brun, that the Germans would have had in hand their projected 5 tons of heavy water by June 1944 [19-22]. If so, a successful German chain-reacting heavy water and natural uranium pile, similar to America's CP-3, would not have been out of the question, in the opinion of most. In contrast to heavy water, uranium metal seems to have been relatively plentiful in German wartime circles, mainly used as a substitute for tungsten in the production of specialized steels. Moreover, a heavy-water reactor requires relatively less uranium than a carbon-moderated reactor [19-23].

On the other hand, a reactor such as CP-3 is a notoriously inefficient producer of weapons-grade plutonium, and the production of enough German plutonium for a single nuclear weapon in a reasonable time frame, assuming a prolongation of the war in Europe by an uncertain amount, seems most unlikely to those who have studied the matter. This seems all the more the case in light of the German reactor team's evident naïveté in reactor *technology*, as opposed to a purely theoretical understanding of the physics of the uranium lattice, coupled with the formidable problems of separating the two plutonium isotopes 239 and 240, among a host of other things. Even a doubling of Vemork's production rate, a not unreasonable scenario in Brun's opinion, albeit at the serious expense of decreased fertilizer production, does not alter the poor outlook for a single hypothetical bomb in German hands [19-24]. Perhaps a more plausible route to a German weapon could have been via separated uranium-235; the records suggest isotope separation was pushed harder than any other portion of the German nuclear power project [19-25].

Despite a qualitative appreciation of atomic bomb physics by Heisenberg and his closest colleagues, not even a conceptual design for a bomb came out of the German effort. For that matter, they did not even try. Hardly anybody in or close to the uranium project believed a bomb, if one could be built at all, could influence the outcome of the ongoing war. About all the notion of a bomb proved good for was as a carrot 'dangling seductively' in certain quarters of the National Socialist German state [19-26]. It kept a minimum effort going, focused on nuclear power, either for submarine propulsion (another carrot) or for unspecified applications in some postwar period.

To blame the failure, if it can be called that, on one or more single technical factors, however, overlooks numerous broader factors plaguing the wartime German project—factors endlessly debated in recent and earlier studies of the uranium project [19-27]. It suffices to note but a few of them: lack of concentration on a particular technology (plutonium versus uranium-235); lack of material (and human) resources in a wartime economy; complacency and predominance of pure over applied science among German scientific, not military, heads of the uranium effort (along with an ingrained conviction of German monopoly in atomic research); inter-service rivalries in the German military hierarchy; lack of priority at the highest level in the German command structure. It goes without saying that, in addition to purely technical, managerial, and bureaucratic considerations, come questions of ideology and science policy—never-ending questions in a virtual cottage industry of books on Nazism and science [19-28].

Then again, considering all the hurdles facing the Uranium Club, it is perhaps remarkable that such a small cadre of German scientists achieved as much as they did. Their series of subcritical pile experiments beat the Americans to the first positive neutron-production coefficient, and came tantalizingly close to a self-sustaining nuclear chain reaction. The irony is that heavy water, on which they staked so much, was a prime factor in these achievements, but, for all the misgivings of the Allied powers, played no significant role in what began as an effort to harness nuclear energy for a weapon of war.

19.2. A few of the personalities

We cannot dwell on the post-WWII careers of the many individuals who's contributions are sketched in the preceding pages. Still, a few concluding remarks seem appropriate for those whose names are closely linked with heavy water, its uses and wartime politics—especially for their less well known lieutenants who figured so importantly in the thread of events.

Starting with the Norwegians, the tragic demise of Leif Tronstad on the eve of Allied victory has been covered at length and needs no further elaboration here. His close associate Jomar Brun returned from Britain to his homeland and the industrial giant he and Tronstad had served so well. Brun rejoined Norsk Hydro as chief engineer, serving at Vemork until 1951 when he left for Trondheim and a professorship in the Institute for Applied Electrochemistry at NTH. Among the heavy-water saboteurs, each went their own way after the war. Knut Haukelid attended the Norwegian Military Academy, and then the Staff College, but retired early as a Lieutenant Colonel to his beloved mountains in 1973. Born to Norwegian parents in New York, on the 40th Anniversary of the Vemork raid in 1983, the US surprised him with an American passport. Jens Poulsson, too, retired from the Royal Norwegian Army, with the rank of Colonel. Joachim Rønneberg became a Norwegian broadcasting station executive. Knut Haugland participated in Thor Heyerdahl's *Kon Tiki* expedition, and became the administrative head of the Resistance Museum when it opened at Akershus Fortress in Oslo. On being queried by her neighbors about her son setting out on such a dangerous Pacific voyage, Haugland's mother is said to have shrugged it off: 'My Knut would never do anything dangerous!' [19-29].

In 1990, ten of the eleven surviving heavy-water saboteurs, all in their 70s, participated in a retracing of a portion of their heroic route through Telemark's perilous terrain, organized by Per Pynten of Norsk Hydro, Rjukan and helping inaugurate a five-mile hiking trail from Rjukan Fjellstue to the Vemork Industrial Museum—the latter housed inside the massive walls of the old Vemork power station [19-30].

On the night of 27–28 February 1993, the 50th anniversary of the *Gunnerside* attack, a memorial stone in honor of the saboteurs was unveiled on the rim of the gorge at Vemork, in the presence of the surviving saboteurs (including Knut-Lier Hansen, who participated in the subsequent action on Lake Tinn) and many invited guests. Cadets from the Norwegian War College repeated the climb through the gorge at the correct hour shortly before midnight, wearing lit helmet lights; this time the saboteurs took the easy route across the suspension bridge over the gorge [19-31].

As for the French, the subsequent career of Joliot has been often told. Following the liberation of France, he persuaded General de Gaulle, president of the provisional government, to create a French Atomic Energy Commission in the vague interest of weaponry and as a vehicle for pushing nuclear energy sources. (Among those who helped sway the general was Raoul Dautry, the former Minister of Armament who had joined the Resistance on the night of the Normandy invasion, and the redoubtable Jacques Allier, hero of the heavy-water coup in 1940.) The

Commissariat à l'Énergie Atomique, established in 1945, was initially headed by Joliot. Under his overall direction, and with Lew Kowarski brought back from Canada in technical charge, France's first atomic pile, Zoé (*zéro énergie, oxide, et eau lourde*) began operation at Fort de Châtillon in December 1948. Once again the man responsible for obtaining the needed heavy water from Norsk Hydro was Allier, later a member of Hydro's board of directors. In May 1946 he contracted Hydro to deliver up to 5 tons of heavy water, or about two years' output [19-32]. Under Joliot's tenure, moreover, a major nuclear research center rose on the plateau of Saclay. In 1950 he was removed from his function as High Commissioner over his political views, replaced by Francis Perrin. Forthwith Joliot divided his time between academia at the Collège de France and in the cause of Communism and world peace. He succeeded Irène Joliot-Curie as head of the Radium Institute upon her death at the age of 58; Joliot himself died at the same early age in 1958.

The year 1944 was a turning point in the affairs of Joliot's two principal assistants, the volatile Halban and the hefty Kowarski. After relinquishing the directorship of the Anglo-Canadian nuclear project in Montreal under pressure, Halban spent two stressful years doing neutron physics at his own expense in Montreal and New York City [19-33]. Despite all their difficulties with him, the British, recognizing his considerable efforts in organizing the reactor team at Montreal, secured him a professorship under F A Lindemann (Lord Cherwell) at Oxford in 1946. Halban left Britain for Paris in 1955, taking charge of a nuclear physics group at the École Normale. Always weak of heart, he died at 56 in 1964.

Halban's exit from Montreal opened the doors to the atomic project for Kowarski, who supervised the design and construction of ZEEP, the first heavy-water pilot reactor outside the United States; ZEEP went critical just as World War II ended, and became the forerunner for Canada's family of CANDU (heavy water) reactors. In 1946, Kowarski returned to Paris in charge of Zoé, a near-repeat performance of ZEEP, and as Director of Scientific Services in Joliot's CEA [19-34]. Beginning in 1952, he redirected his energies to the building of CERN, the joint European high-energy physics facility that came into being on a provisional basis that year. He became Director of Scientific and Technical Services at the fledgling organization in Geneva, and later devoted his talents to developing bubble chamber and computer techniques for analyzing the evanescent tracks signaling the debris from colliding subatomic particles. Kowarski also advised the European Nuclear Energy Agency and other bodies on nuclear policy, and remained a spokesman for heavy-water reactor technology.

Upon returning to Germany from British house arrest in 1946, Heisenberg resumed the directorship of the Max Planck Institute for Physics in Göttingen, formerly his old KWI for Physics in Berlin-Dahlem until it was terminated by Allied bombs. To placate the Western occupying authorities, the Institute had been renamed after someone less identified with militarism. At the same time, Heisenberg became professor of physics at the University of Göttingen, and was soon a leading spokesman for West German science policy. In 1958, he and his Institute relocated to Munich, where he died in 1976.

Paul Harteck, too, returned from Farm Hall to academic life after the war, first as Rector and professor of physical chemistry at the University of Hamburg during 1948–1950, and from 1951 to 1968 as Distinguished Research Professor of Physical Chemistry at Rensselaer Polytechnic Institute in Troy, New York. He retired from RPI in 1982, and died in 1985 at his last home in Santa Barbara, California. On hearing of his death, his old partner Hans Suess, himself by then Professor Emeritus in the Department of Chemistry, University of California at San Diego, wrote as follows:

> I feel greatly indebted to him, as he had been decidedly instrumental in getting me safely through Nazi-times and the war years in Germany. Without his efforts, I would probably have been drafted by the military and may not have survived. [19-35]

Karl Wirtz, too, returned to a distinguished career. He was professor of physics at Göttingen from 1948 to 1957, and subsequently professor in Karlsruhe and active in founding that city's Nuclear Research Center (Kernforschungszentrum Karlsruhe, or KFK). At the KFK he was responsible for the design of research reactor FR2, and later Director Emeritus of the Institute for Neutron Physics and Reactor Technology. Wirtz, like Harteck, became involved in American science as well. He was a visiting professor at the University of Washington, and was the first European to serve as member of the board of directors of the American Nuclear Society.

Paul Rosbaud, incidentally, was smuggled out of Berlin in military uniform in late 1945 by Eric Welsh [19-36]. In the post-war years he helped establish a Springer Verlag affiliate in London, and subsequently another publishing venture, Pergamon Press, snitched away from him by the British publishing magnate Robert Maxwell. The American Institute of Physics awarded Rosbaud its first Tate Medal for his services to scientific publishing in 1961. His reward for his role as master spy was dinner, during the same visit to New York, with General Groves—always suspicious of 'the British agent in Berlin' [19-37]. Rosbaud died two years later, and was buried at sea, as was his wish.

It is a measure of Rosbaud's elusiveness that he was never identified to John S Wilson while 'the Colonel' headed the Scandinavian Section of the SOE. Wilson and his family retained strong ties with Norway after the war, regarding the country as their second native land [19-38]. Wilson died in 1969, though his daughter Margaret still frequently attends periodic Norwegian reunions on his behalf among Home Force veterans. Eric Welsh, who ran his own Norwegian network as head of the Norwegian Section of the SIS, died in 1954. His passing went unnoticed in Britain, but was movingly acknowledged in an obituary by his Norwegian counterpart, Alfred Roscher Lund [19-39].

In the Soviet Union, a rather peripheral waystop on our journey, Kurchatov remained overall scientific director of the nuclear project until his death in 1960. In 1950, Iu B Khariton was appointed director of operations for building the first thermonuclear device at the weapons-production facility Arzamas-16, about 400 kilometers east of Moscow. Khariton's close colleague Ia B Zel'dovich became his deputy (in charge of the theoretical department at Arzamas-16), as did Igor Evgen'evich Tamm (first brought on board by Kurchatov to check on the work of Zel'dovich's group). Abram Alikhanov, in charge of the first Soviet heavy-water reactor, shared the honor of Hero of Socialist Labor with many of his fellow nuclear scientists following the successful first Soviet hydrogen-bomb test in August of 1953. Alikhanov, who with Kurchatov had been among the first cyclotron builders outside Berkeley in the mid-1930s, outlived his old colleague by a decade.

In America, finally, the two chiefly responsible for the isotope that caused such a wartime fuss, left a legacy of peaceful researches. Harold Urey outlived his mentor Gilbert Lewis by nearly two decades. Lewis retired from his official titles as Dean of the College of Chemistry and Chairman of the Department of Chemistry at Berkeley, but never from chemistry. In fact, he died alone on a Saturday morning in his laboratory in Gilman Hall, in the course of performing an experiment on dielectric constants of materials. Over the years he received many honors at home and abroad, but, surprisingly to many, not the Nobel Prize. His contributions to thermodynamics, and his theory of the chemical bond, it is frequently argued, amply deserved that singular honor [19-40]. By the same token, the importance of his devotion to deuterium, short-lived though it was, has been sadly downplayed.

In his 1946 letter to his old mentor, K F Bonhoeffer, H S Taylor explains that he is back at Princeton, at work again, after four years of atomic energy research, including devising a process for tonnage production of heavy water. P.S., he adds, 'G N Lewis in California died a few months ago'. Bonhoeffer himself, incidentally, was in poor health in Berlin at the time, as Urey went to some effort (unsuccessfully, as it turned out) to get the Russian authorities to agree to having him leave the Russian zone for a period of recuperation in Switzerland [19-41].

In his postwar career, Urey turned nearly all his attention to geoscience and cosmic chemistry, minor themes in his researches during the 1930s. At the Fermi Institute for Nuclear Studies of the University of Chicago, where he held forth from 1945 to 1958, he collaborated with Willard Libby, Hans Suess, Harrison Brown, and others in elaborating aspects of the new discipline of cosmochemistry and planetary cosmogony. In 1958 he again struck west and on a new professional path, becoming professor-at-large at the University of California at La Jolla; there he embarked on yet another career as participant in, and adviser to, the fledgling American space science program. Among other things, he participated in the planning of the series of automated lunar exploration spacecraft that preceded the manned Apollo landings. Ironically, returned lunar samples brought evidence contradicting Urey's celebrated theory of the Moon's origin [19-42].

Urey died in 1981, fifty years after the achievement that earned him the Nobel Prize in physics, and set in motion the events chronicled in these pages. In the event, the extreme scarcity of deuterium rendered heavy water somewhat a will-o'-the-wisp in the wartime race for nuclear energy: as the moderator of choice in subcritical pile experiments in German hands, as a backup for graphite in the ultimately victorious Allied nuclear effort, and as the focus of small but dramatic military and clandestine operations in wartime Norway.

Appendix A

SOME PROPERTIES OF HEAVY WATER (D₂O) COMPARED TO WATER (H₂O)

Table A.1.

Property	Value or ratio D_2O/H_2O	Units
Molecular weight	20.028/18.016	^{12}C scale
Density	1.108/1.000	g/cm^3
Freezing point	3.82/0.00	°C
Boiling point	101.42/100.00	°C
Temperature of maximum density	11.23/3.98	°C
Viscosity (55 °C)	1.107/0.890	mPa s
Abundance in ordinary water	~1 part in 5000 ~200 ppm 0.0145%	

Appendix B

A CHRONOLOGY OF HEAVY WATER

Numbers in parentheses refer to the section in which the subject is discussed.

1905	December	Norsk Hydro-Elektrisk Kvælstofaktieselskap established (4.3).
1913	Fall	The concept of isotopes proposed independently by Frederick Soddy and Kasimir Fajans (4.1).
1920	June	Neutral doublet ('neutron') and isotope of hydrogen suggested by Ernest Rutherford (3.1).
1927	May	Norsk Hydro–I G Farben collaboration started (4.4).
1930	June	Penetrating radiation discovered by Walther Bothe and Herbert Becker (3.1).
1931	Summer	An isotope of hydrogen proposed by Raymond Birge and Donald Menzel (4.1).
	Fall	Discovery of deuterium by Harold Urey, Ferdinand Brickwedde, and George Murphy (4.1).
1932	February	Discovery of the neutron by James Chadwick (3.2).
1933	Spring	Odd Hassel proposes Norsk Hydro undertake deuterium production, using electrolyzers at Vemork (4.4).
		Paul Harteck prepares heavy water while serving Fellowship under Rutherford at Cambridge (4.2, 9.1).
	April	Very pure heavy water prepared by prolonged electrolysis by Gilbert Lewis; initiation of heavy-water studies by Lewis and his school (4.2).

	May	K F Bonhoeffer requests lye sample from Vemork (4.4).
	June	Special session on deuterium by American Physical Society (4.2).
	July	Ernst Hochheim requests lye samples from Vemork (4.4). Bonhoeffer's initial analysis of Vemork lye sample encourages Norsk Hydro to undertake deuterium production (4.4).
	Fall	First experiment by Brun and Tronstad (4.4). First experiments with accelerated H^2-nuclei (4.2). First demonstration of nuclear fusion by Mark Oliphant, Harteck, and Rutherford (9.1).
	November	Bonhoeffer's final analysis of Vemork lye sample (4.4).
	December	'Discussion on heavy hydrogen' at the Royal Society (4.2).
1934	January	Discovery of induced radioactivity by Frédéric and Irène Joliot-Curie (5.1).
	March	Rutherford and Urey suggest Norsk Hydro attempt tritium preparation (9.1).
	Spring	First heavy-water production by Norsk Hydro (4.4).
	May	Conference at Rjukan on heavy-water production (4.4).
	October	The effectiveness of slow neutrons in causing fission established by Enrico Fermi *et al* (5.2).
	December	Urey awarded Nobel Prize in Chemistry for discovery of deuterium (4.1).
1935	January	First relatively 'pure' heavy water shipped from Norsk Hydro (4.4).
1936	March	Final attempt by Norsk Hydro to prepare tritium by repeated electrolysis of heavy water (9.1).
1939	January	Publication of the discovery by Hahn and Strassmann that bombardment of uranium with neutrons yields barium (6.1). Frisch and Meitner interpret the result as uranium fission (6.2). Experimental confirmation of fission process by Joliot *et al* in Paris, and by several groups in the United States (6.2).
	February	Neutrons liberated from fission detected by Halban, Joliot, and Kowarski (6.2).

April	The number of neutrons from uranium fission determined by Joliot and colleagues (6.2). Francis Perrin deduces critical mass for uranium (6.3). Homogeneous uranium/moderator pile design by Joliot's group (6.3). Meeting of nuclear physicists in Berlin leads to *Uranverein*; decision to proceed with uranium research under German Army Ordnance, as a result of Paul Harteck's letter to Erich Schumann (9.1).
September	First *Schumann Conference* on exploitation of nuclear physics for military purposes (9.1).
October	KWI for Physics designated seat of German uranium project (9.1).
December	First part of Heisenberg's report to Army Ordnance on exploiting nuclear fission (10.1). Germany, through I G Farben, requests heavy water from Norsk Hydro; Hydro management stalls for time (7.1, 10.1).

1940	January	German Army Ordnance takes renewed interest in Norwegian heavy water (7.1, 10.1).
	March	Last shipment of heavy water to Germany on eve of invasion of Norway (4.4). Jacques Allier meets with Axel Aubert of Norsk Hydro, obtains promise of entire Norwegian heavy-water stocks; Norwegian heavy water spirited to France (7.1). Frisch–Peierls memorandum on atomic bomb (8.1). Plans laid for Soviet heavy-water production at Chirchik (18.2).
	June	Halban and Kowarski reach England with Norwegian heavy water (8.2). Harteck's dry-ice-moderated, neutron-pile experiment (9.2). Ia B Zel'dovich and Iu B Khariton determine conditions for slow-neutron chain reaction in heavy-water-moderated uranium pile (18.2).
	Summer	Karl Wirtz meets with Norsk Hydro to discuss expanded heavy-water production (10.1).
	Fall	Expansion of Hydro production facilities completed (10.1). 'Virus House' completed at Berlin-Dahlem; paraffin-moderated U_3O_8-pile at Leipzig (10.1). B-I and L-I, paraffin-moderated uranium oxide pile experiments at Berlin-Dahlem and Leipzig (12.1). First Norwegian heavy water delivered to Leipzig (12.1).
	December	Halban and Kowarski obtain experimental hint that a chain reaction with heavy water is possible (8.2).

1941   January   Walther Bothe's erroneous neutron absorption measurements in graphite. Subsequently the Germans conclude (on economic, not technical, grounds) that graphite is unsuitable as neutron moderator in uranium pile (9.2).

       February   Major papers on heavy-water production presented at high-level nuclear physics research conference in Berlin (12.1). Karl Wirtz meets with Hydro management at Rjukan (10.1).

       March   Norsk Hydro initiates major heavy-water plant retrofit with 18 high-concentration cells (10.1).

       May   German orders for heavy water raised substantially (10.1).

       October   Meeting of Harteck and Wirtz with Norsk Hydro management (10.1).

       September   Leif Tronstad flees Norway (7.3).

       Fall   L-II: first heavy-water-moderated uranium oxide pile (12.1).

       Winter   Germany concludes nuclear fission irrelevant for near-term prosecution of the war; limited research on nuclear-power applications to continue (12.1).

1942   January   Jomar Brun meets with German heavy-water experts in Berlin; plans made for boosting heavy-water production further by catalytic exchange; heavy-water production planned at Såheim and Notodden as well (10.1). L-III: first heavy-water-moderated pile with uranium metal (12.1). President Roosevelt authorizes American atomic bomb project (11.1).

       March   *Galtesund* affair; Einar Skinnarland trained in England and returns to Rjukan (10.2).

       April   The Uranium Committee under Conant decides to proceed full speed with graphite as moderator (11.2).

       May   L-IV: first pile experiment yielding positive neutron production coefficient (12.1).

       Spring   Moribund Soviet nuclear project revived (18.2).

       June   Albert Speer effectively cancels German atomic bomb project (12.1). Expanded heavy-water plant at Vemork becomes operational (10.3).

       Summer   'Codliver sabotage' at Vemork; Norwegian–German catalyzer tests (10.3).

       August   Anglo-Canadian heavy-water effort ratified (15.1).

	September	Construction starts on US-sponsored heavy-water plant at Cominco in Trail, Canada (11.3).
	October	Jomar Brun flees Norway (10.4).
	November	*Freshman*, the failed commando attack on Vemork (12.2).
1943	February	*Gunnerside* attack on Vemork (12.3).
	March	I V Kurchatov takes charge of Soviet nuclear weapons project (18.2). Japanese committee of experts concludes construction of atomic bomb not feasible in near future; NI Project nevertheless initiated at Riken laboratories in Tokyo (18.2).
	May	Heavy-water production resumed at Vemork (13.2).
	Spring	Urey investigates methods for mass-producing heavy water; settles on catalytic exchange (11.2).
	June	Measurements by Fermi confirm suitability of heavy water as moderator (11.3).
	Summer	F-project initiated at Kyoto; stepped-up heavy-water production at Hungnam (Korea) and Kyushu (Japan) (18.2).
	September	First substantial heavy-water production at Trail (11.3).
	Fall	Hydrogen sulfide heavy-water production scheme developed by Karl-Hermann Geib (14.3). Hydrogen distillation pilot plant by Linde in Munich; heavy-water pilot plant at Farben Leuna Works (14.3).
	November	US Eighth Air Force raid on Vemork and Rjukan (13.2).
	December	Heavy-water production by Norsk Hydro discontinued on German orders (14.2).
1944	January	Walther Gerlach succeeds Abraham Esau as German plenipotentiary for nuclear physics (14.1). Underground bunker at Berlin-Dahlem completed (14.1).
	February	Sinking of ferry on Lake Tinn (14.2).
	April	Halban resigns as director of Anglo-Canadian project, replaced by John Cockcroft (15.1).
	May	CP-3, the world's first heavy-water reactor, goes critical (11.3).
	July	Leuna Works destroyed in bombing raid; I G Farben abandons heavy-water production (14.3).
	August	High-concentration apparatus dismantled at Vemork and Såheim for shipment to Germany (14.3).
	Fall	Heavy-water production begins at Chirchik (18.2).

	November	*Alsos* Mission determines that no German atomic bomb exists (16.1).
1945	March	Heisenberg's team at Haigerloch makes Germany's last attempt to achieve chain reaction; B-VIII fails to reach criticality (16.2). Leif Tronstad killed in action on Hardanger Plateau (17.2).
	September	ZEEP goes critical (15.1).
1946	December	Soviet pile F-1, the first self-sustaining reactor in Europe, goes critical (18.2).
1951	July	JEEP goes critical, ushering Norway into the atomic age (19.1).
1956		Cominco discontinues heavy-water production (19 1).
1960s		Norsk Hydro initiates transition from electrolysis to petrochemicals for ammonia production, and Norwegian heavy-water production is ultimately doomed (19.1).

ABBREVIATIONS

The following abbreviations are used extensively in the notes and bibliography.

Archives and Collections

AHQP	Archive for the History of Quantum Physics
BSC	Bohr Scientific Correspondence, Niels Bohr Archive, Copenhagen
CUL	Cambridge University Library
EOL	E O Lawrence Papers, Bancroft Library, University of California, Berkeley
G-	Captured German reports, National Technical Information Service, Oak Ridge, TN. Copies consulted at the Niels Bohr Library, American Institute of Physics
GD	German documents, Ordnance Research and Development Translation Center, Ft Eustis, VA and Sondersammlungen, Deutsches Museum
GLP	Gilbert Lewis Papers, University of California Archives, College of Chemistry, Berkeley
IMC	David Irving Microfilm Collection, Niels Bohr Library and the Deutsches Museum
KBA	Kr Birkeland Archives, Norsk Teknisk Museum, Oslo
MPG	Archiv zur Geschichte der Max-Planck-Gesellschaft, Berlin
MPI	Max-Planck-Institut für Wissenschaftsgeschichte, Berlin
NHAN	Norsk Hydro Archives, Notodden
NHAO	Norsk Hydro Archives, Oslo
NHAR	Norsk Hydro Archives, Rjukan
NHM	Norges Hjemmefrontmuseum Archives, Oslo
NIAM	Norsk Industriarbeidermuseum Archives, Vemork
PHP	Paul Harteck Papers, RPI, Troy, NY
RO	Riksarkivet, Oslo
SON	Deutsches Museum, Munich, Sondersammlungen

Journals and Periodicals

AHES	*Archive for History of Exact Sciences*
AJP	*American Journal of Physics*
CR	*Comptes Rendus Hebdomadaires des Séances de l'Académie des Sciences*
DSB	*Dictionary of Scientific Biography*
HSPBS	*Historical Studies in the Physical and Biological Sciences*
JPR	*Le Journal de Physique et le Radium*
NA	*Nature*
NW	*Die Naturwissenschaften*
PM	*The London, Edinburgh and Dublin Philosophical Magazine*
PR	*The Physical Review*
PRS	*Proceedings of the Royal Society of London*
PT	*Physics Today*
RSDS	*Rivisita di Storia Della Scienza*
ZP	*Zeitschrift für Physik*

Technical Abbreviations

The following abbreviations are used extensively in the text.

AAC	Academic Assistance Council (Britain)
AECL	Atomic Energy of Canada Ltd
AFHQ	Allied Force Headquarters
CANDU	Canadian Deuterium Uranium reactor program
CNRS	Caisse Nationale de la Recherche Scientifique
CPC	Anglo-American Combined Policy Committee
DTM	Department of Terrestrial Magnetism, Carnegie Institution of Washington, DC
EPCI	École de Physique et de Chimie Industrielle (Paris)
FO4	Forsvarets Overkommandos kontor 4 (Norwegian High Command, London, office 4)
GS	Girdler-sulfide process for producing heavy water
ICI	Imperial Chemical Industries
IMI	British code for heavy water
JEEP	Joint Establishment Experimental Pile (Norway)
KFK	Kernforschungszentrum Karlsruhe
KWI	Kaiser Wilhelm Institute
MAUD	Acronym for sub-committee of the Committee for the Scientific Survey of Air Warfare ('Thomson Committee')
MILORG	Norwegian Military Organization of Resistance
NKVD	People's Commissariat for Internal Affairs (USSR)
NRX	National Research Experimental power reactor (Canada)
NS	National Samling (the Quisling party)

NTH   Norges Tekniske Høyskole (Trondheim)
OSS   United States Office of Strategic Services
PTR   Physikalisch-Technische Reichsanstalt (Berlin)
RAF   Royal Air Force
SIS   Secret Intelligence Service
SNO   Sudbury Neutrino Observatory (Canada)
SOE   Special Operations Executive
STS   Special Training School, SOE
USAAF   United States Army Air Forces
XU   Norwegian Resistance Intelligence Organization
ZEEP   Zero Energy Experimental Pile (Canada)
Zoé   Zéro Énergie, Oxide and Eau Lourde (France)

NOTES

Notes section 1.1

1-1   Moulton J L 1967 *A Study of Warfare in Three Dimensions: The Norwegian Campaign of 1940* (Athens, OH: Ohio University Press)

Notes section 2.1

2-1   Marsden E 1962 Rutherford at Manchester *Rutherford at Manchester* ed J B Birks (London: Heywood) p 11

2-2   Marsden E and Lantsberry W C 1915 The passage of α-particles through hydrogen, II *PM* **30** 240–3, on p 243

2-3   Marsden: see note 2-1, p 12

2-4   A term soon coined by Rutherford on the suggestion of Darwin and F Fowler in memory of William Prout who proposed in 1815 that all elements are made up of the atoms of hydrogen as a primordial substance.

2-5   'Experiments on H-particles', 3 January–18 March 1919, Rutherford archives, Cambridge University Library, CUL, Add 7653, NB25. The first reference to 'disintegration of nitrogen' appears on page 39.

2-6   The reaction, written as $\alpha + N^{14} \rightarrow p + O^{17}$, where p stands for a proton, was only confirmed somewhat later by P M S Blackett in Cambridge. Rutherford's picture of the process, not so far from modern concepts, was as follows. The 14 particles inside the nitrogen nucleus are arranged in three α-particles and two other particles, one of which is a proton. These two particles move in the outer regions of the nucleus and are easily dislodged. Boorse H A and Motz L (ed) 1966 *The World of the Atom* (New York: Basic Books) on p 808.

2-7   Rutherford 1919 Collisions of α particles with light atoms. IV. An anomalous effect in nitrogen *PM* **37** 581–7, on p 587

2-8   Feather N 1940 *Lord Rutherford* (Glasgow: Blackie)

2-9   Rutherford in a 1933 broadcast. Clark R W 1981 *The Greatest Power on Earth: The International Race for Nuclear Supremacy* (New York: Harper and Row), quoting John Cockcroft 1950 *The Development and Future of Nuclear Energy* The Romanes Lecture, Sheldonian Theatre, Oxford, 2 June 1950 (Oxford: Clarendon) p 8.

2-10   *The Times* 12 September 1933. R W Clark feels it is not clear whether Rutherford's famous 'moonshine', a fairly common term then, was in fact Rutherford's or that of *The Times*. Clark: see note 2-9, p 33.

2-11   Wells H G 1914 *The World Set Free: A Story of Mankind* (New York: Dutton)

2-12   Soddy F 1906 The internal energy of elements *Journal of the Proceedings of the Institution of Electrical Engineers, Glasgow* **37** 7; 'Advances in the study of radio-active bodies', two lectures to the Royal Institution on 15 and 18 May 1915, reviewed in *Engineering* **99** (1915) 604. Indeed, the year of his relocation to London, the St Louis Post-Dispatch in America, waxing over a grain of radium to be exhibited in that city, wrote as follows: 'By means of the metal all the arsenals in the world might be destroyed. It could make war impossible by exhausting all the accumulated explosives in the world. ... It is even possible that an instrument might be invented which at the touch of a key would blow up the whole earth and bring about the end of the world'. *St Louis Post-Dispatch* 4 October 1903, cited by Jauncey G E M 1946 The early years of radioactivity *AJP* **14** 226–41, on p 239.

2-13   Weart S R 1979 *Scientists in Power* (Cambridge, MA: Harvard University Press) p 39, quoting Perrin J 1924 *Les Atomes* 2nd revised edn (Paris: Alcan) pp 318–9

Notes section 2.2

2-14   Badash L 1996 The discovery of radioactivity *PT* **49** 21–6

2-15   Eve A S 1939 *Rutherford: Being the Life and Letters of the Rt Hon Lord Rutherford, OM* (New York: Cambridge University Press) p 93. Rutherford adds that he and his wife 'could not help observing that the hands of Professor Curie were in a very inflamed and painful state due to exposure to radium rays'.

2-16   In discussions prior to finalizing the Curies' Nobel nomination in 1903, the Swedish Academy had, on the objection of chemists, altered their original citation from one bestowing the physics prize for 'their discovery of the spontaneous radioactive elements' to a citation worded such as to leave open the possibility of a *second* Nobel Prize to the Curies for the discovery of radium. And so it came to pass in 1911 except that, with Pierre no longer around, the second prize went to Marie alone. McGrayne, Sharon Bertsch 1993 *Nobel Prize Women in Science: Their Lives, Struggles, and Momentous Discoveries* (A Birch Lane Press Book, Secaucus, NJ: Carol Publishing Group) p 26.

2-17   Perrin, Francis 1973 Joliot-Curie, Irène *DSB* **7** 157–9

2-18   Goldsmith, Maurice 1976 *Frédéric Joliot-Curie: A Biography* (London: Lawrence and Wishart) p 19
Perrin, Francis 1973 Joliot, Frédéric *DSB* **7** 151–7

2-19   Goldsmith relates a letter from Frédéric to Irène concerning an invitation she had received from the Physical Society in London, inviting them both to its October conference. 'I am vexed by the phrasing of the letter, which gives the impression that I am at your orders. It is that wretched question of the signature which people interpret badly and which pains me'. Frédéric to Irène, 3 May 1934, cited by Goldsmith: see note 2-18, p 34.

2-20   Goldsmith: see note 2-18, p 31

2-21   Hélène Joliot, like her father, graduated first in her class from EPCI, and became a nuclear physicist. She married a grandson of Paul Langevin, also engaged in research, thus maintaining the family of married physicists begun by Pierre and Marie Curie. So did Pierre Joliot who married a physicist specializing in biophysics, like himself.

Notes section 3.1

3-1   Pais A 1991 *Niels Bohr's Times, in Physics, Philosophy, and Policy* (Oxford: Clarendon; New York: Oxford University Press) p 331

3-2   Rutherford 1920 Nuclear constitution of atoms, Bakerian Lecture *PRS* A **97** 374–400; reprinted in *The Collected Papers of Lord Rutherford of Nelson* vol 3, pp 424–8. The Bakerian Lectures date from 1775, and are named after Henry Baker. This was Rutherford's second of these prestigious lectures; he gave the first on radioactivity, in May 1904.

3-3   Rutherford: see note 3-2, p 396

3-4   Rutherford: see note 3-2

3-5   In the very same lecture, Rutherford anticipated yet another particle, namely one with mass three and two units of charge; in other words, a lighter isotope of helium which was isolated by him not long before he died, working in collaboration with Mark Oliphant. Rutherford 1937 The search for the isotopes of hydrogen and helium of mass 3 *NA* **140** 303–5.

3-6   Massey H and Feather N 1976 James Chadwick *Biographical Memoirs of Fellows of the Royal Society* **22** 11–70; Hendry J 1970 Chadwick, James *DSB* **17** (Supplement II) 143–8

3-7   Chadwick set out to study the relative intensities of the discrete lines observed earlier by Rutherford and Harold R Robinson in radioactive β-ray spectra. Although he did identify a few of the most intense lines, he quickly found something altogether unexpected: a *continuous* spectrum. This was something that could not then be explained theoretically; it was an article of faith that spectral lines, including those of β-rays, are necessarily monochromatic, or at least discrete. The continuous spectrum haunted physics for years, and was only explained when Wolfgang Pauli proposed his neutrino in 1930.

3-8   The Vienna group has been described as the *enfant terrible* of nuclear physics by Otto Frisch, who began his career there under Karl Przibram. Possibly because it employed students to do the scintillation counting, biased toward high numbers to please their supervisors, they found that practically all light nuclei could be disintegrated and gave off more protons than anybody else could observe. Said Przibram to Frisch as he left, 'You will tell the people in Berlin, won't you, that we are not quite as bad as they think?' Frisch, Otto R, transcript of oral history interview with Thomas S Kuhn, 8 May 1963 (OH153), p 4; Frisch O R 1976 The discovery of fission: how it all began *PT* **20** 272–7, on p 273.

3-9   Harkins W D 1920 The nuclei of atoms and the new periodic system *PR* **15** 73–94

3-10   Segrè E 1980 *From X-rays to Quarks, Modern Physicists and Their Discoveries* (San Francisco: Freeman) p 180
Dostrovsky S 1973 Bothe, Walther Wilhelm George *DSB* **2** 337–9

3-11   Chadwick, James, transcript, oral history interview with Charles Weiner, 15–20 April 1969 (OH82), p 70

3-12   Dostrovsky: see note 3-10, p 338

3-13   Bothe W and Becker H 1930 Kunstliche Erregung von Kern-γ-Strahlen *ZP* **66** 239–306

Notes section 3.2

3-14   Curie I and Joliot F 1932 Émission de protons de grande vitesse par les substances hydrogénées sous l'influence des rayons gamma très pénétrants *CR* **194** 273

3-15   F Joliot to D Skobel'tzyn, 2 April 1932, cited by Goldsmith: see note 2-18, p 42

3-16   Chadwick J 1964 Some personal notes on the search for the neutron *Proc. 10th Int. Congress on the History of Science (Ithaca, 1962)* (Paris: Hermann) p 161

3-17   Chadwick: see note 3-16. For a discussion of the compelling kinematics in the argument in favor of uncharged *particles*, see Pais: note 3-1, p 399. We now know that Rutherford's 'neutral doublet' bears little resemblance to Chadwick's neutron. Among other things, the spin is wrong, and so is the mass; however, in the historical context, that is a moot point. Pais: see note 3-1, p 398.

3-18   Chadwick J 1932 On the possible existence of the neutron *NA* **129** 312. A full-length paper, 'The existence of a neutron', was dispatched to the Royal Society three months later together with two other papers reporting cloud chamber studies by Norman Feather and P I Dee; they appeared back-to-back in the June issue of the *Proceedings*.
        At the meeting of the British Association in York that September, a session was devoted to the conservation of energy and the neutron. At Calais, on returning from the meeting, Leprince-Ringuet wrote Joliot that Chadwick had begun his own paper at the session by speaking of the work of the Joliot-Curies as the first experiments to lay a serious foundation for the existence of the neutron. In a subsequent press conference at the meeting, M de Broglie had sought to put the matter straight by insisting 'on the priority of [Chadwick's] experiments and their importance'. Goldsmith: see note 2-18, p 47.

3-19   J Chadwick to N Bohr, 24 February 1932, Bohr Scientific Correspondence, 1930–1945 (BSC), AHQP, Microfilm Reel 18, A.L.S., 2 pp; Chadwick: see note 3-11, p 73.

3-20   N Bohr to J Chadwick, 25 March 1932, BSC, AHQP, Microfilm Reel 18, C, 1 p

3-21   F Joliot to D Skobel'tzyn, 2 April 1932, cited by Goldsmith: see note 2-18, pp 46–7

3-22   Translation from German by Barbara Gamow in Gamow G 1966 *Thirty Years that Shook Physics: The Story of Quantum Theory* (New York: Doubleday) p 213. The reference to Wolfgang Pauli came about as follows. Several years before Chadwick's discovery, Pauli coined the name 'neutron' for his hypothetical neutral and *massless* particle postulated for explaining the apparent violation of energy conservation in radioactive β-decay. The name gave rise to heated discussions, but was never sanctified by appearing in print. Consequently, when Chadwick's heavy neutron was announced, Pauli's name was in jeopardy. Enrico Fermi suggested instead calling Pauli's particle the *neutrino* or 'little neutron' in Italian, and the name stuck.

3-23   Goldsmith: see note 2-18, p 47. In fairness, they may have ignored Rutherford's printed lecture, since the Bakerian Lectures were invariably retrospective in tone.

3-24   Goldsmith: see note 2-18, p 48

Notes section 4.1

4-1   Birge R T and Menzel D H 1931 The relative abundance of the oxygen isotopes and the basis of the atomic weight system *PR* **37** 1669

4-2   Murphy G M 1964 The discovery of deuterium *Isotopic and Cosmic Chemistry, Dedicated to Harold C. Urey on his Seventieth Birthday April 29, 1963* ed H Craig, S L Miller and G J Wasserburg (Amsterdam: North-Holland) pp 1–7
Brickwedde F G 1982 Harold Urey and the discovery of deuterium *PT* **34** 208–13

4-3   The basis of figure 4.1 is an article on regularities among the then known atomic nuclei by Barton, dated 1930. It is discussed by Urey the next year in an article of his own on the subject, which contains the seeds of the thinking that led to the discovery of deuterium.
Barton D H R 1930 *PR* **35** 408
Urey H C 1931 The natural system of atomic nuclei *Journal of the American Chemical Society* **53** 2872–80

4-4   Ruark A E and Urey H C 1930 *Atoms, Molecules and Quanta* (New York: McGraw-Hill)

4-5   Tatarewicz J T 1970 Urey, Harold Clayton *DSB* **18** (Supplement II) 943–8

4-6   Brickwedde: see note 4-2, p 208

4-7   Interview with H A Urey by J L Heilbron, 24 March 1964. AHQP, Ms A4:5, transcript of tape 102a, p 5.

4-8   Cited by Tatarewicz: see note 4-5, p 944

4-9   Brickwedde: see note 4-2, p 208

4-10   Murphy: see note 4-2, p 4

4-11   Urey H C, Brickwedde F G and Murphy G M 1932 Abstract No 34 *PR* **39** 854

4-12   Urey H C, Brickwedde F G and Murphy G M 1932 A hydrogen isotope of mass 2 *PR* **39** 164–5. Urey H C, Brickwedde F G and Murphy G M 1932 A hydrogen isotope of mass 2 and its concentration *PR* **40** 1–15. It may be asked why Birge, at Berkeley, did not follow up on his own prediction of an isotope of hydrogen by joining up with his Berkeley colleague William Giaque. Giaque's laboratory was the only other laboratory, besides the National Bureau of Standards, then in a position to produce liquid hydrogen in liter quantities. Besides, it was Giaque, with Herrick Johnston, who had discovered the two isotopes of oxygen. Brickwedde, asking Birge, somewhat later, was told that Birge was simply too busy with other work. When informed of this, Urey's comment was: 'What in the world could Birge have been working on that was so important?'
Brickwedde: see note 4-2, p 212

4-13   Interview: see note 4-7, tape 102b, p 6

4-14   Brickwedde: see note 4-2, p 212

Notes section 4.2

4-15   Minutes of the Chicago Meeting, June 19–24, 1933 *PR* **44** 313–5. A joint meeting with the AAAS in connection with 'A Century of Progress Exposition', the meeting was declared by the editor of *The Physical Review* to have been 'perhaps the most important scientific session [in the history of the American Physical Society] to date' (p 313).

4-16   Aston F W 1934 Discussion on heavy hydrogen *PRS* A **144** 1–28, on p 9. Wrote Lawrence to Gilbert Lewis after the lively session: 'I suppose you had eye-witness accounts of Urey's temper as displayed at Chicago—stimulated by 'deuton' [by then Lawrence and Lewis's preferred term for the H^2 nucleus]. He was certainly "sore" and I believe the general impression was that he made an ass of himself'. E O Lawrence to G N Lewis, *circa* 7 July 1933; Gilbert Lewis Papers, University of California Archives, College of Chemistry, CU-30, Box 3, Item 425.

4-17   'Discussion...' (note 4-16); Heavy hydrogen: the search for a name. Royal Society and 'diplogen' *The Times* 15 December 1933. Besides Rutherford, the Royal Society discussion featured ten European speakers, each focusing on one or more aspects of heavy hydrogen. Following Rutherford's opening address, dealing mainly with prospects for H^2 nuclei as projectiles for studying the transformation of elements, N V Sidgwick picked up on expected chemical aspects of the new isotopes: equilibrium constants and reaction rates involving H^2 ions, and applications of the isotope in earmarking hydrogen atoms in compounds. Aston, naturally, dwelt on the atomic weight discrepancies that led the Americans to the new isotope. Harteck enlivened the discussion with some clever experimental demonstrations of vapor pressure and melting point differences between heavy and normal hydrogen, and on his own procedure for preparing heavy hydrogen by electrolysis while in Cambridge. Soddy returned to the aforesaid intricacies of extending his definition of 'isotope' to heavy hydrogen. M Polanyi addressed questions of ionization and catalysis involving heavy hydrogen; appropriately, he was followed by E Rideal, who drew attention to the work on the *ortho–para* hydrogen system by the brothers Ladislaus and Adalbert Farkas, newly arrived Cambridge refugee scientists from Fritz Haber's Institute in Berlin, where Polanyi had headed chemical kinetics. The Farkas methodology might well be applicable for studying the hydrogen–deuterium system as well. R H Fowler returned to details of electrolytic separation of hydrogen and deuterium, as did R P Bell. J D Bernal reported on measurements on a single crystal of heavy ice, in itself providing tell-tale indications of physical properties of heavy water; in effect, heavy water is 'more icelike' than is ordinary water.
    Following W Jevons' review of recent work on the spectroscopy of deuterium, Rutherford closed the discussion with a gentle jibe at Soddy for his stubborn reluctance to extend the term 'isotope' from 'chemically indistinguishable' heavy radioactive elements differing but slightly in mass, and thus very difficult to separate, to heavy hydrogen with a mass twice that of ordinary hydrogen.

4-18   Calvin, Melvin 1984 Gilbert Newton Lewis: his influence on physical-organic chemists at Berkeley *Journal of Chemical Education* **61** 14–8
    Branch G E K 1984 Gilbert Newton Lewis, 1875–1946 *Journal of Chemical Education* **61** 18–21

4-19   Gerald Branch, the first of Lewis's Berkeley faculty appointees not trained at Berkeley, has characterized Lewis's chemical thermodynamics as America's rival to W Nernst's European version of the subject. Branch: see note 4-18, p 20.

4-20   Washburn E W and Urey H C 1932 Concentration of the H^2 isotope of hydrogen by the fractional electrolysis of water *Proc. Nat. Acad. Sci.* **18** 496

4-21   G N Lewis to H A Urey, 8 April 1933; GLP, Box 4, Item 1016

4-22 Urey to Lewis, 11 April 1933; GLP, Box 4, Item 1017

4-23 Lewis G N and Macdonald R T 1933 Concentration of H^2 isotope *The Journal of Chemical Physics* **1** 341–4
Lewis to Rutherford, 12 July 1933; GLP, Box 3, Item 860

4-24 Urey to Lewis, 17 April 1933; GLP, Box 4, Item 1020

4-25 Lewis to Urey, 13 May 1933; GLP, Box 4, Item 1022

4-26 Urey to Lewis, 16 May 1933; GLP, Box 4, Item 1023

4-27 Urey to Lewis, 18 May 1933; GLP, Box 4, Item 1024

4-28 Urey to Lewis, 29 May 1933, GLP, Box 4, Item 1025. 'In case a H^3 isotope is discovered [added Urey], it could then be given the name "tritium".'

4-29 The Editor 1934 Biology of heavy water *NA* **133** 620, citing, in particular, Gilbert N Lewis 1934 The biology of heavy water *Science* **79** 151–3
Other studies at the time of Lewis's experiment included those of the biologist Samuel L Meyer at Vanderbilt University on the yield of fungus in a blue mold as a function of heavy water in the nutrient medium, and Edward W Washburn and Edgar R Smith at the Bureau of Standards on tissues in plants grown in soil solutions enriched in heavy water. Meyer S L 1934 Deuterium oxide and aspergillus *Science* **79** 210–1. For post-WWII studies at Berkeley of heavy water and mice, see for example Hughes, Tolbert and Calvin 1958 Production of sterility in mice by deuterium oxide *Science* **127** 1445.

4-30 Rutherford, 'Discussion...' (note 4-16) on p 2; Rutherford 1933 Heavy hydrogen *NA* **132** 955–6, on p 956

4-31 Rutherford to Lewis, 20 June 1933; Lewis to Rutherford, 12 July 1933; GLP, Box 3, Items 859 and 860. Urey to Rutherford, 6 January 1934; GLP, Box 4, Item 1028.

4-32 Rutherford to Lewis, 27 July 1933; GLP, Box 3, Item 861

4-33 Rutherford, 'Discussion...' (note 4-16), p 27

4-34 Rutherford to Lewis, 10 August 1933, GLP, Box 3, Item 862

4-35 Lewis to Rutherford, 5 October 1933, GLP, Box 3, Item 863

4-36 Lewis to R H Fowler, 17 January 1934; GLP, Box 2, Item 209

4-37 Fowler to Lewis, 3 February 1934; GLP, Box 2, Item 210

4-38 Fowler to Lewis, 3 February 1934; GLP, Box 2, Item 210

4-39 C P Snow, who had been a student of Rideal, disliked him and portrayed him accordingly in several of his books, among them as the 'irrepressibly effervescent and slightly bogus' Professor Desmond in 1935 *The Search* (Indianapolis, NY: Bobbs-Merrill)
Brock W H 1970 Rideal, Eric Keightley *DSB* **18** (Supplement II) 738–43

4-40 Brun, Jomar and Ræder, Mathias 1970 Tronstad, Leif Hans Larsen *Norsk Biografisk Leksikon, XVII* (Oslo: Aschehoug)
Rideal E K and Evans U R 1945 Prof. Leif Tronstad, OBE, Obituary *NA* **156** 74
Sørensen N A 1946 Minnetale over Professor Leif Tronstad *Årbok, Det norske Videnskaps-Akademi* pp 113–34
Ræder, Mathias 1946 Leif Tronstad, Minnetale *Forhandlinger, Det kongelige norske Videnskabers Selskab* **26** 95–106

Notes section 4.3

4-41 Friedman R M, Civilization and national honour: the rise of Norwegian geophysical and cosmic science, in J P Collett (ed) 1995 *Making Sense of Space: The History of Norwegian Space Activities* (Oslo: Scandinavian University Press) Chapter 1, p 5

4-42   Særland S 1923 Birkeland *Norsk Biografisk Leksikon* vol I (Kristiania: Aschehoug); Egeland A and Leer E 1989 Professor Kr Birkeland: his life and work *IEEE Transactions on Plasma Science* **PS-14** 666–77; Egeland A 1994 *Kristian Birkeland: Mennesket og Forskeren* (Oslo: Norges Banks Seddeltrykkeri), published on the occasion of the issuance of a new 200-krone note bearing a portrait of Kr Birkeland. Egeland's second cited work contains many additional references to Birkeland and his work.

4-43   Arkiv No 2819: Kristian Birkeland, occupying 14 boxes in all

4-44   Birkeland Kr 1908 and 1913 *The Norwegian Aurora Polaris Expedition 1902–1903* vol 1 (Christiania: Aschehoug)

4-45   Devik O 1971 *Blant Fiskere, Forskere og Andre Folk* (Oslo: Aschehoug) p 42. Olaf Devik, an assistant to Birkeland during the peak of the latter's experimental researches, has much of interest to say about Birkeland in his memoirs.

4-46   Born in Arendal, Norway in 1866, Samuel Eyde was schooled initially in Kristiania and subsequently in Berlin where he graduated as a civil engineer in 1891. Following several construction jobs in Germany, he teamed with a German-American engineer. The two formed an engineering firm, and undertook several prize-winning civil construction projects throughout Scandinavia. In time Eyde's interest shifted in another direction, and the day-to-day construction business was left to his partner. What had diverted Eyde's interest, was the potential of Norway's largely untapped hydroelectric power, as we shall see.
Rygh, Arne 1923 Eyde, Samuel *Norsk Biografisk Leksikon* vol 3 (Kristiania: Aschehoug)
Eyde, S 1939 *Mitt Liv of Mitt Livsverk* (Oslo: Gyldendal Norsk)

4-47   Libæk I and Stenersen Ø 1991 *History of Norway from the Ice Age to the Oil Age* (Oslo: Grøndahl) p 91

4-48   Libæk and Stenersen: see note 4-47

4-49   At the same time, Birkeland entered into negotiations with the Swedish mathematician Gösta Mittag-Leffler to be engaged as the professor of physics at the Technical University in Stockholm. Mittag-Leffler had led Birkeland to believe he would be elected to the Nobel Committee for Physics, which would entail a hefty jump in salary over what he earned in Kristiania, and perhaps laboratory space in the Nobel Institute. In the event, the deal fell through, partly because Norway broke away from Sweden just then; Swedish sentiment turned against hiring yet another Norwegian besides Bjerknes. Friedman: see note 4-41, pp 14–5.

4-50   Anker Olsen Kr 1955 *Norsk Hydro Gjennom 50 År: Et Eventyr fra Realitetens Verden* (Oslo: Norsk Hydro-Elektrisk Kvælstofaktieselskab). A three-volume commemorative work, to be published for the 100th anniversary of the founding of Norsk Hydro in 2005, is already under preparation.

4-51   Taylor-Wilkie D (ed) 1991 *Insight Guides: Norway* (Singapore: APA) p 197

4-52   The Telemark water course—a royal bid for Røgnen Falls, see Anker Olsen, note 4-50, p 86

4-53   The process works as follows. Air is heated in an arc furnace until the nitrogen undergoes combustion with the aid of the oxygen naturally present in the air, forming nitrogen dioxide. This chemical combination, on being cooled in water, is absorbed in the water with the formation of dilute nitric acid (NHO_3). The nitric acid, in turn, is combined (neutralized) with lime, forming calcium nitrate, the desired end product which replaced the scarce Chilean nitrate as fertilizer.

4-54   Dahl, Helge 1983 and 1984 *Rjukan* vol I, to 1920, vol II, from 1920 to

1980 (Tinn kommune). In the 1930s, as the saltpeter production method was phased out in favor of the ammonia process, Notodden entered a period of decline, and in the post-WWII years Rjukan underwent the same fate as Norsk Hydro's operations shifted to Herøya near Porsgrund and elsewhere.

Notes section 4.4

4-55   Trimble R F 1993 Odd Hassel *Nobel Laureates in Chemistry 1901–1992* ed L K James (Washington, DC: American Chemical Society and The Chemical Heritage Foundation) pp 514–9

4-56   Kramish A 1986 *The Griffin* (Boston: Houghton Mifflin) pp 33–4

4-57   Kramish: see note 4-56, p 34

4-58   Correspondence between Hassel and Arne Enger, Rjukan; NIAM archives, Box 4F-D02, 960: Tungt vann, Akt. nr. 704E

4-59   Internal Hydro correspondence, 9 September 1933; NHAO, K03. 50. n. 2.1: Tungt vann, teknisk, Hefte 1. In his letter to Hydro in October, Hassel expresses his appreciation for the electrolysis cell, and hopes to convey some interesting results before long. Hassel to Norsk Hydro, 26 October 1933; NHAO, K03. 50. n. 2.1: Tungt vann, teknisk, Hefte 1.

4-60   K F Bonhoeffer *et al* of I G Farben, Frankfurt-Griesheim, to B F Halvorsen, Oslo, 29 May 1933; NIAM, Box 4F-D02, 960: Tungt vann, Akt. nr. 704G. Internal Hydro correspondence, June–July 1933; NHAO, K03. 50. n. 2.1: Tungt vann, teknisk, Hefte 1.

4-61   Brun, Jomar 1985 *Brennpunkt Vemork: 1940–1945* (Oslo: Universitetsforlaget) p 10

4-62   Ernst Hochheim to B Nilssen, 11 July 1933; NIAM, Box 4F-D02, 960: Tungt vann, Akt. nr. 704D. Also NHAO, K03. 50. n. 2.1: Tungt vann, teknisk, Hefte 1.

4-63   Lunde, Erik 1969 Av tungtvannets saga *Aftenposten* 5–10 April; first installment, 5 April, citing Jomar Brun

4-64   Brun to Norsk Hydro management, dated 8 August and 22 September 1933, respectively. Brun: see note 4-61, p 10. Translation by PFD.

4-65   Bonhoeffer to Hydro, 29 July and 20 November 1933; NHAO, K03. 50. n. 2.1: Tungt vann, teknisk, Hefte 1

4-66   Brun: see note 4-61, p 12. Tronstad to Hydro management, Oslo, 19 December 1933; NIAM, Box 4F-D02, 960: Tungt vann, Akt. nr. 704F.

4-67   Hayes P 1987 *Industry and Ideology: I G Farben in the Nazi Era* (Cambridge: Cambridge University Press) p 290. I G Farben, or Interessengemeinschaft Farbenindustrie Aktiengesellschaft to be exact, remained the world's largest chemical concern from its founding in 1925 until its dissolution by the Allies after World War II. It grew out of a complex merger of German manufacturers of chemicals, pharmaceuticals, and dyestuffs.

4-68   Tronstad, handwritten laboratory notes labelled 'Tungt vand' preserved under the archival heading 'Opprinnelige D_2O arbeider, Leifs notater mest' in NIAM, Box 4F-D02, 960: Tungt vann, Akt. nr. 704F

4-69   NIAM, Box 4F-D02, 954: Tungt vann, Akt. nr. 704B and Box 4F-D02, 960: Tungt vann, Akt. nr. 704F; also NHAO, K03. 50. n. 2.1: Tungt vann, teknisk, Hefte 1.

4-70   Pro Memoria, 25 May 1934, signed by Leif Tronstad and Jomar Brun, and covering letter from Tronstad to Hydro management, Rjukan, 1 June 1934; NIAM, Box 4F-D02, 954: Tungt vann, Akt. nr. 704B

4-71 Tronstad L 1934 Production of large quantities of heavy water *NA* **133** 872

4-72 Tronstad L and Brun J 1934 Über die Elektrolytische Anreicherung des Schweren Wasserstoffisotops im Wasser *Zeitschrift für Elektrochemie und angewandte physikalische Chemie* **40** 556–8

4-73 Forsendelser av tungt vann, 8 August 1934 to 19 March 1940; personal archive of Per Pynten, Rjukan

4-74 Anker Olsen: see note 4-50, p 399

4-75 Larsen E J and Cunliffe Barnes T 1934 Parasitism in heavy water of low concentration *NA* **133** 973–4

4-76 Commented Ernest O Lawrence, Lewis's main heavy-water customer, on Lewis's mouse experiment, according to N P Davis, 'This was the most expensive cocktail that I think mouse or man ever had!' Davis N P 1968 *Lawrence and Oppenheimer* (New York: Simon and Schuster) p 53. Agreed Lewis, 'The experiment was a very costly one and I regret that since it was undertaken solely to ascertain whether the heavy water would be lethal, no preparation was made for a careful clinical study of the effects produced'. Lewis: see note 4-29, p 152. As for Hansen's demonstration, he had fasted 12 hours to give the potion maximum effect; a doctor stood by with emergency apparatus in case he showed signs of collapse. Brun gives 1936 for the demonstration, but newspaper headlines at home and abroad show the date as 26 January 1935. Brun: see note 4-61, p 9. NHAN, 03. 50. n. 1.1: Tungt vann, Hefte 1. For correspondence between Hansen and Paul Harteck on heavy water at about this time, see K Hansen to P Harteck, 3 May 1935; Harteck to Hansen, 21 May 1935. PHP, Acc. 85-22, Box 2, File 2.

4-77 Brun: see note 4-61, p 15. The matter of patents for the production system arose, naturally, at an early stage, though Tronstad and Brun were quick to point out that a similar concentration process was already in print in an American journal article by Lewis and Macdonald. In the end, Hydro's patent application was not supported. It was deemed more appropriate to rely on maintaining a measure of confidentiality with regard to company production methods, as was Norsk Hydro practice on many occasions. Brun: see note 4-61, pp 13–4. Lewis and Macdonald: see note 4-23.

4-78 'Forsendelser . . .': see note 4-73

4-79 Sørensen: see note 4-40, p 119

Notes section 5.1

5-1 Goldsmith: see note 2-18, p 50

5-2 Anderson C D with Anderson H L 1983 Unraveling the particle content of cosmic rays *The Birth of Particle Physics* ed L M Brown and L Hoddeson (New York: Cambridge University Press) pp 131–54, on p 140

5-3 Anderson C D 1932 The apparent existence of easily deflectable positives *Science* **76** 238–9

5-4 Dahl P F 1997 Evanescent rays: a French cottage industry *Flash of the Cathode Rays: A History of J J Thomson's Electron* (Bristol: Institute of Physics Publishing) Chapter 14

5-5 Anderson: see note 5-2, pp 148–9

5-6 Anderson C D, transcript, oral history interview with C Weiner, 30 June 1966 (OH16)

5-7 Goldsmith: see note 2-18, p 51

5-8   Mehra, Jagdish 1975 *The Solvay Conferences on Physics: Aspects of the Development of Physics Since 1911* (Dordrecht: Reidel) pp 209–18

5-9   Weart: see note 2-13, p 44

5-10   Gentner W, transcript, oral history interview with C Weiner, 15 November 1971 (OH31300), p 20

5-11   Goldsmith claims that Joliot thought the term 'artificial radioactivity' poorly chosen. To him, the new radioactivity was just as spontaneous and natural as that of the natural radioactive elements. He would have retained the adjective 'artificial' for the radioisotopes produced, not for their radioactivity. Goldsmith: see note 2-18, p 53. Perhaps 'induced radioactivity', used less frequently, is a better term.

5-12   The method used was to dissolve a thin sheet of irradiated aluminum in hydrochloric acid. The hydrogen liberated was collected in an inverted test tube, and was found to carry the activity with it. A different chemical test proved that the radioactive substance was phosphorus.

5-13   Weart: see note 2-13, p 46

5-14   Goldsmith: see note 2-18, pp 57–8

5-15   Joliot-Curie, Frédéric and Irène 1934 Un noveau type de radioactivité *CR* **198** 254

5-16   Joliot-Curie, Frédéric and Irène 1934 Artificial production of a new type of radioelement *NA* **133** 201

5-17   Cited by Goldsmith: see note 2-18, p 57

5-18   Joliot-Curie: see note 5-15

5-19   Davis: see note 4-76, p 60

5-20   Heilbron J L and Seidel R W 1989 *Lawrence and His Laboratory: A History of the Lawrence Berkeley Laboratory* vol 1 (Berkeley, CA: University of California Press)

5-21   Joliot-Curie: see note 5-15

Notes section 5.2

5-22   Amaldi, cited by Holton G 1974 Striking gold in science: Fermi's group and the recapture of Italy's place in physics *Minerva* **12** 159–98, on p 173

5-23   Amaldi E 1984 Neutron work in Rome in 1934–36 and the discovery of uranium fission *RSDS* **1** 1–24

5-24   Segrè: see note 3-10, p 197. About as important was the new type of particle statistics discovered by Fermi in 1926, and independently by Paul Dirac.

5-25   E Rutherford to E Fermi, 23 April 1934, *Collected Papers* vol I (Chicago: University of Chicago Press) p 641

5-26   Segrè E 1971 Fermi, Enrico *DSB* **4** 576–83
Segrè E 1970 *Enrico Fermi: Physicist* (Chicago: University of Chicago Press)
Fermi, Laura 1954 *Atoms in the Family* (Chicago: University of Chicago Press)

5-27   Segrè (1970): see note 5-26, p 18

5-28   Segrè (1970): see note 5-26, p 20

5-29   Segrè (1970): see note 5-26, p 33

5-30   Fermi, Laura (1954): see note 5-26, p 89

5-31   Rutherford to Fermi, 23 April 1934 (letter, see note 5-25)

5-32   Noddack, Ida 1934 Über das Element 93 *Z. Angewandte Chemie* **47** 653–5

5-33   The academic year in Italy was traditionally closed with a solemn convocation of the Accademia dei Lincei, attended by the King. Corbino was the speaker for 1934, and he dwelt on the neutron work in his Institute and, in particular, on the transuranic element. 'For what my own opinion on the matter is worth, and I have followed the investigation daily, I believe that the production of this new element is certain'. Corbino O M Results and perspectives of modern physics *Conferenze e Discorsi*, p 51, cited by Segrè (1970): see note 5-26, p 76. Fermi was dismayed at the premature stir created in the press by Corbino's well intentioned enthusiasm, thinking that he himself had acted rashly in claiming the new element.

5-34   Fermi E, Amaldi E, d'Agostino O, Rasetti F and Segrè E 1934 Artificial radioactivity produced by neutron bombardment *PRS* A **146** 483–500

5-35   Segrè (1970): see note 5-26, p 77

5-36   Segrè (1970): see note 5-26, p 80; Fermi E 1965 *Collected Papers* vol II *United States 1939–1954* (Chicago: University of Chicago Press) p 927

5-37   Segrè (1970): see note 5-26, p 80

5-38   A neutron entering a hydrogenous material at room temperature with an energy of 5 MeV, provided it does not combine with some nucleus, will be reduced to the energy of a molecule at room temperature, or about 0.025 eV, after some 20 collisions. Neutrons which have thus come into thermal equilibrium with matter are known as *thermal neutrons*.

5-39   Fermi, Laura (1954): see note 5-26, pp 99–100

5-40   Fermi E, Amaldi E, Pontecorvo B, Rasetti F and Segrè E 1962 Influence of hydrogenous substances on the radioactivity produced by neutrons Transl. E Segrè, Fermi E *Collected Papers* vol I (Chicago: University of Chicago Press) pp 761–2

Notes section 5.3

5-41   McGrayne: see note 2-16, p 138

5-42   Curie I, von Halban H and Preiswerk P 1935 Sur la création artificielle d'éléments appartenant à une famille radioactive inconnue, lors de l'irradiation par les neutrons *JPR* **6** 361

5-43   Curie I and Savitch P 1937 Sur les radioéléments formés dans l'uranium irradié par les neutrons *JPR* **8** 385–7

5-44   Curie I and Savitch P 1938 Sur la nature du radioélément de périod 3.5 heures formé dans l'uranium irradié par les neutrons *CR* **206** 1643. Translated as 'Concerning the nature of the radioactive element with 3.5-hour half-life, formed from uranium irradiated by neutrons' in Graetzer H G and Anderson D L 1971 *The Discovery of Nuclear Fission, A Documentary History* (New York: Van Nostrand Reinhold) pp 37–8

5-45   Weart: see note 2-13, p 55

5-46   McGrayne: see note 2-16, p 138

5-47   Weart: see note 2-13, p 55

5-48   Sime, Ruth Lewin 1996 *Lise Meitner: A Life in Physics* (Berkeley, CA: University of California Press) p 180

Notes section 5.4

5-49   Sime: see note 5-48, p 184

5-50   Sime: see note 5-48, p 185

5-51   Sime: see note 5-48, p 185

5-52   Frisch O R 1974 Meitner, Lise *DSB* **9** 260–3

5-53   Badash, Lawrence 1972 Hahn, Otto *DSB* **6** 14–7

5-54   Badash: see note 5-53, p 15

Badash L 1969 *Rutherford and Boltwood: Letters on Radioactivity* (New Haven, CT: Yale University Press) p 81

5-55   Kramish: see note 4-56, p 48

5-56   Kramish gives the date as 16 July; Sime quotes Meitner's diary as 12 July. Kramish: see note 4-56, p 49; Sime: see note 5-48, p 204.

5-57   Sime: see note 5-48, p 205

Sime mentions Georg Graue, a former student of Hahn, as playing a role in delaying the denunciation.

5-58   Born in Graz, Yugoslavia, in 1896, the son of an organ maker and choirmaster he would never know, Rosbaud attended trade school in Graz before enlisting in the Austrian army in 1915. Rising from private to lieutenant during the First World War, he participated in the Austro-German campaigns against Italy on the Isonzo River, was much decorated, and agreeably interned by the Allies in 1918. As he later wrote, 'my first two days as prisoner under British guard were the origins of my long-time anglophilia'. Those feelings would come to the fore somewhat later. In the meanwhile, Rosbaud studied chemistry at Darmstadt Technische Hochschule, served a fellowship to the KWI in Berlin-Dahlem, then earned his doctorate in x-ray studies at the Technische Hochschule in Berlin-Charlottenburg. Not able to find a university position, he obtained instead a job as scientific adviser to the new journal *Metallwirtschaft*.

When Hitler came to power in 1933, the owner of *Metallwirtschaft*, Georg Lüttke, began to show his Nazi colors openly, and Rosbaud resolved to devote himself to combatting the new political order. By chance, he was offered and accepted a similar advisory position to the prestigious publishing firm Springer Verlag, a position which gave him broad access to the scientific and academic communities at home and abroad. The German civil service law promulgated in 1933, and more so the *Anschluss* in 1938, placed Rosbaud in a precarious position, being at the very least 'non-Aryan', but by hook and by crook he stayed out of trouble. His future undercover role in the Allied cause would be significantly abetted by his contact in the early 1930s with Francis Edward Foley, agent for MI6, alias the Secret Intelligence Service, then posted to the British legation in Berlin. Kramish: see note 4-56.

5-59   Hahn O 1968 *My Life* Transl. E Kaiser and E Wilkins (New York: Herder and Herder) p 149

5-60   Coster, the co-discoverer with Hevesy of hafnium (named in honor of Copenhagen), received numerous congratulations for his efforts on behalf of Meitner, including a cable from Wolfgang Pauli which read: 'You have made yourself as famous for the abduction of Lise Meitner as for hafnium!' Sime: see note 5-48, p 205.

5-61   Hendry, John 1970 Frisch, Otto Robert *DSB* **17** (Supplement II) 320–2

5-62   L Meitner to O Hahn, 18 September 1938; cited by Sime: see note 5-48, p 215. Max Junior was von Laue.

5-63   Lindemann judged Meitner's plight 'more contingent than actual', and according to Kramish also 'had an aversion to his mother, brother, and animal protein. He never married'. Kramish: see note 4-56, p 187.

5-64   Fermi, Laura (1954): see note 5-26, p 126

5-65   Segrè (1970): see note 5-26, p 98

Notes section 6.1

6-1   Graetzer and Anderson: see note 5-44, p 39

6-2   Friedrich Wilhelm (Fritz) Strassmann was born in Boppard, Germany in 1902, the ninth and last child of a court clerk. In 1907 his father was transferred to Düsseldorf, where Fritz completed elementary schooling, then attended the municipal Oberrealschule. Shortly before taking his final examination, his father died. All but one of his brothers and sisters having left home, and because of the smallness of his mother's pension, Strassmann had to forego a university education. Instead he chose to study chemistry at the Technische Hochschule in Hanover, matriculating in 1920. He received his diploma as a chemical engineer in 1924, and his doctorate in physical chemistry under Hermann Braune in 1929. He then became a lecture assistant to Braune, and shortly afterwards accepted Hahn's offer of a scholarship by the Notgemeinschaft der Deutschen Wissenschaft to the KWI for Chemistry; there he eagerly took up radiochemistry, the strong card of the institute.

     When his scholarship expired in 1932, Strassmann was allowed to continue without pay under Hahn, but in 1934 Meitner persuaded Hahn to grant him a modest wage out of contingency funds. After Hitler came to power, Strassmann refused more lucrative industrial offers, which would have required him to join the Nazi party. Instead, after joining the working team of Meitner and Hahn, he was employed as Hahn's assistant. Strassmann's anti-Nazi stance forced him to remain under the shadow of Hahn with his liberal political attitude, instead of developing his own field of research. Kraft F 1970 Strassmann, Friedrich Wilhelm (Fritz) *DSB* **18** (Supplement II) 880–7

6-3   Hahn to Meitner, 25 October 1938; copy, Hahn–Meitner correspondence, A. 538, Map 3, Archiv zur Geschichte der Max-Planck-Gesellschaft (MPG), Berlin-Dahlem

6-4   Hahn O and Strassmann F 1938 Über die Entstehung von Radiumisotopen aus Uran beim Bestrahlen mit schnellen und verlangsamten Neutronen *NW* **26** 755–6; submitted 8 November, published 18 November 1938. Translated by Graetzer and Anderson: see note 5-44, pp 41–4.

6-5   Fermi, Laura (1954): see note 5-26, p 156

6-6   Curiously, this meeting between Hahn and Meitner, kept an absolute secret outside Copenhagen at the time, remains largely that to this day. Even in his memoirs, where Hahn recalls his meeting with both Frisch and Bohr on that occasion, nothing is said about this particular meeting with Meitner. Sime: see note 5-48, pp 227–8. As for Jutz Frisch, he was shipped to Dachau, but in the spring of 1939 both he and Gusti received a Swedish visa, and both reached Stockholm safely. Oddly, their belongings from Vienna reached Stockholm before Meitner's from Dahlem; hers did arrive at long last in May of 1939.

6-7   Sime: see note 5-48, p 228

6-8   Hahn O 1962 *A Scientific Autobiography* Transl. and edited by W Ley, introduced by G T Seaborg (New York: Scribner) p 155; original Hahn 1962 *Vom Radiothor zur Uranspaltung* (Braunschweig: Viewet)

6-9   Hahn to Meitner, 19 December 1938; copy, Hahn-Meitner correspondence, A. 538, Map 3 (MPG)

     Kraft F 1981 *Im Schatten der Sensation: Leben und Wirken von Fritz Strassmann* (Weinheim: Chemie), cited by Kraft: see note 6-2, p 883. Original letter reproduced in *DJ* **29** 23–6, and given in rough translation in *DJ* **29** 28.

6-10   Hahn: see note 6-8, p 157

6-11   Hahn O and Strassmann F 1939 Über den Nachweis und das Verhalten der bei der Bestrahlung des Urans mittels Neutronen enstehenden Erdalkalimetalle *NW* **27** 11–5; received 22 December 1938, published 6 January 1939. Translated by Graetzer and Anderson: see note 5-44, p 44–7.

6-12   Meitner to Hahn, 21 December 1938; original, item 04873, Meitner-Hahn correspondence (MPG); cited by Sime: see note 5-48, p 235. Original letter reproduced in *DJ* **29** 26–9, and given in rough translation in *DJ* **29** 29-30.

6-13   Frisch O R 1979 *What little I Remember* (Cambridge: Cambridge University Press) pp 115–6
Frisch, interview by C Weiner, 3 May 1967, pp 33-34 (OH 154)

6-14   Wheeler J A 1967 Mechanism of fission *PT* **20** 278–81

6-15   Evans R D 1955 *The Atomic Nucleus* (New York: McGraw-Hill) pp 365–97

6-16   Subtracting the masses of the known barium isotopes from the mass of the uranium nucleus, Hahn obtained 'masurium' (i.e., technetium, then still undiscovered), or elements in the middle of the platinum group, with chemical characteristics which could presumably be reconciled with those of the transuranics observed earlier. In fact, he should have subtracted the atomic *numbers* (charges), not masses. Meitner and Frisch appreciated that atomic numbers are the relevant parameters, that must balance properly in the reaction, and gave the correct explanation for 'fission' (their term) as

$$_{92}U \rightarrow {}_{56}Ba + {}_{36}Kr$$

where Kr (krypton) is an inert gas. Hahn (1962): see note 6-8, pp 157–8. Meitner L and Frisch O R 1939 Disintegration of uranium by neutrons: a new type of nuclear reaction *NA* **143** 239; communicated 16 January 1939, published 11 February 1939.

6-17   As early as 1936, Strassmann had done some uranium experiments on his own and found what seemed to be barium. When informed of it, Meitner retorted that barium was highly unlikely and told him he might as well throw away his notes. Weart S 1983 The discovery of fission and a nuclear physics paradigm *Otto Hahn and the Rise of Nuclear Physics* ed W R Shea (Dordrecht: Reidel) pp 91–133, on p 112

6-18   Meitner to Hahn, 3 January 1939; original, item 04875, Meitner–Hahn correspondence (MPG); Kraft: see note 6-9, p 271. In assigning credit to the team, Kraft is adamant in insisting that the credit be shared equally among the team partners, Otto Hahn, the organic and nuclear chemist, Fritz Strassmann, the analytical and physical chemist, and Lise Meitner, the theoretical and experimental nuclear physicist. Meitner's claim to collaborator rests, in Kraft's opinion, not just for her correct interpretation, with Frisch, of the Hahn–Strassmann results, but for her original instigation in 1934 of the joint studies of the 'transuranics' that led to the final discovery. More generally, 'the methodology of no one field was dispensable'. Kraft F, Internal and external conditions for the discovery of nuclear fission by the Berlin team, pp 135–65 in Shea: see note 6-17. The apparatus used in the fission discovery has been on permanent display in the chemistry section of the Deutsches Museum, Munich, since 1953. For years the accompanying text failed to mention Meitner in the discovery—a blunder only corrected in 1990. See also Sime: note 5-48, pp 499–500.

Notes section 6.2

6-19   Frisch (1979): see note 6-13, p 116. Stuewer notes that Frisch erroneously placed his dramatic meeting with Bohr on the day Bohr was leaving for America. Stuewer R H 1985 Bringing the news of fission to America *PT* **38** 49–56, on p 51.

6-20   Hartcup G and Allibone T E 1984 *Cockcroft and the Atom* (Bristol: Adam Hilger) p 86

6-21   Meitner and Frisch: see note 6-16

6-22   Frisch (1979): see note 6-13, p 117

6-23   Frisch (1979): see note 6-13, p 117. Characteristically, the doubting Placzek likened prospects for the instability of uranium against fission, in the face of its known α-instabilities, to 'the man who is run over by a motor car and whose autopsy shows that he had a fatal tumor and would have died within a few days anyway'. Frisch O R 1967 The discovery of fission; how it all began *PT* **20** 272–7, on p 276. When Hevesy saw the French paper, he remarked that 'Curie had told him already last fall that she found very light elements from uranium, but she obviously did not trust herself to publish it'. To which Frisch added, 'well, she already has the Nobel Prize, she can be satisfied'. Frisch to Meitner, 8 January 1939, cited by Sime: see note 5-48, p 245.

6-24   Frisch O R 1939 Physical evidence for the division of heavy nuclei under neutron bombardment *NA* **143** 276; submitted 16 January, published 18 February 1939

6-25   Frisch: see note 6-24

6-26   Frisch: see note 6-13, p 104

6-27   Interview with Kowarski by C Weiner, 20 March 1969 to 20 November 1971, transcript, p 74 (OH265)

6-28   Goldsmith: see note 2-18, p 70

6-29   Anderson H L, Booth E T, Duning J R, Fermi E, Glasoe G N and Slack F G 1939 The fission of uranium *PR* **55** 511–2

6-30   Stuewer: see note 6-19, p 54. When Bohr followed Rosenfeld to Princeton, he found to his disappointment no letter from Frisch. He drafted a note on his own, outlining his and Rosenfeld's review of the fission problem *en route* to New York, and sent it, with a cover letter, back to Frisch, asking him to see that it was forwarded to *Nature*. It, along with a follow-up letter a few days later, would cross the Atlantic at the same time that a long overdue letter from Frisch, written on 22 January, went the other way. Stuewer: see note 6-19, p 53. Though the paper by Hahn and Strassmann reached Princeton on 20 January, and the Frisch–Meitner explanation had leaked out, Bohr felt badly about the matter, writing Margrethe that it was all his fault that they heard about the explanation of Frisch and Meitner . . .. Rhodes, Richard 1986 *The Making of the Atomic Bomb* (New York: Simon and Schuster) p 269.

6-31   As noted in section 5.1, the Berkeley group had bombarded a variety of substances with deuterons; the many disintegrations observed were attributed to the breakup of unstable deuterons. Similar observations, in fact, had been made at the Cavendish Laboratory as well. Tuve and his group had difficulty verifying either the Berkeley or the Cambridge results, and became suspicious that something was amiss on the west coast and abroad. After replacing their solid targets with gaseous targets of high purity, the DTM team found that deuteron bombardment produced no disintegrations; however, whenever they introduced a small amount of deuterium gas into a target, numerous disintegrations were observed. Tuve and company finally concluded that what had been observed by their colleagues was actually the result of deuterons colliding with deuterons. Cornell T D 1988 Merle Antony Tuve: pioneer nuclear physicist *PT* **41** 57–64, on p 61.

6-32   Roberts R B, Meyer R C and Hafstad L R 1939 Droplet fission of uranium and thorium nuclei *PR* **55** 416–7

6-33   Fowler R D and Dodson R W 1939 Intensely ionizing particles produced by neutron bombardment of uranium and thorium *PR* **55** 417–8

6-34   Present for the historic occasion were Robert Meyer, Merle Tuve, Enrico Fermi, Richard Roberts, Léon Rosenfeld, Erik Bohr, Niels Bohr, Gregory Breit, and John A Fleming (Director of the DTM). Present at the meeting but absent from the photograph taken on the occasion was Edward Teller.

6-35   Green G K and Alvarez L W 1939 Heavily ionizing particles from uranium *PR* **55** 417

6-36   The three letters were followed by one by Bohr, in which he concluded that it was not the heavy uranium isotope U^{238} that is primarily responsible for fission, but the lighter, rare isotope U^{235}. Stuewer: see note 6-19, p 56.
Bohr N 1939 Resonances in uranium and thorium disintegrations and the phenomenon of nuclear fission *PR* **55** 418–9

6-37   Cornell: see note 6-31. Contrasting Tuve and Lawrence, Cornell notes, for instance, their different attitudes toward the publication of experimental results. Tuve viewed research as the means of establishing scholarly benchmarks, and withheld publication until he was absolutely sure of his results. Lawrence, by contrast, favored fast publication. Cornell: see note 6-31, p 57. A case in point is Lawrence's unfortunate brush with deuterium contamination, and Tuve's patient sleuthing of the problem (note 6-31). In congratulating the Washington researchers on their nuclear physics investigations, Rutherford added: 'I am very pleased to see that the rush period of work on transmutation has come to an end, and as your papers show, results of real value can only be obtained by accurate and long-continued experiment'. Rutherford to Tuve, 17 November 1936; Carnegie Institution Archives, 1530 P Street NW, Washington, DC, under the heading 'Department of Terrestrial Magnetism—Director 1935–1940'.

6-38   Joliot F 1939 Preuve expérimentale de la rupture explosive des noyaux d'uranium et de thorium sous l'action des neutrons *CR* **208** 341–3

6-39   Szilard, Leo 1972 *The Collected Works: Scientific Papers* (Cambridge, MA: MIT Press), p 530. A year later, Szilard took out patents relating to possible nuclear chain reactions. To keep the concept out of the hands of Hitler's scientists, he assigned the patents to the British Admiralty. Bernstein B J 1987 Leo Szilard: Giving peace a chance in the nuclear age *PT* **40** 40–7.

6-40   Joliot 1966 Chemical evidence of the transmutation of elements *Nobel Lectures, including Presentation Speeches and Laureate's Biographies, Chemistry 1922–1941* (Amsterdam: Elsevier) pp 369–73, on p 373

6-41   Frisch (1979): see note 6-13, p 118

6-42   Frisch (1979): see note 6-13, p 118

6-43   Weart: see note 2-13, p 67

6-44   Weart S R 1970 Kowarski, Lew *DSB* **17** (Supplement II) 494–5
Goldschmidt B L 1979 Lew Kowarski *PT* **32** 68–70
Frisch O R 1979 Lew Kowarski *NA* **282** 541

6-45   Weart: see note 2-13, p 66

6-46   Halban, cited by Weart: see note 2-13, p 68. See also transcript of interview with Kowarski (note 6-27), pp 77–8. According to Kowarski, Joliot was the 'teacher', Halban 'in experimental charge' and he himself was 'the bright assistant'.

6-47   Reconstructed from the figure on p 69 of Weart: see note 2-13

6-48   Halban papers, 1939–1940, microfilm collection, Niels Bohr Library at the American Institute of Physics (MI30)

6-49   Halban H von, Joliot F and Kowarski L 1939 Liberation of neutrons in the nuclear explosion of uranium *NA* **143** 470–1. Interestingly, the neutron density distribution curve—the essence of the communication—was published without any caption. This article is followed immediately by the article by Meitner and Frisch 1939 Products of the fission of the uranium nucleus *NA* **143** 471–2, dated 6 March, Copenhagen.

6-50   Heilbron and Seidel: see note 5-20, p 444

6-51   Weart: see note 2-13, p 80

6-52   Anderson H, Fermi E and Hanstein H B 1939 Production of neutrons in uranium bombarded by neutrons *PR* **55** 797–8

6-53   Kowarski, in his interview with C Weiner, spoke of 'Halban-originated' versus 'Kowarski-originated' experiments. 'Incidentally [he continued], Fermi and Szilard at the same time were in exactly the same kind of opposition—Fermi proposing ideas like Halban's and Szilard proposing ideas like mine'. Kowarski: see note 6-27, pp 78–9.

6-54   Joliot F, Halban H von Jr and Kowarski L 1939 Number of neutrons liberated in the nuclear explosion of uranium *NA* **143** 680, dated 7 April. Szilard and Zinn's corresponding paper is dated 16 March at Columbia: Szilard L and Zinn W 1939 Instantaneous emission of fast neutrons in the interaction of slow neutrons with uranium *PR* **55** 799.

6-55   Interview with Kowarski by J J Ermenc, pp 169–8, on p 171; Ermenc J J 1967 *Atomic Bomb Scientists: Memoirs, 1939–1945* (Westport, CT: Meckler). See also the interview with C Weiner: note 6-27, p 76.

6-56   Halban, cited by Weart: see note 2-13, p 72

6-57   Kramish: see note 4-56, p 52

Notes section 6.3

6-58   Weart 1976 Scientists with a secret *PT* **29** 23–30, on p 24

6-59   Szilard to Joliot Curie, 2 February 1939, reproduced by Weart S R and Szilard G W (ed) 1978 *Leo Szilard: His Version of the Facts* (Cambridge, MA: MIT Press) pp 69–70

6-60   Fermi to Joliot, 4 February 1939. Joliot-Curie papers, Radium Institute, Paris; cited by Weart: see note 2-13, p 79.

6-61   Anderson *et al*: see note 6-52

6-62   Halban *et al*: see note 6-49

6-63   Goldsmith: see note 2-18, p 73

6-64   Weisskopf to von Halban, 31 March 1939, reproduced by Weart and Szilard: see note 6-59, p 71

6-65   Weisskopf to Blackett, 31 March 1939, in Weart and Szilard: see note 6-59, p 70

6-66   Weisskopf to Blackett, about 1 April 1939, in Weart and Szilard: see note 6-59, pp 72–3. Blackett and Cockcroft promptly replied that they would support the plea for secrecy.

6-67   Joliot, Halban, and Kowarski to Szilard, 5 April 1939, in Weart and Szilard: see note 6-59, p 73

6-68   Science Service, 24 February 1939, reprinted in *Science News Letter*, 11 March 1939, p 140

6-69   Szilard to Joliot, 6 April 1939, reproduced by Weart and Szilard: see note 6-59, p 73

6-70   Szilard and Zinn: see note 6-54

6-71   Szilard to Joliot, 7 April 1939, reproduced by Weart and Szilard: see note 6-59, p 74

6-72   Joliot to Szilard, 6 April 1939, in Weart and Szilard: see note 6-59, p 74

6-73   Goldsmith: see note 2-18, p 74

6-74   Joliot to Szilard, 19 April 1939, reproduced by Weart and Szilard: see note 6-59, pp 78–9

6-75   See e.g., Joliot, von Halban and Kowarski: note 6-54

The French value for v, the mean number of neutrons per fission, was 3.5 ± 0.7, or slightly in excess of the accepted value of 2.5. Kowarski claims he realized, after publication, that their value was high, and obtained a new value of 2.9. Ermenc: see note 6-55, p 171. In reworking their data in January 1940, Louis Turner at Princeton corrected for a certain error in the French analysis, and came up with $v = 2.6 \pm 0.6$.

Turner L 1940 Secondary neutrons from uranium *PR* **57** 334; Weart: see note 2-13, pp 303–4

6-76   The concept of a critical mass applies equally well for a controlled chain reaction (a reactor), or a divergent reaction (a bomb). In either case it refers to the minimum mass, or size of fissile material for which more neutrons find nuclei to fission than find surface to escape from.

6-77   Rudolf Peierls showed in 1940 that fast-neutron fission in U^{238} (making up 99.28% of natural uranium) was inefficient because of scattering and parasitic effects of capture resonances of that isotope. Bohr had already shown that U^{238} does not fission under slow-neutron bombardment, only the rare isotope U^{235}. These matters are further treated in section 8.1.

6-78   Perrin F 1939 Calcul relatif aux conditions éventuelles de transmutation en chaine de l'uranium *CR* **208** 1394–6; Weart: see note 2-13, p 94

6-79   Clark R W 1961 *The Birth of the Bomb* (New York: Horizon) p 28

6-80   Clark quoting Sengier; Clark: see note 6-79, p 23

6-81   Flügge S 1939 Kann der Energieinhalt der Atomkerne technisch nutzbar gemacht werden? *NW* **27** 402–10

6-82   Walker M 1989 *German National Socialism and the Quest for Nuclear Power 1939–1949* (Cambridge: Cambridge University Press) pp 17–8

6-83   Clark: see note 6-79, p 35

6-84   Clark: see note 6-79, pp 36–7

6-85   Cross section: the effective area of a target nucleus exposed to bombardment by nuclear particles (neutrons, α-particles, etc), used to express the probability of a particular nuclear process or reaction. The problem with water as a moderator is that, after a certain number of collisions with protons, a neutron combines with a proton to form a deuteron, releasing in the process binding energy in the form of γ-rays.

6-86   Kowarski see note 6-27, p 83

6-87   Halban, Joliot, Kowarski and Perrin 1939 Mise en évidence d'une réaction nucléaire en chaine au sein d'une masse uranifère *JPR* **10** 428–9

6-88   Weart: see note 2-13, p 118

Notes section 6.4

6-89 Goldsmith: see note 2-18, p 77

6-90 Adler F and Halban H von 1939 Control of the chain reaction involved in fission of the uranium nucleus *NA* **143** 793

6-91 If *k* exceeds 1, the chain reaction will eventually become uncontrollable. In order for it to remain under control, the multiplication factor must be kept at a constant value of 1 with the use of neutron absorbers. The multiplication factor is frequently referred to as the *reproduction factor*.

6-92 Pli cacheté No 11, 620, opened on 18 August 1948. Joliot-Curie, Frédéric and Irène 1961 *Oeuvres Scientifiques Complètes* (Paris: Presses Universitaires) pp 673–7

6-93 Maurois, André 1940 Raoul Dautry *Revue de Paris* **47** (3) 20–31

6-94 Kowarski in Ermenc: see note 6-55, p 172. Kowarski claims he, himself pushed deuterium as a moderator over Halban's lingering preference for carbon. Ermenc: see note 6-55, pp 172, 174.

6-95 Frisch O, Halban H von and Koch J 1938 *Kgl. Danske Videnskabernes Selskab, Mathematisk-fysiske Meddeleser* **15** No 10. See also interview of Frisch by C Weiner: note 6-13, p 31.

6-96 Goldsmith: see note 2-18, p 80

6-97 The building housing the Banque de Paris was originally prepared for the headquarters of Louis XV's Naval Minister in 1750. The office of the General Director was the site of Napoléon Bonaparte's civil wedding to Marie-Josèphe-Rose Tascher in 1796.

Notes section 7.1

7-1 Goldsmith: see note 2-18, p 85

7-2 Nicolai Stephansen to Brun, in Brun: see note 4-61, p 17. Bacterial cultures for the development of bacterial weapons seemed one possibility to Brun, when asked by Stephansen. Brun: see note 4-61, p 16.

7-3 'Forsendelser...': see note 4-73. NHAO, K03. 50. n. 2.1: Tungt vann, teknisk, Hefte 4.

7-4 Stephansen to Brun, 12 January 1940, in a handwritten note concerning 'a matter I do not wish to broach over the telephone'. Cited by Lunde: note 4-63; NHAO, K03. 50. n. 2.1: Tungt vann, teknisk, Hefte 4.

7-5 Brun: see note 4-61, p 16

7-6 Stephansen to Brun, 3 February 1940; Brun: see note 4-61, p 16

7-7 Anker Olsen: see note 4-50, p 399
Brun: see note 4-61, p 17

7-8 The final say in the matter lay with the Norwegian foreign ministry; NHAO, K03. 50. n. 2.1: Tungt vann, teknisk, Hefte 4.

7-9 In 1940, Hydro's stock of highly concentrated heavy water (99.6–99.9% purity) amounted to 185 kg.

7-10 See e.g., Goldschmidt, Bertrand 1990 *Atomic Rivals* Transl. G M Tenner (New Brunswick: Rutgers University Press) p 77

7-11 Anker Olsen: see note 4-50, p 398

7-12 In 1940, Hydro's Board had seven members, one Swede (Marc Wallenberg, Chairman), two Norwegians, three French, and one German.

7-13 Together with the holding company I G Chemie registered in Basel, Switzerland, I G Farbenindustrie held 25% of Hydro's joint capital.

7-14 Allier, 'Affaire de l'eau lourde', report for Dautry, February 1945, cited by Goldsmith: see note 2-18, pp 86–7

7-15   Goldsmith: see note 2-18, p 87

7-16   Anker Olsen: see note 4-50, p 399; also cited by Goldschmidt: see note 7-10, p 77, and Clark: see note 6-79, p 71. In fact, Aubert, who died in 1943, was not bothered by the Germans when they occupied Norway. However, Eriksen, his successor as General Manager to Norsk Hydro, was arrested by the Nazi authorities in 1943 for refusing production for the Germans. Allier, by the way, was elected to the Board of Directors of Norsk Hydro in 1955.

7-17   Brun: see note 4-61, p 16. Clark disagrees with Brun, who ought to know, claiming the canisters were manufactured outside Oslo, and sent up to Rjukan in advance. Clark: see note 6-79, p 71.

7-18   Kramish: see note 4-56, p 84

7-19   Among other things, he brought a request for Aubert to allow Jomar Brun to spend two weeks in Paris, orienting the French about all aspects of heavy-water production, should production in France be deemed practical. Brun: see note 4-61, p 19.

Notes section 7.2

7-20   French A P and Kennedy P J (ed) 1985 *Niels Bohr: A Centenary Volume* (Cambridge, MA: Harvard University Press) p 256. Among the physicists awaiting Bohr's return at his Institute at Blegdamsvej 17 that morning was Lise Meitner. She had arrived from Sweden the day before to use the cyclotron at the Institute, since Siegbahn's cyclotron was not yet operational. It took Bohr until the end of April before he was able to slip Meitner back across Øresund.

7-21   Moulton: see note 1-1, pp 63–4

7-22   Moulton: see note 1-1, p 93

7-23   Elting J R and the Editors of Time-Life Books 1981 *Battles for Scandinavia* (Alexandria, VA: Time-Life Books) p 50

7-24   The Germans attained their other Norwegian objectives on 9 April, albeit at considerable cost to Admiral Ræder's Naval forces. The squadrons bound for Trondheim and Narvik had slipped out of Wilhelmshaven on 7 April, escorted by the *Gneisenau* and *Scharnhorst*. They eluded the British in the North Sea, who thought the German warships were attempting to break out into the Atlantic. The British should, in fact, have been tipped off by the sinking in the Skagerrak by a Polish submarine of the *Rio de Janeiro*, a ship of the Sea Transport Echelon *en route* to Bergen. German survivors blurted out their destination, but the news, discussed by the Storting and broadcast over the evening news, was dismissed as nonsense by everybody. Narvik fell according to Ræder's plans, with Norway's two ancient coastal defense ships *Norge* and *Eidsvoll* knocked out by the destroyers of Group 1. (Subsequently, the British would wreak revenge at Narvik, destroying fully half of Ræder's total destroyer force.) Group 2, led by the heavy cruiser *Hipper*, took Trondheim without difficulty. Group 3 for Bergen, led by the light cruisers *Köln* and *Königsberg*, fared less well. *Königsberg* and the auxiliary *Bremse* were damaged from Kvarven fortress in the approaches to Bergen; the *Königsberg* was subsequently sunk in Bergen harbor by Fleet Air Arm bombers. Group 4, advancing toward Kristiansand in line ahead led by the cruiser *Karlsruhe*, battled token resistance in heavy fog by two island forts, but by 3 p.m. that city was in German hands as well. The *Karlsruhe* was, however, torpedoed and sunk on returning from Kristiansand.

7-25   The heroic and amazing evacuation of the Norwegian gold reserves, valued at 100 million dollars, is well told in *Operation: Gold Transport*, book I in Adamson H C and Klem P 1964 *Blood on the Midnight Sun* (New York: Norton), pp 15–109. It is also told by one of the participants in that venture, the well known Norwegian poet Nordahl Grieg, who later perished in an RAF bomber over Berlin. Grieg N 1945 *Flagget* (Oslo: Gyldendal).

7-26   Hambro C J 1940 *I Saw it Happen in Norway* (New York: Appleton-Century) pp 25–30. Carl Hambro was president of the Storting in 1940.

7-27   Kramish: see note 4-56, p 82

7-28   Moulton: see note 1-1, pp 151–2

Notes section 7.3

7-29   'War and occupation' in Anker Olsen: see note 4-50, pp 397–423, on p 397 Fjeldbu, Sigmund 1980 *Et Lite Sted på Verdenskartet: Rjukan 1940–1950* (Oslo: Tiden). Similar defenses were put up by Hydro in the winter of 1939–1940 at its facilities at Herøya to the south outside Porsgrunn. What role they actually played, if any, when the crunch came is less clear from *Norsk Hydro*.

7-30   Anker Olsen: see note 4-50, p 398
DEN = Deutschland-England-Norwegen

7-31   Anker Olsen: see note 4-50, p 400

7-32   Kramish: see note 4-56, p 83

7-33   Kramish: see note 4-56, p 83

7-34   'Tyskernes besettelse av Rjukan i Mai 1940'. Rjukan, 28 July 1945, originally written fall 1940; NIAM, Vemork.

7-35   'Military events 9 April–2 May 1940' in Anker Olsen: see note 4-50, pp 400–2, on pp 401–2

7-36   Powers T 1993 *Heisenberg's War: The Secret History of the German Bomb* (New York: Knopf) p 156

7-37   Sørensen: see note 4-40, p 120

7-38   Bjørn Rørholt, one of the leaders of the illegal group in Trondheim, hatched up a scheme by which Quisling was to be kidnapped and exhibited in Piccadilly Circus for a shilling a look; the money was to go to the Norwegian Fighter Fund. However, it was the Germans who struck first, catching the telegraphist on the way from Bymarka into town. Rørholt escaped once again. Haukelid K 1989 *Skis Against the Atom* (Minot, ND: North American Heritage Press) pp 29–30. Haukelid's book first appeared in English in 1954.

7-39   The immediate cause of Tronstad's departure had to do with the physical chemist Jan Herman Reimers, who also had smuggled in a transmitter, which he operated out of Oslo, and which Tronstad also made good use of. For unrelated reasons, Reimers had found it necessary to go into hiding in Oslo. In so doing, he left word with his mother that he was off on a trip to visit Tronstad in Trondheim, with whom he had, indeed, collaborated on various technical matters in times past. Not long after he had left, his mother came across certain ration cards of his, which she promptly mailed to Tronstad, along with a greeting. By pure chance, 'forgotten ration cards' was the warning of danger in Tronstad's group; hence Tronstad left for Oslo in all haste—just in time to escape Gestapo agents, who happened to be closing in on him. Sørensen: see note 4-40, p 122.

7-40   Dagbok 1, Dagbøker 1941–1945, Leif Tronstad, Riksarkivet, privatarkiv 354 (gift of Edla Tronstad)

7-41   Gallagher T 1975 *Assault in Norway: Sabotaging the Nazi Nuclear Bomb* (New York: Harcourt Brace Jovanovich) p 11

Notes section 8.1

8-1   Rozental S (ed) 1967 The forties and the fifties *Niels Bohr, His Life and Work as Seen by his Friends and Colleagues* (Amsterdam: North-Holland; New York: Wiley) pp 149–90

8-2   Meitner to O W Richardson, 15 May 1940; cited, for example, by Hartcup and Allibone: see note 6-20, p 121

8-3   Clark: see note 6-79, p 77. Hartcup and Allibone (see note 6-20, p 122) reports it as a rather inept contraction from the cumbersome *MAP Uranium Development Committee*, where MAP stands for Ministry of Aircraft Production.

8-4   Tizard to William Elliot, 4 April 1940, cited by Weart: see note 2-13, p 148; also by Gowing M 1965 *Britain and Atomic Energy, 1939–1945* (New York: St Martin's) Chapter 1

8-5   Frisch (1979): see note 6-13, p 123

8-6   Walker: see note 6-82, pp 29–30; Clusius K and Dickel G 1939 Das Trennrohr *Zeitschrift für Physikalische Chemie* **44** 459

8-7   Rudolf Peierls was born in 1907, the son of a Jewish industrialist, Heinrich Peierls, who rose in the administrative ranks of the Allgemeine Elektrizitäts-Gesellschaft (AEG). Young Peierls entered the University of Berlin in 1925, intending to take up experimental physics on the advice of Fritz Haber, who knew his father somewhat. He received his introduction to quantum physics from Walther Bothe, not from Planck whose lectures 'were the worst he ... ever attended'. After less than a year at the University of Berlin, he gravitated to Munich (1926, under Sommerfeld), Leipzig (1928, Heisenberg), and finally Zürich in 1929, where he completed his PhD under Pauli. Arriving in England in 1933, he remained restless, moving from Cambridge to Manchester, back to Cambridge, and finally settling at Birmingham in 1937.
Peierls R 1985 *Bird of Paradise* (Princeton: Princeton University Press)

8-8   Frisch (1979): see note 6-13, p 126

8-9   The technical part is reproduced in Gowing: see note 8-4, pp 389–93. The second part is apparently no longer extant, but a copy exists in the Tizard papers and is reproduced in Clark R 1965 *Tizard* (Cambridge, MA: MIT Press) pp 215–8.

8-10   Peierls: see note 8-7, p 155

8-11   Peierls: see note 8-7, p 154

8-12   Peierls: see note 8-7, pp 155–6

8-13   Franz (Francis) E Simon (1893–1956) was, like Peierls, the son of a well-to-do Jewish family in Berlin, who also began studies at the University of Berlin. In Simon's case, his studies were interrupted by World War I, in which he was seriously wounded in action. Recovering from a second injury sustained on the eve of the Armistice, he resumed his studies with Walther Nernst, under whom he received his doctorate in 1921 and stayed on in Nernst's Institute as his 'Assistant Extraordinary'. When Nernst became president of the Physikalisch-Technische Reichsanstalt in nearby Charlottenburg, Simon kept his low-temperature program going, before filling a professorship vacated by Nernst's early assistant Arnold

Eucken at the Technische Hochschule in Breslau in 1931. When Hitler came to power, remaining in Germany was anathema to Simon even though, as a front line fighter in the Great War and holder of the Iron Cross, First Class, he was exempt from the decree dismissing Jews from academic positions. Before long he was the senior member of a team of crack Continental refugee scientists assembled by Lindemann at Oxford's languishing Clarendon Laboratory, on a par with those of Cambridge's Cavendish Laboratory. In 1936 Simon was also appointed University Reader in Thermodynamics at Balliol College. Many years later he succeeded Lord Cherwell, née Lindemann, as Dr Lee's Professor of Experimental Philosophy—albeit a month before his untimely death.

8-14   Simon F, Estimate of the size of an actual separation plant, rewritten, under a different title, in Gowing: see note 8-4, p 416

Notes section 8.2

8-15   Wernick R 1977 *Blitzkrieg* (Alexandria, VA: Time-Life Books)
        Churchill W S 1949 *The Second World War: Their Finest Hour* (Cambridge, MA: Houghton Mifflin)
8-16   Weart: see note 2-13, p 137
8-17   Weart: see note 2-13, p 137, citing Dautry *Le Monde* 15 August 1945
8-18   Weart: see note 2-13, pp 139–40. Reports on this research were deposited as sealed notes in the archives of the Académie des Sciences, opened after the war. *Ibid.*, p 140.
8-19   Goldsmith: see note 2-18, p 90
8-20   Clark: see note 6-79, pp 95–6
8-21   Goldschmidt: see note 7-10, p 79
8-22   Kowarski: see note 6-27, p 104
8-23   Kowarski: see note 6-27, p 104
8-24   Goldsmith: see note 2-18, p 92
8-25   Bichelonne would choose Vichy France and collaboration, becoming state secretary for industrial production in the Laval government. Subsequently he followed Pétain to Germany and died in 1944 under somewhat murky circumstances. Goldschmidt: see note 7-10, p 81.
8-26   Kramish: see note 4-56, p 88
8-27   On 6 June 1944—D-day by happenstance—Irène and the two children hiked over the Jura massif into Switzerland, Irène lugging a heavy physics tome in her knapsack. McGrayne: see note 2-16, pp 140–1.
8-28   Certainly the name Curie was extremely popular in American circles, and his wife's sister Eve was on very friendly terms with the Roosevelt family. Goldsmith: see note 2-18, p 95. Goldsmith has even speculated that, due to Joliot's reputation in Europe and the US, de Gaulle's influence might have been correspondingly greater with respect to Roosevelt and Churchill. *Ibid.*

Notes section 9.1

9-1   Attending, besides Esau, were Georg Joos and Wilhelm Hanle of Göttingen, Hans Geiger, then chairman of physics at the Technische Hochschule in Berlin, Josef Mattauch of the KWI, Walther Bothe of Heidelberg, Gerhard Hoffmann of Leipzig, and William Dames.

9-2   Hutton R S 1964 *Recollections of a Technologist* (London: Sir Isaac Pitman & Sons Ltd) p 180. Rosbaud considered Mattauch among the 'trustworthy' German scientists. Others in that category on Rosbaud's list included Hahn, Strassmann, and Max von Laue. Sime: see note 5-48, p 486, citing Rosbaud to S Goudsmit, 18 July 1958.

9-3   Kramish: see note 4-56, p 53

9-4   Harteck P and Groth W to Army Ordnance, 24 April 1939; PHP, Acc. 85-22, Box 2 (copy, incl. English translation). The original of this letter has proven hard to locate; Harteck's own original copy was lost when the British authorities emptied his desk in his Hamburg Institute in 1945. Equally influential in setting the stage for the Reich Ministry meeting was a letter from Georg Joos to the Education Minister, proposing that the Reich study the employment of uranium fission for the production of energy. Walker: see note 6-82, p 17.

9-5   Irving, David 1967 *The German Atomic Bomb: The History of Nuclear Research in Nazi Germany* (New York: Simon and Schuster) p 37; also issued in 1967 as *The Virus House. Germany's Atomic Research and Allied Countermeasures* (London: William Kimber)

9-6   *The Development of Nuclear Energy in Germany, 1939–1945*, edited typescript of taped interview with Harteck by J J Ermenc, 6 June 1967 (PHP, Acc. 85-22, Box 2), published in Ermenc: see note 6-55, p 85

9-7   Ermenc: see note 6-55, p 91

9-8   In June 1932, to Harteck's everlasting pride, Rutherford ended his reply to Haber's letter of inquiry on the following note: '. . . we shall be glad to welcome such a distinguished investigator from your Laboratory as Dr Harteck'. Rutherford to F Haber, 17 June 1932; PHP, Acc. MC 17, Box 2, Folder 2.

It took until the next February for the deal to be consummated, when the Rockefeller Foundation agreed to commit not more than '$150/month for a period of not more than twelve months'. W E Tisdale to F Haber, 20 February 1933; PHP, Acc. 85-22, Box 1, Folder 3.

9-9   Oliphant M L E, Harteck P and Lord Rutherford 1934 Transmutation effects observed with heavy hydrogen *PRS* **144** 692–703, summarized in 1934 *NA* **133** 413

A nuclear fusion reaction, in which light nuclei combine to form a nucleus of a higher mass number, can be said to represent the opposite of nuclear *fission*. Fusion is essentially a thermal reaction, requiring no critical mass and is therefore potentially unlimited. Moreover, deuterium, the efforts of Urey and Tronstad notwithstanding, is more easily separated from hydrogen than is U^{235} from natural uranium. Irving rates the Oliphant–Harteck–Rutherford paper as having been 'of little less moment than Hahn's and Strassmann's [sic] 1939 paper on the fission of the uranium nucleus'. Irving: see note 9-5, p 48.

9-10   A third variety of hydrogen? Experiments at Cambridge *The Times* London, 22 March 1934

9-11   Rutherford to the Director, Norsk Hydro, 19 March 1935. NHAN, K03. 50. n. 1.1: Tungt vann, general, Hefte 1.

9-12   Urey to Dr S B Bronne [sic], 28 February and 22 March 1935. NIAM, Box 4F-D02, 960: Tungt vann, Akt. nr. 704C.

9-13   Axel Aubert to Rutherford, 23 March 1935; C Kahrs Kielland, Chief, Teknisk Avdeling, to Hydro management, Rjukan, 28 March 1935. NHAN, K03. 50. n. 1.1: Tungt vann, general, Hefte 1.

9-14   Leif Tronstad to Hydro management, Oslo, 10 May 1935. NIAM, Box 4F-D02, 960: Tungt vann, Akt. nr. 704F.

Speculations on an appropriate method for concentrating T_2O, and musings on his stay in London and Cambridge, are also found in Tronstad's journal for 1932–1935, NHAR, kindly brought to my attention by Per Pynten at Rjukan.

9-15   Rutherford to the Director, Norsk Hydro, 3 April 1936. NHAO, K03. 50. n. 2.1: Tungt vann, Hefte 4.

Brun J, Report on work in order to produce triterium concentrate by electrolyzing down heavy water, 6 March 1936; Hydro archives, *ibid*.

Rutherford 1937 The search for the isotopes of hydrogen and helium of mass 3 *NA* **140** 303–5

9-16   Lunde: see note 4-63, first installment, 5 April 1969

In 1935 a group of Princeton scientists, including H S Taylor and Walker Bleakney, had concentrated 1/2 cc of heavy water and claimed to detect spectroscopic traces of tritium in it. Doubt was subsequently cast on that finding, when tritium was found to be radioactive. Rutherford, whose own heavy-water sample appeared devoid of tritium, had lectured Taylor for having arrived at a 'foolish conclusion'. The finding much later that Rutherford's own sample was highly radioactive meant belated vindication of Taylor's own work. 'I am certainly glad to learn that Rutherford's sample was lousy with tritium', he is said to have remarked. Laurence W L 1950 Cosmic rays are found to produce tritium, vital for hydrogen bomb *New York Times* Friday, 15 September 1950.

9-17   Rutherford to G N Lewis, 30 October 1933. GLP, Box 3, Item 865. 'When I came back [from the 1933 Solvay Conference], I found [Harteck] had got a sufficient concentration to be useful for a good deal of our work'.

9-18   Harteck P 'Bericht über Meine Arbeit in Cambridge und Erfahrungen augemeiner Art' 1 June 1934; PHP, Acc. MC 17, Box 2, Folder 2.

'Fortunately for me and the Cavendish', begins Harteck's report, within a week of his arrival at the Laboratory, there arose a need for heavy water in many experiments—'just the right task for a physical chemist'. Harteck P 1934 The preparation of heavy hydrogen *PRS* **46** 277–80.

News of Harteck's heavy water traveled fast; the experiments were barely over when the Deutsches Museum, Munich, requested a sample of the precious liquid for their collection. Director, Deutsches Museum, to Harteck, 5 December 1933; PHP, Acc. MC 17, Box 2, Folder 2.

9-19   P Harteck to G N Lewis, 4 June 1933; G N Lewis to P Harteck, 23 June 1933. GLP, Box 3, Items 328 and 329. One of the batches of heavy water was hand-carried by John Cockcroft, who had stopped in Berkeley during his American tour in 1933. He purchased 10 dollars worth from Lewis, but only with difficulty got it past British customs officials, who could not understand why a liquid resembling water should be brought into the UK. On top of that, Rutherford scolded him for having spent so much on the heavy water without authorization. Hartcup and Allibone: see note 6-20, p 61.

9-20   One of these demonstrations, illustrating the difference in vapor pressure, consisted of a U-shaped tube containing mercury. On one side was a little heavy water and on the other side a little normal water. At room temperature there was a difference of 3 mm in the water levels, that increased with increasing temperature. Harteck: see note 4-17, p 9–11 (on p 9). Another difference between normal and heavy water gave rise to an April joke in Rjukan. The freezing point of heavy water is 4 °C, compared to 0 °C for normal water, leading the *Rjukan Dagblad* to report on plans for a more durable skating rink at Rjukan, flooded with heavy water. Brun: see note 4-61, p 8.

9-21 Wilson D 1983 *Rutherford: Simple Genius* (Cambridge, MA: MIT Press) p 569. Harteck, cited by Ermenc: see note 6-55, p 92.

9-22 Among other things, Professor Rideal, Tronstad's mentor shortly before, took a liking to Harteck as well, and had him lend a hand to the Farkas brothers in their experiments on ortho- and para-deuterium during their short refugee stay at Cambridge. Harteck: see note 9-18, p 1.
Farkas A and L 1933 Some experiments on heavy hydrogen *NA* **132** 894; 1934 Chemical separation of diplogen from hydrogen *NA* **133** 139.

9-23 Irving: see note 9-5, p 42

9-24 Walker: see note 6-82, p 18
The Reich Research Council (*Reichsforschungsrat*), within the Ministry of Education, only acquired some clout when Hitler put Goering in charge of it in May 1942.

9-25 Goudsmit, Samuel A 1986 *Alsos* (Los Angeles: Tomash) p 166

9-26 Irving: see note 9-5, p 42; Powers: see note 7-36, p 14

9-27 At Farm Hall in 1945 Heisenberg gruffly referred to Diebner as 'not really a physicist... more a kind of *Verwaltungsmann* (Administrator)'. Heisenberg, cited by Bernstein, Jeremy 1996 *Hitler's Uranium Club: The Secret Recordings at Farm Hall* (Woodbury, NY: AIP) p 259.

9-28 Rittner, Major T H, cited by Bernstein: see note 9-27, p 72

9-29 Bernstein: see note 9-27, p 73

9-30 IMC, DJ 29-85 is a facsimile of Otto Hahn's diary for 10–16 September 1939

9-31 Farm Hall Report 4, see note 9-27, p 145

9-32 Harteck to D Irving, 19 April 1966; PHP, Acc. 85-22, Box 1, Folder 1

9-33 Irving: see note 9-5, p 42

9-34 Harteck, cited by Ermenc: see note 6-55, p 128

9-35 Turner L A 1940 The missing heavy nuclei *PR* **57** 950

9-36 McMillan E and Abelson P 1940 Radioactive element 93 *PR* **57** 1185–6
Starke K 1942 Anreichung des Künstlich Radioaktiven Uran-Isotops U^{239} und Seines Folgeproduktes 93^{239} (element 93) *NW* **30** 577–82
Hahn O and Strassmann F 1942 Zur Folge nach der Entstehung des 2.3 Tage-Isotopes des Elements 93 aus Uran *NW* **30** 256–60
It should be emphasized that the Germans appreciated the potential of element 94, at least as early as 1942: that it could be produced in a 'uranium machine', and appeared to be a nuclear explosive fully as effective as U-235. Moreover, element 94 was easier to manufacture than U-235, since it could be separated from uranium by chemical means.

9-37 Goudsmit: see note 9-25, p xii

Notes section 9.2

9-38 Bothe W, Lebensbeschreibung, in Drossel R 1975 *Walther Bothe, Bemerkungen zu seinen kernphysikalischen Arbeiten auf Grund der Durchsicht seiner Laborbucher* (Max-Planck-Institut für Kernphysik, Heidelberg, unpublished); cited by Rhodes: see note 6-30, p 345

9-39 Bothe W 1940 Die Diffusionslänge für thermische Neutronen in Kohle, G-12 (7 June 1940)
Bothe W and Jensen P 1941 Die Absorption thermischers Neutronen in Elektrographit, G-71 (20 January 1941), published 1944 *ZP* **122** 749–55

Fermi's value for the absorption cross section was not published at the time, on the insistence of Szilard and Pegram, at the cost of Fermi, for once, nearly losing his temper. Anderson H and Fermi E 'Production and absorption of slow neutrons by carbon', see Fermi: note 5-36, pp 32-40. Weart and Szilard: see note 6-59, p 116

9-40   Walker: see note 6-82, p 26
9-41   Kramish: see note 4-56, pp 174–5
9-42   Halban H and Kowarski L 'Discussion of the composition and constitution of uranium-containing systems in view of producing a divergent nuclear reaction chain', IMC, DJ 31-877; also British Atomic Energy Authority Files, Report BR-94, AERE Library, Harwell.
9-43   Clark: see note 6-79, p 102
9-44   Harteck, cited by Ermenc: see note 6-55, p 102
9-45   Harteck, cited by Ermenc: see note 6-55, pp 103–4
9-46   I G Farben to Harteck, 3 and 5 May 1940; PHP, Acc. MC 17, Box 2, Folder 6
9-47   Heisenberg to Harteck, 4 May 1940; PHP, Acc. 85-22, Box 2, Folder on correspondence 1935–1943
9-48   Harteck to OKH, 3 June 1940, reporting CO_2 experiments have begun, IMC, DJ 29-648
9-49   Harteck P, Jensen J, Knauer F and Suess H 1940 Über die Bremsung, die Diffusion und Einfang von Neutronen in fester Kohlensäure und über ihren Einfang in Uran, G-36, 19 August 1940
9-50   Heisenberg, cited by Ermenc: see note 6-55, p 29

Notes section 9.3

9-51   Goldsmith: see note 2-18, p 98
9-52   Gentner W 1980 *Discussions with Frédéric Joliot-Curie in Occupied Paris during the Years 1940–1942* (Heidelberg: Max-Planck-Institut für Kernphysik). Shortly before his death in September 1980, Gentner recounted his experiences in occupied Paris during 1940–1942. His memoirs were printed by the Max-Planck Institute, as of interest to his friends and colleagues. See also transcript of oral history interview with C Weiner: note 5-10, pp 104–5.
9-53   Schumann suspected the uranium had been shipped to North Africa. In fact, part of the stock of uranium was hidden near Toulouse, where it was confiscated by the Americans after the Liberation. Some uranium, in a less pure state, was at Ivry. Goldsmith: see note 2-18, p 99.
9-54   Five European cyclotrons were completed by the start of WWII. Those at Cambridge, Liverpool, and Copenhagen obtained their first beam in 1938, those at Paris and Stockholm in 1939. The Heidelberg machine would obtain its first beam in 1943. Heilbron and Seidel: see note 5-20, p 321.
9-55   Goldsmith: see note 2-18, p 101
9-56   Allier, report on 1940 heavy water discussions at Vichy, 1945, Jacques Allier Papers, Paris; cited by Weart: see note 2-13, p 315
9-57   In October 1943 Allier was called to Oslo for a general dressing-down by the German authorities, who were enraged at the destruction of the Hydro light-metal works at Herøya by American bombers, and unhappy with French interests in Norsk Hydro. In the event, the Swedish Consul

General in Paris saw to it that Allier's visa was denied at the eleventh hour. Anker Olsen: see note 4-50, p 409.

9-58   Gentner: see note 9-52

Notes section 10.1

10-1   'Military events 9 April–2 May 1940' in Anker Olsen: see note 4-50, pp 400–2, on p 402

Norsk Hydro's situation in France remained precarious from the onset of the occupation. When the German's marched into Paris, Horace Finaly, Director General of the Banque de Paris and a Jew, fled to America. He was replaced by H Oster of I G Farben, who subsequently became the second German member of Hydro's Board of Directors, and played a role in Hydro's wartime activities in Norway. 'The Germans bypass the French' in Anker Olsen: see note 4-50, pp 406–9, on p 406.

10-2   'The Herøya Installation', pp 409–11; 'The Attack on Herøya', pp 411–2, in Anker Olsen: note 4-50

10-3   Heisenberg W 1939 'Die Möglichkeit der Technischen Energiewinnung aus der Uranspaltung', G-39 (6 December 1939)

10-4   Harteck P 'A layer arrangement of uranium and heavy water, to avoid resonance absorption in uranium-238', cited by Irving: see note 9-5, p 49

10-5   Harteck to Heisenberg, 15 January 1940; Heisenberg to Harteck, 18 January 1940. 'Unterlagen betreffend SH.200' (Documents concerning SH.200). Erik Bagge Papers, Kiel, copy in PHP, Acc. 85-22, Box 1, Folder 1, letters reproduced in IMC, DJ 29-625. 'Unterlagen' gives a succinct summary of essential milestones in the German heavy-water effort, from January 1940 to December 1942.

10-6   Harteck–Bonhoeffer correspondence, January–March 1940; PHP, Acc. MC 17, Box 1, Folder 6. Bonhoeffer to I G Farben 28 February 1940, *ibid.* and IMC, DJ 29-632.

10-7   Harteck to Army Ordnance, 24 January 1940; Army Ordnance to Harteck, 1 February 1940. Erik Bagge Papers, Kiel: see note 10-5, PHP, Acc. 85-22, Box 1, Folder 1.

10-8   Army Ordnance to Harteck, 1 February 1940. Erik Bagge Papers, Kiel: see note 10-5, PHP, Acc. 85-22, Box 1, Folder 1; see also Walker: note 6-82, p 27.

10-9   Harteck to Bonhoeffer, 14 April 1940; PHP, Acc. MC 17, Box 1, Folder 6

10-10   Irving citing Harteck: see note 9-5, p 66; see also Erik Bagge Papers, Kiel: note 10-5, PHP, Acc. 85-22, Box 1, Folder 1, for February 1940

10-11   Brun: see note 4-61, p 20; NHAO, K03. 50. n. 2.1: Tungt vann, Hefte 5

10-12   Walker: see note 6-82, pp 41–2; IMC, DJ 29-1179

10-13   Note from meeting with General Director, 18 February 1941. NHAO, K03. 50. n. 2.1: Tungt vann, Hefte 5.

10-14   Brun: see note 4-61, p 21

10-15   Brun: see note 4-61, p 21

10-16   Döpel R, Döpel K and Heisenberg W 1940 Bestimmung der Diffusionslänge thermischer Neutronen in schwerem Wasser, G-23, 7 August 1940

10-17   Visit with Hydro management of Dr Tzschirner, head of *Wehrwirtschaftstab Norwegen*, according to 'note concerning heavy water' by E Hv Eide, 27 August 1941; NHAO, K03. 50. n. 2.1: Tungt vann, Hefte 5. For a good technical description of the electrolytic process at Vemork, see 'History of process' Benedict M and Pigford T H 1957 *Nuclear Chemical Engineering* (New York: McGraw-Hill) pp 431–4.

10-18 Brun: see note 4-61, p 22. Harteck to D Irving, 19 April 1966; PHP, Acc. 85-22, Box 1, Folder 1. In an effort to clear his name of 'numerous falsehoods' raised in connection with his heavy-water involvement on behalf of *Wehrwirtschaftstab Norwegen*, Schoepke wrote a lengthy letter to, perhaps among others, Paul Harteck in 1951. In it, he goes over the whole affair, maintaining that he acted solely in the interest of Norway and its people, and enclosing as well declarations on his behalf by then *Fregattenkapitän* Frederich Ebeling, also attached to WW, and Ludwig Angerer, a colleague of Harteck accompanying him to Norway on occasion. Schoepke to Harteck, 30 October, 9 December 1951; PHP, Acc. 85-22, Box 1, Folder 'Schoepke'. I have not found Harteck's reply, if any.

10-19 Report from meeting with *Wehrwirtschaftstab Norwegen*, 2 and 6 October 1941; Ergebnisse der Besprechung zwischen den Herren Ing. Brun, Prof. Harteck und Dr. Wirtz, 4 October 1941. NHAO, K03. 50. n. 2.1: Tungt vann, Hefte 5.

10-20 Protokoll über die Sitzung in Heereswaffenamt am 22/11/41; PHP, Acc. 85-22, Box 1, Folder 1; see Erik Bagge Papers, Kiel: note 10-5, PHP, Acc. 85-22, Box 1, Folder 1. Among those present were Bonhoeffer, Clusius, Harteck, Suess, and Wirtz.

10-21 Internal Hydro correspondence concerning regeneration of heavy water, December 1941; NHAO, K03. 50. n. 2.1: Tungt vann, Hefte 6.

10-22 For more on the Harteck–Suess exchange process, see Benedict and Pigford: note 10-17, pp 440–3
Murphy G M (ed) 1955 *Production of Heavy Water* (New York: McGraw-Hill)
The advantages of the Harteck–Suess process are discussed by Harteck in his interview with Ermenc: see note 6-55, p 109.
The pros and cons of the catalytic exchange process were also discussed by Harteck with Hahn at Farm Hall on 23 September 1945; Operation 'Epsilon', Report 9' in Bernstein: note 9-27, pp 280–3.
See also Harteck P 'Die Produktion von schwerem Wasser' G-86 (December 1941); Wirtz K 'Die elektrolytische Schwerwassergewinnung in Norwegen' G-198 (26–28 February 1941).

10-23 Suess H 1968 Virus House: comments and reminiscences *Bulletin of the Atomic Scientists* **24** 36–9, on p 38. Suess claims, in his article, and so informed Brun after the war, that a few hundred grams of platinum would have sufficed as a catalyst.

10-24 Brun: see note 4-61, p 24

10-25 Brun: see note 4-61, p 25

10-26 Brun: see note 4-61, p 16. Brun's wartime memoirs are also recorded in his typewritten notes entitled 'Some impressions from my work with Z [i.e., heavy water], especially regarding the "co-operation" with the Germans', stamped MOST SECRET and dated 30 November 1942. They were thus prepared about three weeks after he arrived in London, in time for assisting the Allied authorities in preparing for *Freshman* (see section 12.2). Copy courtesy of NHM, consulted at NIAM. NHM, Box 10, Mappe SCS, Sak: Brun. The particular discussion with Tzschirner and Wirtz is related on p 1.

10-27 Anker Olsen: see note 4-50, p 415

10-28 Haukelid: see note 7-38, p 174. Haukelid's production figures for 1942, obtained from Norsk Hydro after the war, differ somewhat (by up to

30%) from the figures reported by Brun to the British authorities in January 1943. 'Memorandum submitted by Dr. Sverre Hagen on the methods for manufacture of D_2O of the N.H.', London, 29 January 1943, signed Sverre Hagen (alias Jomar Brun). Copy obtained from NIAM, courtesy NHM.

10-29 Colonel Tzschirner to Norsk Hydro, 15 January 1942, reproduced in Brun: see note 4-61, p 23

10-30 Brun to N Stephansen, Berlin, 22 January 1942; Brun, 28 January 1942, Report on visit in Berlin, 20–25 January 1942. NHAO, K03. 50. n. 2.1: Tungt vann, Hefte 6. Brun, 'Journey to Germany', pp 3–5 in 'Some impressions from my work with Z . . .': see note 10-26, NIAM.

10-31 Brun: see note 4-61, p 26

10-32 Brun: see note 4-61, p 27

10-33 Brun: see note 4-61, p 28

10-34 Anker Olsen: see note 4-50, p 416

Notes section 10.2

10-35 Tronstad L, Dagbok 2 og 3, Dagbøker 1941–1945, RO, privatarkiv 354

10-36 Sørensen tells of a meeting between German military and industrial experts recorded during the fall of 1942. Experiments involving long-range rockets which could reach English targets from northern France were discussed, and a site named Peenemünde came up several times in the conversation. The news reached Tronstad, Peenemünde was subsequently plastered by a major RAF raid, and the V1 and V2 attacks on England were delayed by at least half a year. Sørensen: see note 4-40, pp 128–9.

10-37 Howarth D 1955 and 1974 *The Shetland Bus* (London: Fontana–Collins)

10-38 Adamson and Klem: see note 7-25, pp 116–36

Notes section 10.3

10-39 Irving: see note 9-5, p 115; translated from *Månedsrapport for drift av Vannstoffabrikken Vemork* in IMC, DJ 31-456

10-40 Harteck P 'Die Produktion von schwerem Wasser' (see note 10-22). Moreover, in the event of a failure to divert industrial capacity to atomic weapons development, notes Suess in a retrospective account, Hitler certainly would have had its sponsors executed. Suess: see note 10-23, p 37.

10-41 Brun: see note 4-61, p 21

10-42 Larsen, like Brun and Tronstad, came from NTH in Trondheim, where he had been a student of Tronstad. He joined the Vemork facility in 1936, where he worked initially on the production of pilot amounts of heavy paraffin and other deuterium derivatives, and transferred to the hydrogen-production plant in 1939. He fled to England in early 1944. It was he who packed the famous 185 kg of heavy water at Vemork on 9 March 1940.

10-43 IMC, DJ 31-446

10-44 *Månedsrapport for drift av Vannstoffabrikken Vemork* in IMC, DJ 31-446, 447

10-45 Gallagher: see note 7-41, p 16

10-46 Irving: see note 9-5, p 115

10-47 NHAO, K03. 50. n. 2.1: Tungt vann, Hefte 8. Anker Olsen: see note 4-50, p 352; Harteck to OKH, 23 March 1942, IMC, DJ 29-709.

10-48 'Memorandum submitted by Dr. Sverre Hagen . . . ': see note 10-28

10-49 *Månedsrapport for drift av Vannstoffabrikken Vemork*, June 1942, IMC, DJ 31-446

Harteck P 1953 Die elektrolytische Deuteriumgewinnung in Norwegen *Naturforschung und Medizin in Deutschland 1939–1945, Band 14, Kernphysik und Kosmische Strahlen* teil II, ed W Bothe and S Flügge (Weinheim: Chemie) pp 182–3

10-50 NHAO, K03. 50. n. 2.1: Tungt vann, Hefte 8

10-51 Wirtz, 'Die elektrolytische Schwerwassergewinnung in Norwegen': see note 10-22, p 2

10-52 C Kielland, 8 April 1942, internal correspondence, NHAO, K03. 50. n. 2.1: Tungt vann, Hefte 7.

10-53 Brun, 22 September 1942, 'Bericht über die ersten Versuchs-und Betriebsergebnisse der als Versuchsanlage Ausgebauten "Halben 6. Stufe" von J. Brun und H. Suess', NHAO, K03. 50. n. 2.1: Tungt vann, Hefte 9.

In his subsequent report on these tests to the Allied authorities, in 'Some impressions from my work with Z . . . ' (see note 10-26), NIAM, Brun, for reasons of his own, refers to a Dr X, since 'his real name being of no importance, I prefer not to mention it'. Brun: see note 10-26, p 5. Clearly Dr X was Hans Suess. See also G-337, G-338, G-339.

10-54 Brun: see note 4-61, p 31

10-55 'Niederschrift über die Besprechungen am 24. und 25. 7. 42 in Rjukan und Vemork', Oslo, 27 July 1942; NHAO, K03. 50. n. 2.1: Tungt vann, Hefte 8. Minutes of correspondence reproduced in IMC, DJ 29-728-730; also G-188.

The agenda of the conference covered the following principal subjects: (1) catalysts; (2) expansion of stage 6; (3) S.H. 200-Såheim Plant; (4) high-concentration plant in Germany; (5) settlement of materials for Vemork plant expansion; (6) Summary of conclusions.

10-56 Suess: see note 10-23, p 37

10-57 Powers: see note 7-36, pp 158–60

For a thorough discussion of Heisenberg's famous 1941 trip to Denmark, both to see Bohr and to attend an astrophysics conference at the German Cultural Institute in Copenhagen, see Walker: note 6-82, pp 223–8.

10-58 Kramish: see note 4-56, p 131

10-59 XU ('unknown undercover agent') was the largest and best organized intelligence organization in Norway, perhaps anywhere, in World War II. It functioned under the Norwegian High Command (FO2) in London in conjunction with a branch office (Militærkontoret) with the Norwegian military attaché (Mi2) in Stockholm. Brunulf Ottar was one of its founders, in charge of the Stockholm office at 32 Skeppargatan. Wergeland was active in XU as well.

Sæter, Einar and Svein 1995 *XU—i Hemmeleg Teneste 1940–1945* (Oslo: Det Norske Samlaget)

10-60 Hinsley F H 1981 *British Intelligence in the Second World War* vol 2 (London: Her Majesty's Stationery Office) p 126

10-61 Testimonial by L Tronstad on behalf of H Wergeland, 13 March 1939; addendum to Wergeland letters, Bohr Scientific Correspondence, 1930–1945, AHQP, microfilm reel 26

10-62 Brun: see note 4-61, p 95

10-63 Sæter: see note 10-59, p 152

Writing to Bohr in May 1945, Wergeland notes that he is still in military service, and still recovering from the constant anxiety and excitement as of late. 'Now it should be ended, no more preoccupation with weapons in the forest or in dark cellars in fear of detection . . .' Wergeland to N Bohr, 15 May 1945. BSC, AHQP, microfilm reel 26, A.L.S., 2 pp.

10-64 Kramish: see note 4-56, p 189; Sime: see note 5-48, p 305

10-65 See e.g., Hole to Tronstad, 8 October 1942, reporting on comings and goings at the Berlin-Dahlem institutes, according to his 'German sources'. In March 1943 Tronstad warns Hole to be guarded in his dealings with the German visitors. The same goes for his interaction with Lise Meitner. Should circumstances warrant it, London is prepared to fly Hole to England at the highest priority and at their discretion; however, for now he is to 'remain calm and do his job'. Tronstad to Hole, 6 March 1943. NHM, Box 10, Mappe SCS, copy consulted at NIAM.

Wergeland's anti-bomb sentiment may have been a factor in his post-WWII crusade against nuclear energy for industrial purposes in Norway (in connection with the Halden reactor), ostensibly on economic and environmental grounds. Njål Hole became another outspoken critic. *Verdens Gang* 20 January 1956; *Bergens Tidende* 21 May 1956.

Notes section 10.4

10-66 Brun: see note 4-61, p 28

10-67 Brun: see note 4-61, pp 31, 35

10-68 Excerpt of report sent by courier to London, spring 1942.
Brun: see note 4-61, p 29

10-69 Excerpt of report sent by courier to London, summer 1942.
Brun: see note 4-61, p 32

10-70 *Månedsrapport for drift av Vannstoffabrikken Vemork*, September 1942, IMC, DJ 31-447
Brun: see note 4-61, p 31
Brun 'The creation and running of the pilot plant' pp 5-6 of 'Some impressions from my work with Z . . .': see note 10-26, NIAM.
Nickel catalyst No 5233 had been settled upon as the catalyst of choice at the Rjukan–Vemork conference the previous July. Brun: see note 4-61, p 31. The apparatus belonging to the exchange plant is also indicated in figure 10.3 (left center), marked in red on the original drawing.

10-71 Brun: see note 4-61, pp 35–6
Brun T J (Mrs Brun), Recollections from the Second World War, February 1995, Jomar Bruns Arkiv, Hydro Arkiv No 00006.010, Rjukan

10-72 In Tronstad's personal journal for 4 November is noted: 'Theodor [Eric Welsh] explained that WC [Churchill] himself wanted Master here as fast as possible. This is an interesting fact'. Tronstad, Dagbok 4, Dagbøker 1941–1945, RO, Privatarkiv 354.

10-73 Brun: see note 4-61, pp 36–7
Brun T J: see note 10-71. Thus Brun's disappearance led to some unpleasantness for relatives left behind, including interrogations and searches, but nobody was imprisoned or otherwise punished. Thor Abrahamsen, who helped Brun prepare for his escape, and his wife lost their lives during the American bombing of Vemork in November 1943.

10-74 Tronstad, Dagbok 4, Dagbøker 1941–1945, RO, Privatarkiv 354

10-75 Brun: see note 4-61, p 37

Notes section 11.1

11-1   Fermi's use of the term 'pile' for a chain-reacting system, a term (*pila* in Italian) usually associated with the Voltaic battery, is of doubtful significance; Fermi used the term in the American sense of a 'heap' of uranium and graphite. Segrè (1970): see note 5-26, p 116.

11-2   Fermi, Laura (1954): see note 5-26, p 164

11-3   Clark: see note 6-79, p 166

11-4   Clark: see note 6-79, p 162. The magnetron, it was soon apparent, was superior to the American *klystron* then also under development. The Naval Research Laboratory, across town in Washington, DC, had a klystron transmitter operating at 10 watts; the British device gave a thousand times as much power at the same wavelength.

11-5   Gosling F G 1994 *The Manhattan Project: Making the Atomic Bomb* (Washington, DC: US Department of Energy Report DOE/HR-0096) p 9. The MAUD Report, building on the theoretical foundation of Frisch and Peierls, dismissed plutonium production, thermal diffusion, electromagnetic separation, and the centrifuge; a critical mass of 10 kg of uranium-235 from gaseous diffusion should lead to a practical bomb in approximately two years.

11-6   Hewlett R G and Anderson O E Jr 1962 *The New World, 1939–1946, Volume I, A History of the United States Atomic Energy Commission* (University Park, PA: Pennsylvania State University Press) p 49

Notes section 11.2

11-7   Szilard to Fermi, 8 July 1939: see Szilard, *Collected Works*, note 6-39, pp 195–6

11-8   Fermi to Szilard, 9 July 1939: see Szilard, *Collected Works*, note 6-39, p 196

11-9   Szilard to Fermi, 3 July 1939: see Szilard, *Collected Works*, note 6-39, p 194

11-10   Fermi to Szilard, 9 July 1939: see Szilard, *Collected Works*, note 6-39, p 197

11-11   Fermi E 1955 Physics at Columbia University, the genesis of the nuclear energy project *PT* **8** 282–6

11-12   Anderson H L and Fermi E 1965 Production and absorption of slow neutrons by carbon *E. Fermi Collected Papers* vol I, see note 5-40, pp 32–40

Frisch, Koch, and Halban had determined an upper limit for the absorption cross section in 1937 of 1×10^{-26} cm^2; Halban's own value in December of 1939 was 0.6×10^{-26} cm^2. Fermi and Anderson's value, as noted, was half as large, or 0.3×10^{-26} cm^2.

11-13   Hartcup and Allibone: see note 6-20, p 102

11-14   Kowarski, cited by Ermenc: see note 6-55, p 179

Halban H and Kowarski L 1940 Evidence for a potentially divergent nuclear reaction chain in a system, below the critical size, containing U and D, December 1940, OSRD, Report BR-4

11-15   Cockcroft to R H Fowler, 28 December 1940, Public Records Office, London, cited by Weart: see note 2-13, p 176

11-16   Urey to Conant, 2 April 1943, EOL, film 2248, roll 041, frame 091042. In a homogeneous reactor, the fuel, moderator, and coolant are mixed together, in a true solution or mechanically in a structure small compared to the distance traveled by neutrons between collisions; in a heterogeneous reactor, the three essential components are in separate structures of dimensions large compared with the neutron path length.

An example of the latter type is one in which rods of natural (essentially unseparated) uranium metal are surrounded by heavy-water moderator and coolant.

11-17   Briggs to Lawrence, 4 November 1940; Lawrence to Briggs, 8 November 1940, EOL, film 2248, roll 041, frames 090082, 090083. For the definition of 'thermal neutrons', see note 5-38.

11-18   Briggs to Lawrence, 1 March 1941, EOL, frame 090094

11-19   Urey to Conant, 2 April 1943, EOL, film 2248, roll 041, frame 091043

11-20   Briggs to Urey, 18 March 1941; cited by Weart: see note 2-13, p 181

11-21   Weart: see note 2-13, p 181

11-22   Chadwick J 'Report on the Experiments of Halban and Kowarski' 6 May 1941; cited by Weart: see note 2-13, p 182
Brown A 1997 *The Neutron and the Bomb: A Biography of Sir James Chadwick* (Oxford: Oxford University Press) p 209
Thomson and Cockcroft thought his report sufficiently important for him to bring it to the United States in person. Chadwick replied that it would be awkward, but was willing to go if they insisted. In the event, the Americans were happy to accept his word on the subject. Brown (*ibid.*); Gowing: see note 8-4, p 64.

11-23   In a memorandum of a conference between Fermi and Urey on 6–8 March 1943, the Halban–Kowarski paper of December 1940 (note 11-14), then more than two years old, was still described as 'most relevant for our discussion'. The following conclusions concerning the paper are reiterated: 'The comparison between the experiments of Halban and the theoretical expectation for the case of a hydrogeneous mixture indicates a very serious discrepancy.... Unfortunately neither theory nor experiment are free from serious objections, so that we are not in a position to reach a conclusion that may be trusted.... If new experiments should confirm Halban and his co-workers, a considerable shift in the emphasis on the heavy water program might result'. See 'The utilization of heavy hydrogen in nuclear chain reactions' in Fermi: note 5-36, pp 315–9, on p 319.

11-24   Urey 1942 Report on the general situation of the heavy water project, 25 July 1942, EOL, film 2248, roll 041, frame 091103, 2

11-25   Urey to Conant, 2 April 1943, EOL, film 2248, roll 041, frame 091045, 4

11-26   E V Murphree to Conant, 7 May 1942, EOL, frame 090330; Conant to Murphree, 15 May 1942; Gowing: see note 8-4, Chapter 4

11-27   Clark: see note 6-79, p 182

11-28   Conant to Urey, 7 April 1943, EOL, film 2248, roll 041, frame 091048, 2

Notes section 11.3

11-29   Hewlett and Anderson: see note 11-6, p 67

11-30   Excerpt of report to London, Summer 1942; Brun: see note 4-61, p 33

11-31   Until 1943 and the Allied sabotage operation, 3267 kg, or about 3 tons, of heavy water were produced at Vemork, compared with approximately 55 tons by the American effort (32 tons of D_2O by catalytic exchange and electrolysis at Trail and 23 tons by water distillation in the du Pont plants).

11-32   Maloney J O, Quinn G F and Ray H S 1955 *Production of Heavy Water: Part I, Commercial Production of Heavy Water*; Eidinoff M L, Joris G G, Taylor E, Taylor H S and Urey H C 1955 *Production of Heavy Water: Part II, Laboratory Studies for Separation Processes* (New York: McGraw-Hill)

11-33   Imperial Chemical Industries (ICI) was founded in 1926, as an amalgam of other large British chemical companies. It quickly rose to become a major competitor to I G Farben, the German concern only a year younger. ICI had been interested in producing small quantities of heavy water (5 grams/day at 30% purity) from c. 1934, and began looking into the possibilities of producing heavy water for wartime purposes from the end of 1941. During 1942–1943 they concentrated on alternatives to power-intensive electrolysis—mainly chemical exchange of hydrogen and water, and distillation. However, the alternatives soon appeared more and more unpromising, and the heavy-water effort was dropped.
Editor 1934 News and views *NA* **133** 604; Gowing: see note 8-4, p 229

11-34   Brun: see note 4-61, p 64

11-35   Experiments with the catalytic exchange process in early 1942 at Princeton and Louisiana showed that exchange with water would be too slow to produce heavy water in the quantities needed; however, Taylor found that using steam at atmospheric pressure instead of water was much more promising. Hewlett and Anderson: see note 11-6, p 67.
For more on the primary electrolysis and steam-hydrogen exchange plant at Trail, see pp 443–53 of Benedict and Pigford: note 10-17. The secondary Trail plant is described on pp 431–4.

11-36   Urey to Compton, 2 February 1942, EOL, film 2248, roll 041, frame 090951

11-37   'The development of the first chain-reacting pile' Appendix 4 of Segrè (1970): see note 5-26, pp 231–9

11-38   Wattenberg A 1993 The birth of the nuclear age *PT* **46** 44–51, on p 50

11-39   Monthly Progress Reports, December 1942 through May 1943, EOL, film 2248, roll 041

11-40   Conant to Urey, 7 April 1943, EOL, film 2248, roll 041, frame 091049, 3

11-41   Murphree to Conant, 20 April 1943; 1 July 1943, EOL, film 2248, roll 041, frames 090723 and 090755

11-42   Urey to Murphree, 3 September 1943, EOL, film 2248, roll 041, frame 091072

11-43   Summary of experimental research activities: P-9 experiments, Fermi: see note 5-36, p 303. 'P-9' was the American code word for heavy water.

11-44   Urey to Conant, 21 June 1943, EOL, film 2248, roll 041, frame 091060

11-45   Urey to Conant, 2 April 1943, EOL, film 2248, roll 041, frame 091045

11-46   Conant to Urey, 7 April 1943, EOL, film 2248, roll 041, frame 091047

11-47   Urey to Conant, 16 April 1943, EOL, film 2248, roll 041, frame 091050

11-48   Urey to Conant, 21 June 1943, EOL, film 2248, roll 041, frame 091060

11-49   Conant to Urey, 29 June 1943, EOL, film 2248, roll 041, frame 091053

11-50   Urey to Murphree, 3 September 1943, EOL, film 2248, roll 041, frame 091072

11-51   *E.I. Du Pont de Nemours & Co. Study Report: Production of HDO Water by Direct Method* 4 December 1942 (EOL, film 2248, roll 041, frame 091204 *et seq*). 'Water-distillation process for the isolation of heavy water as employed by E.I. Du Pont de Nemours & Co., Inc. at Morgantown, W.V.', Chapter 3 in *Production of Heavy Water: Part I*: note 11-32.

11-52   *Production of Heavy Water: Part I*: see note 11-32, p 55
For more on the water-distillation plants of the Manhattan District, see Benedict and Pigford: note 10-17, pp 416–20

11-53   Zinn W *et al* 'Design and description of ANL reactors (CP-3, CP-3' and CP-5)' *Int. Conf. on the Peaceful Uses of Atomic Energy (Geneva, August 1955)* conf. paper No P/861

Weinberg A M and Huffman J F 1956 U.S. research reactors *Progress in Nuclear Energy, Series II, Reactors* vol 1, ed R A Charpie, D J Hughes, D J Littler and M Trochers (London: Pergamon) pp 79–96

Notes section 12.1

12-1 Walker: see note 6-82, p 43

12-2 The decision by Army Ordnance was somewhat of a 'non-decision' in Walker's opinion. The research continued with its goals largely intact, and the Army continued its financial support of nuclear power. Walker: see note 6-82, p 51.

12-3 Wirtz see note 10-22. Abstracts of the papers presented on 26–28 February are also included, curiously, as a subset of abstracts in a three-volume 'Accession list of German documents pertaining to guided missiles' prepared by the Ordnance Research and Technology Translation Center, Ft Eustis, Virginia in 1946. The particular abstracts in question appear in Volume I, are mostly dated March 1942, and are numbered sequentially from GD610.29.1 to GD610.29.26, with the first item a summary of the rest. (Copies of the volumes are located at the Deutsches Museum, Munich.) For convenience, they are listed as GD-documents, numbered GD-1 to GD-26, in these notes. Wirtz's abstract is GD-15.

12-4 Suess H 1942 Die Gewinnung von Deuterium durch Austausch bei zwei verschiedenen Temperaturen, G-194 (23 March 1942); also GD-21 (28 April 1942)

12-5 Clusius K and Starke K, Zur Gewinnung von schwerem Wasser, G-134 (24 February 1942); Hydrogen-distillation process for the isolation of heavy water' Chapter 4 in *Production of Heavy Water: Part I*: see note 11-32. The method relies on the difference in vapor pressure between normal and heavy hydrogen. In the production of heavy water, the first stage of the concentration, to 5 or 10%, was always the most difficult, due to the large amounts of water or hydrogen needed. In the German version developed by Klaus Clusius with Linde's Ice Machinery Factory near Munich, about 5% enriched heavy hydrogen would be produced by fractional distillation of hydrogen, already liquefied, and the remaining concentration performed by normal electrolytic means. Among the disadvantages of the method for mass production are the need for invoking large-scale, low-temperature apparatus and the possibility of argon and nitrogen contamination of commercial hydrogen. For more on the method, see Benedict and Pigford: note 10-17, pp 410–3. As described in Chapter 11, the Americans settled on the fractional distillation of water instead for their wartime project. For more on this method, see Benedict and Pigford: note 10-17, pp 413–6. The Germans discarded this relatively simple method for the absurd reason that 'it could be easily copied abroad'. Irving: see note 9-5, p 114. See also GD-16 (March 1942).

12-6 See e.g., Himmler to Minister of Education Bernard Rust, 12 February 1942 (Goudsmit Papers, Box 27, Series IV)

12-7 The agenda of popular lectures was impressive, covering a broad range of issues, from pure nuclear physics to military and industrial applications of nuclear fission. In order of presentation, the lectures were as follows: Schumann, 'Nuclear physics as a weapon'; Hahn, 'The fission of the uranium nucleus'; Heisenberg, 'The theoretical basis for energy production from uranium fission'; Bothe, 'Result of the energy-producing designs examined up until this time'; Geiger, 'The necessity of general basic research'; Clusius, 'Enrichment of the uranium isotopes'; Harteck, 'The production of heavy water'; Esau, 'The extension of the research group "nuclear physics" through cooperation with other Reich departments and industry'. Heisenberg's presentation, in particular, was exemplary, explaining neutron behavior in uranium in terms of human population behavior, with the fission process analogous to 'child-bearing marriage' and neutron capture analogous to 'death'. On a more sober level, Heisenberg explained that the neutron 'mortality rate' could be eliminated by the use of uranium-235, potentially 'an explosive material of unimaginable force'. A big part of the work by Army Ordnance was devoted to the problem of production of this explosive, and, he warned the audience, 'American research also appears to be oriented in this direction, with considerable emphasis'. Element 94, a comparable explosive, was easier to produce (by chemical separation from uranium), but, he also warned, presupposed a self-sustaining chain reaction in uranium. (Germany's only, unsuccessful attempt to produce element 94 was by cyclotron irradiation in Paris.)

Heisenberg's talk is reprinted in translation as Appendix 1 in Bernstein: see note 9-27, pp 373–84.

12-8   American concern over expanded uranium research at the KWI can be traced to an article in *The New York Times* reporting on Debye's disclosure to an American journalist, upon his arrival in America, of the circumstances of his departure and requisitioning of his laboratory for 'other purposes'. 'Report of new sources of power' *The New York Times* 5 May 1940.

12-9   Walker: see note 6-82, p 95

12-10   Heisenberg and Wirtz 'Grossversuche zur Vorbereitung der Konstruktion eines Uranbrenner' in Bothe and Flügge: see note 10-49, pp 143–62, on p 152

12-11   Heisenberg and Wirtz: see note 12-10, pp 149–50

12-12   Brun: see note 4-61, p 31
'On memorandum of February 6th 1943 submitted by N. Stephansen on the production of D_2O of the Norsk Hydro', London, 24 February 1943, signed Sverre Hagen (code name for Jomar Brun). NHM, Box 10, Mappe SCS (copy consulted at NIAM).

12-13   Status of research problems in experimental nuclear physics, in Fermi: see note 5-36, pp 203–5

12-14   Heisenberg and Wirtz: see note 12-10, pp 150–3
Döpel and Heisenberg 1942 Die Neutronenvermehrung in einem D_2O-38-Metallschichtensystem, G-373 (March 1942). See also GD-1 (March 1942, dealing with LII and LIII) and GD-2 (March 1942, dealing with BIII and BIV).

12-15   Heisenberg and Wirtz: see note 12-10, pp 150–3
Döpel and Heisenberg 1942 Der experimentelle Nachweis der effektiven Neutronenvermehrung in einem Kugel-Schichten-System aus D_2O und Uran-Metall, G-136 (July 1942)
Fermi: see note 12-13

12-16   Heisenberg, in Ermenc: see note 6-55, pp 20–1; Kramish: see note 4-56, p 100; Powers: see note 7-36, p 154
What had happened was that water had leaked into the outer sphere of uranium metal, reacting with the uranium to produce an explosive hydrogen-gas mixture. The advantage of cast-metal uranium over powder was not lost on Döpel in his report of this incident, the second and more serious of two such mishaps involving pyrophorous

uranium in L-IV. Döpel 1942 Bericht über zwei Unfälle beim Umgang mit Uranmetall, G-135 (9 July 1942).

12-17 Speer A 1962 *Inside the Third Reich* (New York: Macmillan) p 226; Heisenberg, in Ermenc: see note 6-55, p 33

12-18 Speer: see note 12-17; Irving: see note 9-5, p 121

12-19 Only three days after the meeting with Speer, Harteck, somewhat disturbed over recent de-emphasis on uranium separation in light of L-IV and its forerunners, wrote Army Ordnance. There were two routes to nuclear power, he reminded officials. 'Machine type I', as he put it, was predicated on 5 tons of natural uranium, and the same amount of heavy water. 'Machine type II', one not emphasized in Germany thus far, but possibly pursued by the Americans, consisted of smaller amounts of enriched uranium metal and less heavy water or even normal water. Harteck to K Diebner, 26 June 1942; PHP, Acc. 85-22, Box 1, Folder 1.

12-20 Heisenberg, in Ermenc: see note 6-55, p 34

Notes section 12.2

12-21 Bøhn, Per Anders 1946 *IMI, Norsk Innsats i Kampen om Atomkraften* (Trondheim: Bruns Bokhandels Forlag)
Adamson and Klem: see note 7-25
Gallagher: see note 7-41
Haukelid: see note 7-38
Among the several popular films of the heavy-water operations of World War II, the best was no doubt the French–Norwegian 'Battle of Heavy Water' ('La Bataille de l'Eau Lourde', or 'Kampen om Tungtvannet'); screenplay by Jean Marin, supervised by Jean Duville, directed by Titus Vibe Müller, 1948, which starred many of the original participants, from Joliot-Curie to some of the actual saboteurs, in a reenactment of the events.

12-22 Kramish: see note 4-56, p 160

12-23 Kramish: see note 4-56, p 129

12-24 Powers: see note 7-36, pp 106–7
Kramish: see note 4-56, p 163

12-25 Wiggan R 1986 *Operation Freshman: The Rjukan Heavy Water Raid 1942* (London: William Kimber). Landing two 'Catalina' flying boats on a lake on the Vidda was discussed but dismissed early in the planning stage, since there was not much time; after the next moon, the lakes were frozen. *Ibid.*, p 25.

12-26 The leader of the group was Sub-Lieutenant Jens Anton Poulsson, whose parents, two sisters, and a younger brother were still living in Rjukan. Second in command was Sub-Lieutenant Knut Haugland, another Rjukan man, and radio operator, whose name is also associated with the *Kon-Tiki* expedition of 1947. Completing the party were Sergeants Claus Helberg, former classmate of Poulsson in Rjukan's public school, and Arne Kjelstrup, a plumber from Oslo but experienced at skiing in the Rjukan region of Telemark. Kjelstrup had fought his way out of Norway with a German bullet in the hip under heroic conditions in 1940. Poulsson J A 1982 *Aksjon Vemork: Vinterkrig på Hardangervidda* (Oslo: Gyldendal Norsk).

12-27 Two units were to be spared, to ensure a minimal amount of nitrogen fertilizer for Norwegian agriculture. Brun: see note 4-61, p 67.

12-28 The notes 'Some impressions from my work with Z...' (note 10-26) were prepared by Brun in London in connection with planning for *Freshman*.

'Some impressions...' is dated 30 November 1942. Brun's subsequent, fuller report on heavy-water production at Vemork, 'Memorandum submitted by Dr. Sverre Hagen...' (note 10-28), is dated 29 January 1943.

12-29 Haukelid: see note 7-38, p 59

12-30 The well known Norwegian meteorologist Sverre Petterssen, then a Lieutenant Colonel, was at Wick helping to prepare forecasts for the operation. On the afternoon of the 19th he expressed concern to his forecasting colleagues over the arctic air pattern building up for that particular night. Wiggan: see note 12-25, p 52.

12-31 Hitler's *Führerbefehl* of 18 October 1942 was issued in Norway on 26 October. The commandos executed at Trandum were buried there, those at Egersund in an unmarked grave at Ogna near Egersund. In 1945 the bodies at Trandum were exhumed and reburied with full British and Norwegian military honors in the military section of Vestre Gravlund Cemetery in Oslo. The fourteen bodies from Ogna were reburied in Egenes Cemetery on the outskirts of Stavanger. Their execution was the subject of Nuremberg exhibit US-545.

Fehlis, head of the Gestapo in Oslo, who signed the execution order for the men at Trandum, committed suicide a few days before advance parties of 1st British Airborne Division Royal Engineers landed in Norway in May 1945. Major-General Karl Marie von Behren, then commanding the *Ortskommandantur* at Stavanger, went on trial before a military court at Hamburg in 1946; he was found not guilty to the charge of killing 14 British POWs at Sletten. General Nikolaus von Falkenhorst appeared before a military court at Brunswick in 1946. He was sentenced to be shot, but the sentence was commuted to life imprisonment. He was released in 1953 as an act of clemency, suffering from ill health. He died in 1968.

Of those acting under General von Behren, three were tried before a military court in Oslo. *Stabsarzt* Werner Fritz Seeling, the doctor who administered the poison, was sentenced to death and shot at Akershus in January 1946. *Hauptscharführer* Eric Hoffmann, the NCO in charge of motor transport who shot one of the prisoners in the back of the head while being ordered down some cellar steps at Gestapo headquarters, was hanged in Germany later in the year. *Unterscharführer* Fritz Feuerlein, a brutish gaoler in the Gestapo prison in Stavanger, was sentenced to imprisonment for life, but subsequently handed over to the Russians for alleged atrocities against Russian POWs. Wiggan: see note 12-25, pp 136–58.

12-32 The fact that many of the commandos wore civilian ski-clothing under their battle dress, and some had no unit insignia, gave rise to a dispute between the *Wehrmacht* and the Gestapo. The Army claimed the British were soldiers in uniform and hence *they* had jurisdiction; the Gestapo that they were saboteurs in civilian dress, and hence technically spies. The Gestapo won the argument, still peeved about the Army's precipitous execution of the British in Egersund; Hitler's commando order did not preclude thorough interrogation before execution.

12-33 Among those arrested was Torstein Skinnarland, the only local aware of the Norwegian ski party in the area. He was shipped to Gestapo headquarters in Oslo, did not talk, and survived the war in Grini concentration camp.

12-34 On 15 December, SS-General Wilhelm Rediess, chief of the secret police in Norway, warned Berlin that there were 'several indications that the

British placed great importance on the destruction of Vemork's heavy water installation'. Gallagher: see note 7-41, p 63; Irving: see note 9-5, pp 141, 312, citing General Rediess's Oslo Headquarter's *Meldungen aus Norwegen Nr. 49*, dated Oslo, 15 December 1942.

General Rediess's full title, *Höherer SS-und Polizeiführer beim Reichskommissar für die besetzten Norwegischen Gebiete, SS-Obergruppenführer und General der Polizei* may appear somewhat comic, but his organization was hardly a laughing matter for Milorg and other Norwegian Resistance elements.

12-35 Brun, 'Memorandum submitted by Dr. Sverre Hagen . . . ': see note 10-28, p 15

Notes section 12.3

12-36 Brun: see note 4-61, p 70. The decision on whether or not the attack should be repeated rested with R V Jones, Eric Welsh's boss in the Intelligence Service, according to Jones. After sleeping on the decision overnight, he told Welsh that another attempt should be made. He adds that 'I am unable to say whether my decision was the vital one that led to the next raid because I suspect that Welsh was beginning to use me, as he ultimately used others in more eminent positions, as a puppet'. Jones R V 1978 *Most Secret War* (London: Hamish Hamilton) p 308.

12-37 This famous company, attached to Special Forces, England, had been formed in Britain by Martin Linge (1894–1941), a well known Norwegian actor and army reserve officer. It was made up of Norwegians who found their way over to England by one way or another. Captain Linge himself won a legendary reputation among the Norwegian forces abroad. He fell in the much-publicized commando raid on Vaagsøy in 1941. The name 'Linge Company' dates from 1945; during the war it was simply 'Norwegian Independent Company No. One'.

12-38 Operational Instructions No 1, 15 December 1942; reprinted in Kraglund I and Moland A 1987 *Norge i Krig: Vol. 6, Hjemmefront* (Oslo: Aschehoug) p 208

12-39 Uniforms were not so much for their own protection; Hitler's proclamation of 19 October 1942 called for all commandos to be summarily shot, in uniform or not, as attested to by the *Freshman* aftermath. However, it was hoped that, in the event of capture, it would persuade the Germans that the men were not underground saboteurs, and thereby reduce chances of reprisals against civilians. Adamson and Klem: see note 7-25, p 157.

12-40 Operational Instructions: see note 12-38

12-41 Farm Hall, as the manor was called, received further fame as the place where Heisenberg, Diebner, Hahn, Harteck, Wirtz, and a few other German scientists were interned in 1945, as we shall relate in due course.

12-42 Ukerapport, Rjukan, 27 February 1993. NHAO, K03. 50. n. 2.1: Tungt vann, Hefte 11.

12-43 One reason why the Germans reacted so casually might have been that they were inured from frequent small explosions from burners on the open balconies of the electrolysis building, used for regaining deuterium from hydrogen gas by combustion. Gallagher: see note 7-41, p 147.

12-44 The armored shears have never been located, according to Per Pynten. Perhaps they remain rusting away somewhere along the south face of the gorge.

12-45   Tronstad, Dagbok 5, Dagbøker 1941–1945, RO, Privatarkiv 354
12-46   Haukelid: see note 7-38, p 121

Notes section 13.1

13-1   'The running of the plant was satisfactory up to the 28th February at 1:15 a.m., when the high-concentration plant was put out of action by sabotage'; *Månedsrapport for drift av Vannstoffabrikken Vemork*, February 1943, IMC, DJ 31-446.   Bjarne Nilssen, 'Rapport om sabotasje i vanstoffabrikken Vemork Söndag den 28. Febr. 1943', Rjukan, 6 March 1943, NIAM, Vemork; Preliminary report on 'Krigskade i vår vannstoffabrikk på Vemork, anlegg for tungt vann, i vannstoffabrikkens kjeller, Söndag den 28 Febr.' 10 March 1943, NHAO, K03. 50. n. 2.1: Tungt vann, Hefte 11.

13-2   Larsen's recollections, IMC, DJ 31-449. A set of ruined electrolytic cells are on exhibit at Norsk Industriarbeidermuseum, Vemork, though it is unlikely they are the original set (Per Pynten, private information, 1998).   An intact heavy-water cell is also on exhibit at Norges Hjemmefrontmuseum, Oslo, including reconstructed charges as placed in position during the *Gunnerside* action on 28 February.

13-3   Brun: see note 4-61, p 73; Anker Olsen: see note 4-50, p 417

13-4   Brun himself, whose parents had been interrogated following his disappearance, later expressed appreciation for Falkenhorst's decision that the operation be considered a purely military one, and that, accordingly, no reprisals should be taken against the local population. Gallagher: see note 7-41, p 163. Johansen, the night watchman, was after the war praised for his calm behavior during the attack, and later promoted to foreman. Brun: see note 4-61, p 73.

13-5   Gallagher: see note 7-41, pp 162–3

13-6   Anker Olsen: see note 4-50, p 417

13-7   DSO = Distinguished Service Order; MM = Military Medal; MC = Medal for Courage; OBE = Order of the British Empire.

13-8   Berkei F *et al* 1942 'Bericht über einem Würfelversuch mit Uranoxyd und Paraffin', G-125 (before 26 November 1942)
Heisenberg and Wirtz 'Grossversuche zur Vorbereitung der Konstruktion eines Uranbrenner' in Bothe and Flügge: see note 10-49, pp 148–9
Walker: see note 6-82, pp 95–6

13-9   Heisenberg to Baertel van der Waerden, cited by Powers: see note 7-36, p 323

13-10   Diebner *et al* 1943 'Bericht über einen Versuch mit Würfeln aus Uran-Metall und schwerem Eis' G-212 (July 1943)
Heisenberg and Wirtz 'Grossversuche zur Vorbereitung der Konstruktion eines Uranbrenner' in Bothe and Flügge: see note 10-49, pp 148–9
Walker: see note 6-82, pp 99–100

13-11   Irving: see note 9-5, p 172. 'Unterlagen betreffend SH.200' (note 10-5), PHP, Acc. 85-22, Box 1, Folder 1, for December 1942. Brun, in his comprehensive report to the British authorities, identified another Italian electrolysis plant at Nera Montoro (Tivoli), two plants of comparable capacity in France (at Lannemezan and Pierrefitte) and at least one in Sweden (at Ljunga), as well as lesser plants in Germany, Poland, Denmark, and Norway (Sandar Fabrikker in Sandefjord and De No Fa in Fredrikstad). Brun, 'Memorandum submitted by Dr. Sverre Hagen...': see note 10-28, pp 13–4

Notes section 13.2

13-12   Internal Hydro correspondence, March 1943, NHAO, K03. 50. n. 2.1: Tungt vann, Hefte 11.
13-13   Brun: see note 4-61, p 74
13-14   Anker Olsen: see note 4-50, p 417
13-15   The two-pronged raid attacked Herøya and the German U-boat installations in Trondheim harbor—the latter target calling for a 1900 mile round trip lasting 12 hours. Hammel E 1994 *America's Air War Against Germany in Europe and North Africa; Chronology 1942–1945* (Pacifica, CA: Pacifica) p 157
Jablonski E 1965 *Flying Fortresses: The Illustrated Biography of the B-17s and the Men Who Flew Them* (New York: Doubleday) p 177
13-16   Anker Olsen, sections on Herøya: see note 4-50. The raid on Herøya was also a blatant repudiation of SOE's own plans, which included an attack on the light-metal facility, but by means more sparing of plant workers and the surrounding community. To that end, Kjell Nielsen, the chief engineer at the plant, had kept FO4 and SOE informed about progress on its construction. As late as May 1943 a courier smuggled into Herøya had learned that the start of light-metal production could not be expected during 1943. Any sabotage, he was informed, would be most profitably carried out *after* the plant had become operational, with maximum German investment at stake. The attack in July thus came as a complete surprise.
13-17   Eriksen B 'Notat fra konferanse hos Regierungsrat Albrecht, 4 August, 1943', NHAO, K03. 50. n. 2.1: Tungt vann, Hefte 12.
13-18   Eriksen's arrest, on 23 August 1943, coincided with a German round-up of former Norwegian Army officers; thus the exact reason for his arrest remains unclear.
13-19   Brun: see note 4-61, p 77
13-20   Brun: see note 4-61, p 77
13-21   Major John M Bennett to his father, cited in Jablonski: see note 13-15, p 203
13-22   Major John M Bennett to his father, cited in Jablonski: see note 13-15, p 204
13-23   Ingebretsen, 'Report on bomber attack on Vemork, Tuesday November 16, 1943', dated 21 December 1943, cited by Brun: see note 4-61, p 78
13-24   'The bombing attack on Rjukan' in Anker Olsen: see note 4-50, pp 418–9
13-25   Brun: see note 4-61, p 79
13-26   Tronstad, Dagbok 7, Dagbøker 1941–1945, RO, Privatarkiv 354
13-27   Anker Olsen: see note 4-50, p 420
13-28   Anker Olsen: see note 4-50, p 420. Major Bennett, who led the 100th Bombardment Group over Vemork, described the electrolysis plant as being used by the Germans 'to manufacture high explosives'. Jablonski: see note 13-15, p 203.

Notes section 14.1

14-1   Irving: see note 9-5, p 200
14-2   Walker: see note 6-82, p 101
14-3   Diebner K *et al* 'Über die Neutronenvermehrung einer Anordnung aus Uranwürfeln und schwerem Wasser (G-111)', G-210

Heisenberg and Wirtz 'Grossversuche zur Vorbereitung der Konstruktion eines Uranbrenner' in Bothe and Flügge: see note 10-49, pp 151

14-4   Heisenberg's *S*-matrix theory was published in 1943.

14-5   Walker: see note 6-82, p 104

14-6   Heisenberg, cited by Ermenc: see note 6-55, p 14

14-7   Piekalkiewicz J 1985 *The Air War: 1939–1945* (Poole: Blandford) p 434
Webster, Sir Charles and Noble, Frankland 1961 'The Battle of Berlin, November 1943–March 1944' *The Strategic Air Offensive Against Germany 1939–1945, Vol II: Endeavour* Part 4 (London: Her Majesty's Stationery Office) pp 190–211

14-8   Heilbron J L 1986 *The Dilemmas of an Upright Man: Max Planck as a Spokesman for German Science* (Berkeley, CA: University of California Press) p 193

14-9   Max von Laue to Meitner, 20 February 1944, cited by Sime: see note 5-48, p 298

14-10   Max von Laue to Theodore H von Laue, 26 May 1945. Microfilm copy, SON, HS Standnummer 1976-20

14-11   Heisenberg and Wirtz 'Grossversuche zur Vorbereitung der Konstruktion eines Uranbrenner' in Bothe and Flügge: see note 10-49, pp 156–7

14-12   Bothe to Heisenberg, 14 September 1942, 30 July 1943, cited by Walker: see note 6-82, p 256

14-13   Bopp F *et al* 1945 Bericht über einen Versuch mit 1.5 to D_2O und U und 40 cm Kohlerückstreumantel. (B7), G-300, 3 January 1945
Heisenberg and Wirtz 'Grossversuche zur Vorbereitung der Konstruktion eines Uranbrenner' in Bothe and Flügge: see note 10-49, pp 155–7

14-14   Vögler to Heisenberg, 31 October 1944, cited by Walker: see note 6-82, p 151

14-15   Whiting C and the Editors of Time-Life Books 1982 *The Home Front: Germany* (New York: Time-Life Books) p 144, citing Bielenberg C 1968 *The Past is Myself: The Experiences of an Englishwoman in Wartime Germany* (London: Chatto and Windus)

14-16   Walker: see note 6-82, p 137

Notes section 14.2

14-17   Tronstad to Skinnarland, 29 January 1944
Brun: see note 4-61, p 84
Haukelid: see note 7-38, p 181

14-18   Internal correspondence, January 1944; NHAO, K03, 50. n. 2.1: Tungt vann, Hefte 12.

14-19   Adamson and Klem: see note 7-25, p 179. Hans Suess was among the visitors evaluating the possibility of selecting and securing the contents of the electrolyzers sufficiently enriched that it could be further processed in Germany. Suess: see note 10-23, p 39.

14-20   Skinnarland to London, 3 February 1944; Brun: see note 4-61, p 84; Gallagher: see note 7-41, p 209

14-21   Skinnarland to London, 6 February 1944; Brun: see note 4-61, p 84

14-22   Brun: see note 4-61, p 84. Tronstad, Dagbok 8, Dagbøker 1941–1945, RO, Privatarkiv 354. While Home Secretary and Minister of Home Security 1939–1940, Sir Anderson had supervised the development of the 'Anderson shelters' in which thousands of British families took refuge from German bombs.

14-23   Welsh to Tronstad, 8 February 1944; Brun: see note 4-61, p 84

14-24 Breakdown of the heavy water consignment is given in DJ 31-432; also in Haukelid: see note 7-38, pp 197–8
Irving gives a slightly different breakdown, and quotes 613.68 liters as the equivalent of 100% heavy water. Irving: see note 9-5, p 207.

14-25 A German Gottow Army physicist, Lieutenant Werner Czulius, had been placed in personal charge of the transport by Kurt Diebner. Himmler had even dispatched a special Air Group to operate Fieseler Fi156 Storchs from a landing strip near the north end of Lake Tinn. Irving: see note 9-5, p 206.

14-26 Skinnarland to London, 9 February 1944; Haukelid: see note 7-38, p 183; Bøhn: see note 12-21, p 62

14-27 London to Skinnarland, 10 February 1944; Gallagher: see note 7-41, p 215; Bøhn: see note 12-21, p 64

14-28 Skinnarland to London, 16 February 1944; Brun: see note 4-61, p 85

14-29 London to Skinnarland, 16 February 1944; Brun: see note 4-61, p 85; Gallagher: see note 7-41, p 211

14-30 Haukelid: see note 7-38, p 197; Irving: see note 9-5, p 205

14-31 Haukelid: see note 7-38, p 193

14-32 Haukelid: see note 7-38, p 194

14-33 Gallagher: see note 7-41, p 227

14-34 Accounts differ as to the exact loss of lives. Norsk Hydro's commemorative volume of 1955, as well as Jomar Brun and Per Bøhn in their respective accounts, gives 18 lost and 29 saved. Haukelid, too, reports 18 lost, but claims the Captain later reckoned the German loss alone to be over 20. Irving gives 26 lost, based on records of the German 20th Army in Norway. Anker Olsen: see note 4-50, p 421; Brun: see note 4-61, p 86; Bøhn: see note 12-21, p 74; Haukelid: see note 7-38, p 197; Irving: see note 9-5, p 315.

14-35 Brun: see note 4-61, p 86. Two disputes still surround the *Hydro* sinking. One has to do with the justification for the operation, the other with Tronstad's role in it. Some argue it was the decisive action that ended heavy-water production in the German cause once and for all, and with it any hope for a German self-sustaining pile before the war ended. In point of fact, the Germans did not give up on heavy water quite yet, as we shall see. Countering that argument, others contend that by the time of the ferry sinking, any hope for a German nuclear explosive was long gone, though this was far from obvious to those who authorized or carried out the attack. Walker: see note 6-82, p 138.
As for Tronstad's concurrence in the sinking, he assured Brun, shortly after the operation, that he had no choice in the matter. Had the Allies not struck the ferry, they would surely have struck the Vemork plant once again, or the rail transport somewhere else, with loss of innocent lives nearly as certain. Brun: see note 4-61, p 86.

14-36 Brun: see note 4-61, p 87

14-37 Irving: see note 9-5, p 212

14-38 Internal correspondence, 8 March 1944; NHAR, courtesy of Per Pynten

Notes section 14.3

14-39 Harteck P 'Die Planung einer Deuteriumgewinnung in Deutschland' pp 183–86 in Bothe and Flügge: see note 10-49
For more on dual-temperature deuterium processes, see Benedict and Pigford: note 10-17, pp 453–62

14-40 Irving: see note 9-5, p 115

14-41 Harteck, in Bernstein: see note 9-27, p 172

14-42 Hedwig Geib (Mrs Geib) to Mrs K F Bonhoeffer, 15 April 1952. Geib–Bonhoeffer correspondence, Abteilung III, Repositur 23 (MPG).

14-43 Harteck, in Ermenc: see note 6-55, p 110

14-44 The GS process was named after the Girdler Corporation at Dana, Indiana, where the first pilot plant based on it was operated. It was adopted as the method of choice in the United States in 1950, and in Canada in the mid-sixties to supply the CANDU (Canada Deuterium Uranium) power-reactor program. It was also adopted in India. 1965 *Proc. Third Int. Conf. on the Peaceful Uses of Atomic Energy (Geneva, 31 August–9 September 1964) vol 12: Nuclear Fuels—III. Raw Materials* (New York: United Nations).

Rae H K (ed) 1978 Selecting heavy water processes *Separation of Hydrogen Isotopes* (Washington, DC: American Chemical Society) pp 1–26

14-45 Harteck 'Die Planung einer Deuteriumgewinnung in Deutschland' in Bothe and Flügge: see note 10-49, p 186

14-46 Nøkleby, Berit *Holdningskamp* vol 4, p 137, in *Norge i Krig* (note 12-38).

14-47 Hassel O 1943 The cyclohexane problem *Tidsskr. Kjemi. Berves. Metall* **3** 32–4

With the German occupation of Norway, Hassel stopped publishing in German journals, which left him with relatively obscure Norwegian journals of very limited circulation. The 1943 paper was only abstracted abroad in 1945.

14-48 Kramish: see note 4-56, p 182

14-49 Rosbaud to Harteck, 14 January 1944, 8 February 944; Harteck to Rosbaud, 17 January 1944, 4 February 1944; cited by Walker: see note 6-82, p 140

14-50 Harteck 'Die Planung einer Deuteriumgewinnung in Deutschland' in Bothe and Flügge: see note 10-49, p 183

14-51 Gerlach W 'Bericht über die Arbeiten auf kernphysikalischem Gebiet vom 1. Febr. bis 31 März 1944, III. Schweres Wasser', written 30 May 1944; PHP, Acc. 85-22, Box 1, Folder 1

Walker: see note 6-82, p 144

14-52 Gerlach: see note 14-51. In addition to the two initiatives with Lüde and Linde, there still remained a possible column at Leuna, and a further concentration process still to be decided at Schering.

14-53 Walker: see note 6-82, p 145

14-54 One disquieting feature of that day's aerial operation was the first encounter with *Luftwaffe's* long-feared Me-163 jet fighters.

14-55 Item 15, 'Leuna', in *Extracts from a Notebook Marked 'Ab Juli 1944' Kept by Walther Gerlach*, with commentary interspersed by David Irving; PHP, Acc. 85-22, Box 1

Harteck, in Ermenc: see note 6-55, p 110

14-56 In effect, the decision had already been made for them at a higher level, by Goering's deputy, Dr Fritz Görnnert. When Gerlach paid a visit on him, probably the very day before the USAAF raid, Görnnert declared that a heavy-water plant at Leuna was out of the question, among other things due to the threat of air raids(!). Gerlach: see note 14-51.

14-57 Brun: see note 4-61, p 88

14-58 Gerlach: see note 14-51. Harteck to D Irving, 19 April 1966; PHP, Acc. 85-22, Box 1, Folder 1

14-59 Harteck to Irving, *ibid.*

Notes section 15.1

15-1   Two senior Canadian scientists on the project from the start were George C Laurence, radiologist with the National Research Council, and George M Volkoff of the University of British Columbia.

15-2   Reflected Kowarski in later years, 'I have always considered it a major tragedy of my life because [a stint at Chicago] didn't come about'. Ermenc: see note 6-55, p 190.
    As Goldschmidt observed dryly about Halban's offer to come to Chicago, Vannevar Bush 'ran no risk whatever [of a foreign team establishing itself in the United States]. He was practically certain of Halban's refusal, seeing that Halban was too attached to his position as group leader'. Goldschmidt: see note 7-10, p 145.

15-3   Ermenc: see note 6-55, p 172

15-4   When, after World War II, Joliot requested Cockcroft to locate the 185 kg of heavy water, 130 kg were recovered in Canada, and another 16.7 kg at Harwell, Britain's Atomic Energy Research Establishment. Adds Lunde in his wartime account of heavy water, the 'Norwegian' heavy water had long since been diluted with material from other sources. Lunde: see note 4-63, 4th installment. France's 'repayment' for the heavy water, borrowed from Norsk Hydro in 1940 free of charge, took the form of a 5 ton heavy-water order for Zoé, the first French reactor after the war.

15-5   Ermenc: see note 6-55, p 191

15-6   Eggleston W 1965 *Canada's Nuclear Story* (Toronto: Clarke, Irwin) p 107. A good account of the early history of nuclear research in Canada, Eggleston's work suffers from the lack of any scholarly apparatus—notes, references, all but the simplest form of bibliography.

15-7   Goldschmidt: see note 7-10, p 179. Halban happened to be in Chicago the very day Fermi's pile went critical. However, not being one of the limited circle of individuals crowding the squash court balcony on that historic occasion, he was not even informed of the event at the time. *Ibid.* p 153.

15-8   Goldschmidt 'From school to laboratory' Chapter 1 in *Atomic Rivals*: see note 7-10, pp 3–19

15-9   Gowing: see note 8-4, p 193

15-10   Goldschmidt 'Conant's letter' in *Atomic Rivals*: see note 7-10, pp 176–9 Eggleston: see note 15-6, p 64

15-11   Weart: see note 2-13, p 197. Any doubt Roosevelt may have harbored about the wisdom of a restrictive policy vanished when he became aware of an Anglo–Russian agreement on the exchange of new weapons, though the British point out that the agreement excluded such weapons as the atomic bomb still on the drawing board. Eggleston: see note 15-6, p 68.

15-12   Among other things, the Cambridge experiment of Halban and Kowarski was repeated at Montreal by the British physicist Fred W Fenning, who had worked with the two Frenchmen at the Cavendish Laboratory.

15-13   Weart: see note 2-13, p 202

15-14   The Combined Policy Committee was conceived as a bi-partite vehicle for sorting out exchange of information between the United States, the United Kingdom, and Canada, and as an attempt to restore Anglo–American collaboration.

15-15   Eggleston: see note 15-6, p 90

15-16   Eggleston: see note 15-6, p 92

15-17   Brown: see note 11-22, p 225

15-18   Brown: see note 11-22, p 250

15-19   Atomic Energy of Canada Limited 1997 *Canada Enters the Nuclear Age: A Technical History of Atomic Energy of Canada Limited* (Montreal: McGill–Queen's University Press) p 5

15-20   NRX, the third reactor to become operational outside the United States, went critical in July of 1947. (The second reactor was the graphite-moderated F-1 built by Kurchatov's team near Moscow.) Designed for some tens of megawatts, NRX began at two watts, and was gradually stepped up to 40 MW. In addition to producing plutonium, it was the world's first 'breeder' reactor, converting thorium pellets into U-233 under radiation bombardment.

Notes section 16.1

16-1   Groves L R 1962 *Now It Can Be Told* (New York: Harper and Row) p 191

16-2   Lieutenant Colonel Boris T Pash, a former school teacher turned Army intelligence officer, was the son of the senior Bishop of the Eastern Orthodox Church in North America. The young Pash made his reputation by his forceful interrogation of Robert Oppenheimer, and other staff members in Lawrence's Berkeley laboratory, about their Communist affiliations, using unorthodox security methods. Groves chose him to head *Alsos*, because 'his thorough competence and great drive had made a lasting impression on me'. Groves: see note 16-1, p 193.

16-3   Powers: see note 7-36, pp 300–1

16-4   Amaldi had inherited Fermi's chair in Rome, keeping Italian physics going as best he could during difficult times. Gian Carlo Wick, born in 1909, had obtained his doctorate at Turin, and worked with Born at Göttingen and Friedrich Hund at Leipzig, where he met Amaldi. He also got to know Heisenberg rather well. In 1932 Wick came to Rome, where he remained until Emilio Segrè arranged for his appointment as professor at Palermo.

16-5   Rome fell on 5 June 1944. The very same day Pash, back in Italy, met up with Amaldi and Wick in the city. A minor tug-of-war ensued over Amaldi between Pash and Morris ('Moe') Berg, baseball player, dabbler in physics, and more recently an agent of the OSS. Amaldi, it turned out, had done no further work on fission since very early in the war, when he and Wick had purposely decided to abandon such work for the duration of the war. The Germans had made no attempt to exploit Italian scientists for war research. All Amaldi could add was that he was sure 'the Germans were working on [fission]'. Wick was better informed. He had studied with Heisenberg at Leipzig, and the two exchanged letters during the war. Thus he could inform Berg that Heisenberg had succeeded Peter Debye at the KWI for Physics in Berlin, and even that, as of June 1944, Heisenberg's team was somewhere in southern Germany; just where, Wick did not know, or was not saying. Powers: see note 7-36, p 308.

16-6   Peierls: see note 8-7, p 168

16-7   Irving: see note 9-5, p 224. 'TA project: enemy intelligence. Excerpts from the joint Anglo–US report to the Chancellor of the Exchequer and Major General LR Groves, 28 Nov. 1944 (Reproduced from CAB 126/244)', Appendix 29, pp 931–44 in Hinsley F H 1989 *British Intelligence* *in the Second World War, its Influence on Strategy and Operations* vol 3, part II (London: Her Majesty's Stationery Office).

16-8   Goldberg S 1990 Goudsmit, Samuel Abraham *DSB* **17** (Supplement II) 362–68

16-9   Goudsmit: see note 9-25, p 15

16-10 Goudsmit: see note 9-25, p 18

16-11 Goldsmith: see note 2-18, p 109

16-12 Schoenbrun D 1980 *Soldiers of the Night: The Story of the French Resistance* (New York: New American Library) pp 448–9

16-13 Goudsmit: see note 9-25, p 71

16-14 Goudsmit: see note 9-25, p 96

Notes section 16.2

16-15 Cassidy D 1992 *Uncertainty: The Life and Science of Werner Heisenberg* (New York: Freeman) p 488

16-16 For the complex details of relocating to Hechingen under problematic circumstances, see the correspondence between Julius Hiby, entrusted with moving the KWI for Physics south, and his old friend Karl Wirtz, during July 1943–November 1944. Abteilung I, Repositur 34 (MPG).

16-17 Irving: see note 9-5, p 266

16-18 Irving: see note 9-5, p 267; Powers: see note 7-36, p 407

16-19 Irving: see note 9-5, p 268

16-20 Groves: see note 16-1, p 216

16-21 Winnacker K and Wirtz K 1979 *Nuclear Energy in Germany* Transl. from German by D Goodman (La Grange Park, IL: American Nuclear Society) p 26. This book deals mainly with post-WW II nuclear energy developments in Germany.

16-22 Heisenberg and Wirtz 'Grossversuche zur Vorbereitung der Konstruktion eines Uranbrenner' in Bothe and Flügge: see note 10-49, pp 158–61

16-23 Plotting the reciprocal neutron counting rate as a function of some parameter that controls the pile reactivity (e.g., fuel loading or disposition of the moderator) is a standard procedure for estimating the onset of criticality. Keepin G R 1965 Neutron multiplication measurements *Physics of Nuclear Kinetics* (Reading, MA: Addison Wesley) pp 223–35

Notes section 17.1

17-1   Goudsmit: see note 9-25, p 77
Pash B T 1970 *The Alsos Mission* (New York: Award) p 173

17-2   Goudsmit: see note 9-25, p 88

17-3   Irving: see note 9-5, p 275

17-4   Pash: see note 17-1, p 201

17-5   Irving: see note 9-5, p 282; Pash: see note 17-1, p 206

17-6   Goudsmit: see note 9-25, p 109

17-7   Bernstein: see note 9-27, p 49

17-8   Powers: see note 7-36, p 426

17-9   Goudsmit: see note 9-25, p 122

17-10 Harteck, in Ermenc: see note 6-55, p 125. 'When I was taken in custody by the Allied Military', recalled Harteck, 'I was very happy. I had no family cares and I was in best humor. I was in high spirits. All the dangers were over. Looking back I am sometimes astonished that I survived all the dangers of the Third Reich and the war'.

Notes section 17.2

17-11 Hauge J Chr 1985 *Frigjøringen* (Oslo: Gyldendal Norsk) p 9
Eriksen K E and Halvorsen T 1987 *Norge i Krig: Vol. 8, Frigjøring* (Oslo: Aschehoug) p 9
Gjelsvig T 1979 *Norwegian Resistance 1940–1945* (London: Hurst)

17-12 Arranged against these formidable forces on Norwegian soil were some 20 000 armed members of Milorg, the fighting arm of the Norwegian underground, subsequently called the Home Forces, plus a contingent of Linge Company soldiers. In addition, Red Army troops held liberated Norwegian territory far north in eastern Finnmark, but did not appear disposed to move on short notice west of the Tana River. Standing by in Sweden were 14 000 Norwegian police troops, trained to preserve order in the event of a German collapse. The Norwegian Brigade in Scotland numbered about 4300, and the Norwegian Navy abroad disposed of some 51 vessels and about 7500 men.

17-13 An organization within the overall rubric of the Norwegian Home Forces, for dealing with this problem, already existed: B-org., or *Bedriftsorganisasjonen*, charged with 'quiet sabotage' of plants important to the German war effort, and protection of same against German sabotage. *Norge i Krig: Vol. 6, Hjemmefront*: see note 12-38, p 229.

17-14 Brun: see note 4-61, p 104

17-15 *Sunshine* encompassed three operational areas in Telemark, under the respective designations 'Lamplight', 'Moonlight', and 'Starlight'. 'Moonlight' covered Rjukan, Tinn, and Rauland, and was headed by Jens Anton Poulsson and, after Tronstad fell in battle, Poulsson assumed overall control of *Sunshine* as well. Poulsson: see note 12-26, pp 115–6.

17-16 His insistence in this regard has been seen as a combination of an ingrained idealism and psychic pressure sooner or later affecting those who dispose of troops in mortal combat from the safety of an office desk—a wish to share in the action. Sørensen: see note 4-40, p 133. That the Allied High Command acceded in his proposal to participate in the operation was clearly a mistaken judgment on their part, yet retrospective reflections on this decision tend to conclude that it was inevitable. Sørensen: see note 4-40, p 132.

17-17 Hauge J Chr 1985 Milorg under Okkupasjonen 1940–1945, pamphlet, Norsk Hjemmefrontmuseum, Oslo

17-18 Tronstad to his wife, London, 27 August 1944, NHAR; Tronstad, Dagbok 10, Dagbøker 1941–1945, RO, Privatarkiv 354

17-19 Brun: see note 4-61, pp 107–8

17-20 Poulsson: see note 12-26, p 129; Bøhn: see note 12-21, p 91

17-21 Bøhn: see note 12-21, p 95

17-22 For B-org., see note 17-13

17-23 Poulsson: see note 12-26, pp 134–6; Bøhn: see note 12-21, p 96

17-24 Bøhn: see note 12-21, pp 103–10; Brun: see note 4-61, pp 108–11

17-25 Brun: see note 4-61, p 109

17-26 Bøhn: see note 12-21, p 108

17-27 Haukelid: see note 7-38, p 222

17-28 The tragedy at Syrebekkstøylen had a sad aftermath as well. When members of the Home Forces came to arrest Johans Lognvik, he was working in the field along with his 79-year-old father. Johans at first refused to surrender, and the old man drew a knife and came to his aid. One of the arresting party fired his pistol, aiming for the hand with the knife, but hit the old man in the stomach and killed him instead. In the subsequent trial of the sheriff and his brother, the sheriff received 5 years in custody, his brother 10 years minus 864 days already served.

17-29 Sørensen: see note 4-40, p 134

17-30 Brun: see note 4-61, p 113. Also after the war, friends and colleagues of Tronstad established a student fund in his memory. At his birthplace on the outskirts of Oslo, a monument stands in his honor in a central spot named after him. Streets in Rjukan and Trondheim also bear his name.

Notes section 18.1

18-1 Discussed on page 213.

18-2 Anker Olsen: see note 4-50, p 425

18-3 Holloway D 1994 *Stalin and the Bomb: The Soviet Union and Atomic Energy 1939–1956* (New Haven, CT: Yale University Press) p 109

18-4 Goudsmit: see note 9-25, p 125; Irving: see note 9-5, p 291
The bunker, long since filled in, was located at the foot of the still standing 'Turm der Blitze' (Lightning Tower), then housing a high-tension cascade generator and more recently housing the archives of the Free University of Berlin. Henning E and Kazemi M 1998 *Dahlem— Domain of Science: A Walking Tour of the Berlin Institutes of the Kaiser Wilhelm/Max Planck Society in the 'German Oxford'* (Munich: Max Planck Society). Then again, it was not scientists or equipment that the Soviets were primarily after, but nuclear materials—above all uranium. They realized full well the Americans had bombed the Auer Company at Oranienburg, destined for the Soviet zone north of Berlin, to deny them the Auer works. Some uranium stores were also captured at Stassfurt from under the nose of the Russians; however, considerable uranium stores fell to them at Stadtilm and Rhensberg. More important in the long run for the Soviet nuclear program was its confiscation of extensive uranium deposits in South-West Saxony in the future East Germany. Holloway: see note 18-3, pp 111–2.

18-5 Powers: see note 7-36, p 428

18-6 Groves: see note 16-1, 'The Germans hear the news', Chapter 24, p 303

18-7 Bernstein: see note 9-27
A completely unedited publication of the British version appeared as Frank, Sir Charles 1993 *Operation Epsilon: The Farm Hall Transcripts* (Bristol: Institute of Physics Publishing). Excerpts of Bernstein's book were published in Bernstein J and Cassidy D 1995 Bomb apologetics: Farm Hall, August 1945 *PT* **48** 32–6

18-8 Cassidy in Bernstein: see note 9-27, p xvi

18-9 Rittner in Bernstein: see note 9-27, p 126

18-10 Rittner in Bernstein: see note 9-27, p 129

18-11 Rittner in Bernstein: see note 9-27, p 130

18-12 Rittner in Bernstein: see note 9-27, p 131

18-13 Bernstein and Cassidy: see note 18-7, p 35

Notes section 18.2

18-14 Holloway: see note 18-3

18-15 Reed T and Kramish A 1996 Trinity at Dubna *PT* **49** 30–5, on p 32

18-16 Holloway: see note 18-3, pp 50–1. Georgii Flerov and Lev Russinov reported 3 ± 1 neutrons per fission, compared to Halban and Kowarski's estimate of 3.5 on the average; the correct figure is about 2.5. *Ibid.*

18-17   Zel'dovich Ia B and Khariton Iu B 1939 O tsepnom raspade urana pod deistviem medlennykh neitronov *Zhurnal Eksperimental'noi i Teoreticheskoi Fiziki* **1** 29–36. Holloway: see note 18-3, p 53.

18-18   Zel'dovich Ia B and Khariton Iu B 1940 Delenie i tsepnoi raspad urana *Uspekhi Fizicheskikh Nauk* **4** 353. Holloway: see note 18-3, p 64.

18-19   Kurchatov I V 1941 Delenie tiazhelykh iader *Uspekhi Fizicheskikh Nauk* **2** 159–69. Holloway: see note 18-3, p 65.

18-20   Holloway: see note 18-3, p 64

18-21   Kramish A 1959 *Atomic Energy in the Soviet Union* (Stanford: Stanford University Press) p 27. The use of Bamag-electrolyzers was foreseen for the plant. Brun, 'Memorandum submitted by Dr. Sverre Hagen . . .': see note 10-28, p 14. The industrial city of Chirchik was created in 1935 from several villages surrounding the construction site for the Chirchik hydroelectric power station and the electrochemical works powered by it.

18-22   Kramish: see note 18-21, p 30

18-23   The notebook was captured in April 1942 on the southern shore of Taganrog Bay on the Sea of Azow, and contained a list of materials needed for building an atomic bomb, plus calculations of the amount of energy released by a critical mass of U^{235}. Holloway: see note 18-3, p 85.

18-24   Golovin I N 1968 *I. V. Kurchatov: A Socialist–Realist Biography of the Soviet Nuclear Scientist* Transl. from Russian by W H Dougherty (Bloomington, IN: Selbstverlag) p 41

18-25   Holloway: see note 18-3, p 92

18-26   Holloway: see note 18-3, p 97

18-27   Kramish: see note 18-21, p 83

18-28   Holloway: see note 18-3, p 101. Washington was more generous with uranium compounds, but not the metal. Uranium oxide and uranium nitride were transshipped in 100 kg lots on several occasions, but whether it ended up in the hands of Soviet atomic scientists or metallurgists for steel alloy production is not so clear. *Ibid.*

18-29   Holloway: see note 18-3, p 189

18-30   'Fizicheskii pervyi uranovyi kotel', or, roughly, 'first physics uranium pile'.

18-31   Coleman S K 1990 Riken from 1945 to 1948: the reorganization of Japan's physical and chemical research institute under the American occupation *Technology and Culture* **31** 228–50

18-32   Heilbron and Seidel: see note 5-20, p 317

18-33   Shapley D 1978 Nuclear weapons history: Japan's wartime bomb projects revealed *Science* **199** 152–7, on p 153
Dover J W 1978 Science, society, and the Japanese atomic-bomb project during World War Two *Bulletin of Concerned Asian Scholars* **10** 41–54, on p 47
Dover J W 1993 'NI' and 'F': Japan's wartime atomic bomb research *Japan in War and Peace: Selected Essays* (New York: New Press) Chapter 3, pp 55–100, on p 74

18-34   Dover, 'Japan in War and Peace . . .': see note 18-33, p 76

18-35   Dover, 'Japan in War and Peace . . .': see note 18-33, p 77. NI (*Ni-go Kenkyu*), is understood to have been short for Nishina Project.

18-36   Dover, 'Japan in War and Peace . . .': see note 18-33, p 79

18-37   Dover, 'Japan in War and Peace . . .': see note 18-33, p 80

18-38   Dover, 'Japan in War and Peace . . .': see note 18-33, p 81. In F Project (*F-go Kenkyu*), the 'F' stood for 'fission'.

18-39   Shapley: see note 18-33, p 155

18-40   Wilcox R K 1995 *Japan's Secret War: Japan's Race Against Time to Build Its Own Atomic Bomb* (New York: Marlowe) p 66

18-41   Interestingly, in his 1943 report to British authorities, Jomar Brun warns that 'one had better take account ... that the Japanese, possessing at least one large electrolysis plant, may be able to produce a considerable quantity of D_2O'. Brun, 'Memorandum submitted by Dr. Sverre Hagen ...': see note 10-28, p 14.

18-42   Wilcox: see note 18-40, p 122. In overall command of the *Atomic Bomb Mission*, also known as 'Group Three', was Brig. Gen. T F Farrell, Groves' deputy.

18-43   Rogers J D 'The Conversion of Commander Rogers', unpublished, private communication from Thomas Northrop

Notes section 19.1

19-1   Guhnfeldt C 1998 USA konfiskerte norsk tungt vann i 1945 *Aftenposten*, 2 March 1998, p 4, col. 1

See also correspondence between S Hagen (alias J Brun) and Col. Wilson, SOE, and Commander Welsh, SIS, during spring–summer 1945 concerning Norsk Hydro apparatus spirited to Germany in 1944. In the event, the British authorities advised against taking exceptional action to locate the material which appeared to be 'widely distributed in Germany', some of it undoubtedly in Russian, some in American hands. NHAO, KO3. 50. n. 21: Tungt vann, Hefte 12.

19-2   'Giant black market with heavy water. Dangerous gang uncovered by Munich criminal police' *Süddeutsche Zeitung* Nr. 17M, 1948. Anker Olsen speaks of the lady in question possessing 'large quantities' of heavy water. Anker Olsen: see note 4-50, p 422.

19-3   Brun J 'Produksjon av tungt vann i Norge etter krigen' 4 August 1944; NHM, Box 10, Mappe SCS

19-4   'Joint Establishment Experimental Pile'

19-5   During WWII, the Dutch kept a large quantity of uranium oxide, diverted from the French stockpile, hidden from the Germans in several railroad freight cars. As a result of discussions between Gunnar Randers, co-designer of JEEP with Odd Dahl, H A Kramers, nominally in charge of the material, and John Cockcroft, England acquired the uranium oxide, in exchange for the modest amount of metallic uranium needed for the Kjeller reactor.

19-6   Anker Olsen: see note 4-50, p 527; Andersen K G and Yttri G 1997 *Et Forsøk Verdt: Forskning og Utvikling i Norsk Hydro Gjennom 90 År* (Oslo: Universitetsforlaget) p 155. In 1955, Hydro provided an additional 7 tons of heavy water for the second Norwegian reactor, at Halden.

19-7   Stokes R G 1988 *Divide and Prosper: The Heirs of I.G. Farben under Allied Authority 1945–1951* (Berkeley: University of California Press)

19-8   Andersen and Yttri: see note 19-6, p 191

19-9   Rae H K (ed) 1978 Selecting heavy water processes *Separation of Hydrogen Isotopes* (Washington, DC: American Chemical Society) Chapter 1, pp 1–26, on p 2

19-10   Egeland E 1992 *Bedriftsliv og Humanitet: Glimt av Hydros Kultur og Kunst* (Oslo: Labyrinth) p 36

19-11   Per Pynten, private information, 1998

19-12   ZEEP, it may be recalled, was essentially a zero-energy experimental device. Rae H K 'Heavy water' *Canada Enters the Nuclear Age . . .*: see note 15-19, Chapter 17; Waltham C 1993 A brief history of heavy water *Physics in Canada* **49** 81–6 and 'Addendum to "A brief history of heavy water" ' pp 356–7.

19-13   Rae, 'Heavy water': see note 19-12, p 333

19-14   Waltham, 'A brief history . . .': see note 19-12, p 84

19-15   For more on the water–H_2S exchange process, see Benedict and Pigford: note 10-17, p 369 and pp 459–62

19-16   Rae, 'Heavy water': see note 19-12, p 333

19-17   Pynten, Per 'Tungtvannets historie i korte trekk' unpublished

19-18   Norway is a case in point, starting with the controversial sale of 20 tons of heavy water from Norsk Hydro to Israel in 1959, ostensibly for the Dimona reactor, and more recently 28 tons of heavy water earmarked for European customers, but which ended up as a backbone of India's nuclear program. Njølstad O 1996 Tungtvannssaken i enda en ny tapning *Aftenposten*, 23 October 1996; Magnus J 1998 Atom-makt med norsk tungt-vann *Verdens Gang*, 13 May 1998.

19-19   Waltham, 'A brief history . . .': see note 19-12, pp 85–6; McDonald A 1998 Sudbury—new neutrino observatory *CERN Courier* **38** 1–4. In the longer run, deuterium's singular, large-scale importance as a moderator may be overshadowed by its promise as fuel for energy production by thermonuclear fusion. It has been estimated that the entire annual energy requirement for the US in the year 2020 could be supplied by the d–d reactions of Chapter 9 by the amount of deuterium contained in something like 5000 tons of D_2O. Katz J J 1992 'Heavy water' *McGraw-Hill Encyclopedia of Science and Technology* vol 8 (New York: McGraw-Hill) p 365

19-20   Brun: see note 4-61, p 97

19-21   Maloney: see note 11-32

19-22   Brun: see note 4-61, p 92

19-23   Walker: see note 6-82, p 169

19-24   Brun: see note 4-61, p 100

19-25   Walker: see note 6-82, p 163. If a single factor is to be blamed for thwarting German isotope separation, it was probably the failure of the Clusius–Dickel tube. *Ibid.*, p 173.

19-26   Walker: see note 6-82, p 59

19-27   See e.g., Irving: note 9-5; Walker: note 6-82; Powers: note 7-36; Goudsmit: note 9-25; Bernstein: note 9-27

19-28   Hopper T 1996 Recent books on Nazism and science *HSPBS* **27** (part 1) 163–76

Notes section 19.2

19-29   Hauge J Chr 1988 *Manuskriper* (Oslo: Tiden Norsk) p 136

19-30   Stokker K 1991 In the footsteps of sabotage *The Norseman* **31** 14–7

19-31   Payton G 1993 Verdig 50-års markering av tungtvannsaksjonene *Varden* 24 February 1993

19-32   Weart: see note 2-13, p 225. The uranium for Zoé came in part from a lot of 9 metric tons of sodium uranate, lent from the Union Minière in 1940 and which sat undisturbed in a freight car on a railway siding in Le Havre all during the war. *Ibid.*, p 224.

19-33   Weart: see note 2-13, p 206

19-34   Weart S R 1990 Kowarski, Lew *DSB* **17** (Supplement II) 494–5

19-35   Hans Suess to Robert R Reeves, Department of Chemistry, RPI, 14 May 1985; PHP, Acc. 85-22, Box 1, Folder 'Suess'

19-36   Kramish: see note 4-56, p 250

19-37   Kramish: see note 4-56, p 177

19-38   Hauge: see note 19-29, p 77.   Wilson sent copies of his short wartime memoir, 'The "heavy water" operations in Norway, 1942–1944', originally broadcast over the BBC, as a 'souvenir' to each of the members of the *'Grouse/Swallow'* and *'Gunnerside'* teams. Wilson, Col. J S 1945 'The "heavy water" operations in Norway, 1942–1944' (London: unpublished).

19-39   Kramish: see note 4-56, p 251

19-40   Branch: see note 4-18, p 21

19-41   Taylor to Bonhoeffer, 12 August 1946; Urey to Fritz Karson, Office of Military Government, 7 February 1947; Urey to Bonhoeffer, 5 April 1947; Bonhoeffer to Urey, 9 March 1947. Abteilung III, Repositur 23 (MPG).

19-42   According to Urey, the Moon was a primordial object formed under relatively cold conditions and later captured by the Earth. Tatarewicz J N 1970 Urey, Harold Clayton *DSB* **18** (Supplement II) pp 943–8, on p 945.

SELECT BIBLIOGRAPHY

The following bibliography lists many, but not all, books, journal articles, and other sources cited in the notes. Thus, bibliographical entries in the *Dictionary of Scientific Biography* are not included. Nor are the German Reports on Atomic Energy, commonly known as the 'G-reports'. Abbreviations are those used in the notes.

Adamson H C and Klem P 1964 *Blood on the Midnight Sun* (New York: Norton)

Amaldi E 1984 Neutron work in Rome in 1934–36 and the discovery of uranium fission *RSDS* **1** 1–24

Andersen K G and Yttri G 1997 *Et Forsøk Verdt: Forskning og Utvikling i Norsk Hydro Gjennom 90 År* (Oslo: Universitetsforlaget)

Anderson C D and Anderson H L 1983 Unraveling the particle content of cosmic rays *The Birth of Particle Physics* ed L M Brown and L Hoddeson (New York: Cambridge University Press)

Anderson H L and Fermi E 1965 Production and absorption of slow neutrons by carbon *E. Fermi, Collected Papers, Vol III, United States 1939–1954* (Chicago: University of Chicago Press)

Anderson H L, Fermi E and Hanstein H B 1939 Production of neutrons in uranium bombarded by neutrons *PR* **55** 797–8

Aston F W 1934 Discussion of heavy water *PRS* **144** 1–28

Badash L 1969 *Rutherford and Boltwood: Letters on Radioactivity* (New Haven, CT: Yale University Press)

Benedict M and Pigford T H 1957 *Nuclear Chemical Engineering* (New York: McGraw-Hill)

Bernstein B J 1987 Leo Szilard: giving peace a chance in the nuclear age *PT* **40** 40–7

Bernstein J 1996 *Hitler's Uranium Club: The Secret Recordings at Farm Hall* (Woodbury, NY: AIP)

Bernstein J and Cassidy D 1995 Bomb apologetics: Farm Hall, August 1945 *PT* **48** 32–6

Beyerchen A 1977 *Scientists Under Hitler: Politics and the Physics Community in the Third Reich* (New Haven, CT: Yale University Press)

Birge R T and Menzel D H 1931 The relative abundance of the oxygen isotopes and the basis of the atomic weight system *PR* **37** 1669

Birkeland K 1908 *The Norwegian Aurora Polaris Expedition 1902–1903* vol 1 (Christiania: Aschehoug)

Bøhn P A 1946 *IMI, Norsk Innsats i Kampen om Atomkraften* (Trondheim: Bruns Bokhandels)

Boorse H A and Motz L (ed) 1966 *The World of the Atom* (New York: Basic Books)

Bothe W and Becker H 1930 Kunstliche Erregung von Kern-γ-Strahlen *ZP* **66** 239–306

Bothe W and Flügge S (ed) 1953 *Naturforschung und Medizin in Deutschland 1939–1945, Band 14, Kernphysik und Kosmische Strahlen* Teil II (Weinheim: Chemie)

Bothe W and Jensen P 1944 Die Absorption thermischers Neutronen in Elektrographit *ZP* **122** 749–55

Branch G E K 1984 Gilbert Newton Lewis, 1875–1946 *Journal of Chemical Education* **61** 18–21

Brickwedde F G 1982 Harold Urey and the discovery of deuterium *PT* **34** 208–13

Brown A 1997 *The Neutron and the Bomb: A Biography of Sir James Chadwick* (Oxford: Oxford University Press)

Brun J 1985 *Brennpunkt Vemork: 1940–1945* (Oslo: Universitetsforlaget)

Brun J and Ræder M 1970 Tronstad, Leif Hans Larsen *Norsk Biografisk Leksikon, XVII* (Oslo: Aschehoug)

Calvin M 1984 Gilbert Newton Lewis: his influence on physical-organic chemistry at Berkeley *Journal of Chemical Education* **61** 14–8

Cassidy D 1992 *Uncertainty: The Life and Science of Werner Heisenberg* (New York: Freeman)

Chadwick J 1932 On the possible existence of the neutron *NA* **129** 312

—— 1932 The existence of a neutron *PRS* **136** 692–708

—— 1964 Some personal notes on the search for the neutron *Proc. 10th Int. Congress on the History of Science (Ithaca, 1962)* (Paris: Hermann)

Clark R W 1961 *The Birth of the Bomb* (New York: Horizon)

—— 1965 *Tizard* (Cambridge, MA: MIT Press)

Cockcroft J 1950 *The Development and Future of Nuclear Energy* (Oxford: Clarendon)

Coleman S K 1990 Riken from 1945 to 1948: the reorganization of Japan's Physical and Chemical Research Institute under the American occupation *Technology and Culture* **31** 228–50

Cornell T D 1988 Merle Antony Tuve: pioneer nuclear physicist *PT* **41** 57–64

Curie I *See* Joliot-Curie

Dahl P F 1997 *Flash of the Cathode Rays: A History of J J Thomson's Electron* (Bristol: Institute of Physics)

Devik O 1971 *Blant Fiskere, Forskere og Andre Folk* (Oslo: Aschehoug)

Dover J W 1978 Science, society, and the Japanese atomic-bomb project during World War Two *Bulletin of Concerned Asian Scholars* **10** 41–54

—— 1993 *Japan in War and Peace: Selected Essays* (New York: New Press)

Egeland A 1994 *Kristian Birkeland: Mennesket of Forskeren* (Oslo: Norges Banks Seddeltrykkeri)

Egeland A and Leer E 1989 Professor Kr. Birkeland: his life and work *IEEE Transactions on Plasma Science* **PS-14** 666–77

Egeland E 1992 *Bedriftsliv og Humanitet: Glimt av Hydros Kultur og Kunst* (Oslo: Labyrinth)

Eggleston W 1965 *Canada's Nuclear Story* (Toronto: Clarke Irwin)

Ermenc J J 1967 *Atomic Bomb Scientists: Memoirs, 1939–1945* (Westport, CT: Meckler)

Eve A S 1939 *Rutherford: Being the Life and Letters of the Rt Hon Lord Rutherford, OM* (New York: Cambridge University Press)

Eyde S 1939 *Mitt Liv og Mitt Livsverk* (Oslo: Gyldendal Norsk)

Feather N 1940 *Lord Rutherford* (Glasgow: Blackie)

Fermi E 1955 Physics at Columbia University: the genesis of the nuclear energy project *PT* **8** 282–6

—— 1962–1965 *Collected Papers* 2 vols, ed E Segrè *et al* (Chicago: University of Chicago Press)

Fermi L 1954 *Atoms in the Family: My Life with Enrico Fermi* (Chicago: University of Chicago Press)

Fjeldbu S 1980 *Et Lite Sted på Verdenskartet: Rjukan 1940–1950* (Oslo: Tiden Norsk)

Flügge S 1939 Kann der Energieinhalt der Atomkerne technisch nutzbar gemacht werden? *NW* **27** 402–10

Friedman R M 1995 Civilization and national honour: the rise of Norwegian geophysical and cosmic science *Making Sense of Space: The History of Norwegian Space Activities* ed J P Collett (Oslo: Scandinavian University Press) Chapter 1

Frisch O R 1967 The discovery of fission: how it all began *PT* **20** 272–7

—— 1979 Lew Kowarski *NA* **282** 541

—— 1979 *What Little I Remember* (Cambridge: Cambridge University Press)

Gallagher T 1975 *Assault in Norway: Sabotaging the Nazi Nuclear Bomb* (New York: Harcourt Brace Jovanovich)

Gamow G 1966 *Thirty Years that Shook Physics: The Story of Quantum Theory* (New York: Doubleday)

Gentner W 1980 *Discussions with Frédéric Joliot-Curie in Occupied Paris During the Years 1940–1942* (Heidelberg: Max-Planck-Institut für Kernphysik)

Gjelsvig T 1979 *Norwegian Resistance 1940–1945* (London: Hurst)

Goldschmidt B 1990 *Atomic Rivals* Transl. G M Tenner (New Brunswick: Rutgers University Press)

Goldsmith M 1976 *Frédéric Joliot-Curie: A Biography* (London: Lawrence and Wishart)

Golovin I N 1968 *I.V. Kurchatov: A Socialist–Realist Biography of the Soviet Nuclear Scientist* Transl. W H Dougherty (Bloomington, IN: Selbstverlag)

Gosling F G 1994 *The Manhattan Project: Making the Atomic Bomb* (Washington, DC: US Department of Energy Report DOE/HR-0096)

Goudsmit S A 1986 *Alsos* (Los Angeles: Tomash)

Gowing M 1965 *Britain and Atomic Energy, 1939–1945* (New York: St Martin's)

Graetzer H G and Anderson D L 1971 *The Discovery of Nuclear Fission: A Documentary History* (New York: Van Nostrand Reinhold)

Groves L R 1962 *Now It Can Be Told: The Story of the Manhattan Project* (New York: Harper and Row)

Hahn O 1962 *A Scientific Autobiography* Transl. and edited by W Ley (New York: Scribner)

—— 1968 *My Life* Transl. E Kaiser and E Wilkins (New York: Herder and Herder)

Hahn O and Strassmann F 1938 Über die Entstehung von Radiumisotopen aus Uran beim Bestrahlen mit schnellen und verlangtsamten Neutronen *Naturwissenschaften* **26** 755–6

—— 1939 Über den Nachweis und das Verhalten der bei Bestrahlung des Urans mittels Neutronen entstehenden Erdalkalimetalle *Naturwissenschaften* **27** 11–5

Halban H von, Joliot F and Kowarski L 1939 Liberation of neutrons in the nuclear explosion of uranium *NA* **143** 470–1

Halban H von and Perrin F 1939 Mise en évidence d'une réaction nucléaire en chaine au sein d'une masse uranifére *JPR* **10** 428–9

Hambro C J 1940 *I Saw it Happen in Norway* (New York: Appleton-Century)

Hammel E 1994 *America's Air War Against Germany in Europe and North Africa; Chronology 1942–1945* (Pacifica, CA: Pacifica)

Harkins W D 1920 The nuclei of atoms and the new periodic system *PR* **15** 73–94

Hartcup G and Allibone T E 1984 *Cockcroft and the Atom* (Bristol: Adam Hilger)

Hauge J C 1985 *Frigjøringen* (Oslo: Gyldendal Norsk)

—— 1988 *Manuskripter* (Oslo: Tiden Norsk)

Haukelid K 1989 *Skis Against the Atom* (Minot, ND: North American Heritage Press)

Hayes P 1987 *Industry and Ideology: I G Farben in the Nazi Era* (Cambridge: Cambridge University Press)

Heilbron J L 1986 *The Dilemmas of an Upright Man: Max Planck as a Spokesman for German Science* (Berkeley: University of California Press)

Heilbron J L and Seidel R W 1989 *Lawrence and His Laboratory: A History of the Lawrence Berkeley Laboratory* vol 1 (Berkeley: University of California Press)

Henning H and Kazemi M 1998 *Dahlem—Domain of Science: A Walking Tour of the Berlin Institutes of the Kaiser Wilhelm/Max Planck Society in the 'German Oxford'* (Munich: Max Planck Society)

Hewlett R G and Anderson O E Jr 1962 *The New World, 1939–1946. A History of the United States Atomic Energy Commission* vol I (University Park, PA: Pennsylvania State University Press)

Hinsley F H 1979–1984 *British Intelligence in the Second World War, its Influence on Strategy and Operations* 3 vols (London: Her Majesty's Stationery Office)

Holloway D 1994 *Stalin and the Bomb: The Soviet Union and Atomic Energy 1939–1956* (New Haven, CT: Yale University Press)

Howarth D 1955 *The Shetland Bus* (London: Fontana–Collins)

Hutton R S 1964 *Recollections of a Technologist* (London: Pitman)

Irving D 1967 *The German Atomic Bomb: The History of Nuclear Research in Nazi Germany* (New York: Simon and Schuster). Also published as *The Virus House: Germany's Atomic Research and Allied Countermeasures* (London: William Kimber)

Jablonski E 1965 *Flying Fortresses: The Illustrated Biography of the B-17s and the Men Who Flew Them* (New York: Doubleday)

Jauncey G E M 1946 The early years of radioactivity *AJP* **14** 226–41

Jensen T B 1995 *Norsk Lokalhistorisk Krigshistorie om 2. Verdenskrig 1940–1945: En Bibliographi* (Oslo: Universitetsbiblioteket)

Joliot-Curie F 1966 Chemical evidence of the transmutation of elements *Nobel Lectures, Including Presentation Speeches and Laureate's Biographies: Chemistry 1922–1941* (Amsterdam: Elsevier) pp 369–73

Joliot-Curie F, Halban H von Jr and Kowarski L 1939 Number of neutrons liberated in the nuclear explosion of uranium *NA* **143** 680

Joliot-Curie F and I 1934 Un noveau type de radioactivité *CR* **198** 254

—— 1934 Artificial production of a new type of radioelement *NA* **133** 201

—— 1961 *Oeuvres Scientifiques Complètes* (Paris: Presses Universitaires)

Joliot-Curie I 1951 La découverte de la radioactivité artificielle *Atomes* No 58, 9–12

Joliot-Curie I and F 1932 Émission de protons de grande vitesse par les substances hydrogénées sous l'influence des rayons gamma très pénétrants *CR* **194** 273

Joliot-Curie I, Halban H von and Preiswerk P 1935 Sur la création artficielle d'élémente appartenant à une familie radioactive inconnue, lors de l'irradiation du thorium par les neutrons *JPR* **6** 361

Joliot-Curie I and Savitch P 1937 Sur les radioéléments formés dans l'uranium irradié par les neutrons *JPR* **8** 385–7

—— 1938 Sur la nature du radioélément de périod 3.5 heures formé dans l'uranium irradié par les neutrons *CR* **206** 1643

Jones R V 1967 Thicker than heavy water *Chemistry and Industry* August 26, 1419–24

—— 1978 *Most Secret War* (London: Hamish Hamilton)

Keepin G R 1965 *Physics of Nuclear Kinetics* (Reading, MA: Addison-Wesley)

Kraft F 1981 *Im Schatten der Sensation: Leben und Wirken von Fritz Strassmann* (Weinheim: Chemie)

Kraft F 1983 Internal and external conditions for the discovery of nuclear fission by the Berlin team *Otto Hahn and the Rise of Nuclear Physics* ed W R Shea (Dordrecht: Reidel) pp 135–65

Kraglund I and Moland A 1987 *Norge i Krig: vol 6, Hjemmefront* (Oslo: Aschehoug)

Kramish A 1959 *Atomic Energy in the Soviet Union* (Stanford, CA: Stanford University Press)

—— 1986 *The Griffin* (Boston: Houghton Mifflin)

Lewis G N 1934 The biology of heavy water *Science* **79** 151–3

Lewis G N and Macdonald R T 1933 Concentration of H^2 isotope *The Journal of Chemical Physics* **1** 341–4

Libæk I and Stenersen Ø 1991 *History of Norway from the Ice Age to the Oil Age* (Oslo: Grøndahl and Søn)

Lunde E 1969 Av tungtvannets saga *Aftenposten* 5–10 April

Maloney J O *et al* 1955 *Production of Heavy Water* (New York: McGraw-Hill)

Marsden E 1962 Rutherford at Manchester *Rutherford at Manchester* ed J B Birks (London: Heywood)

Marsden E and Lantsberry W C 1915 The passage of α-particles through hydrogen, II *PM* **30** 240–3

Massey H and Feather N 1976 James Chadwick *Biographical Memoirs of Fellows of the Royal Society* **22** 11–70

McGrayne S B 1993 *Nobel Prize Women in Science: Their Lives, Struggles, and Momentous Discoveries* (Secaucus, NJ: Carol)

Mehra J 1975 *The Solvay Conferences on Physics: Aspects of the Development of Physics Since 1911* (Dordrecht: Reidel)

Meitner L and Frisch O R 1939 Disintegration of uranium by neutrons: a new type of nuclear reaction *NA* **143** 239

—— 1939 Products of the fission of the uranium nucleus *NA* **143** 471–2

Moulton J L 1967 *A Study of Warfare in Three Dimensions: The Norwegian Campaign of 1940* (Athens, OH: Ohio University Press)

Murphy G M (ed) 1955 *Production of Heavy Water* (New York: McGraw-Hill)

—— 1964 The discovery of deuterium *Isotopic and Cosmic Chemistry, dedicated to Harold C. Urey on His Seventieth Birthday April 29, 1963* ed H Craig, S L Miller and G J Wasserburg (Amsterdam: North-Holland)

Oliphant M L E, Harteck P and Lord Rutherford 1934 Transmutation effects observed with heavy hydrogen *PRS* **144** 692–703

Olsen K A 1955 *Norsk Hydro Gjennom 50 År: Et Eventyr fra Realitetens Verden* (Oslo: Norsk Hydro-Elektrisk Kvælstofaktieselskab)

Pais A 1991 *Niels Bohr's Times, in Physics, Philosophy, and Policy* (Oxford: Clarendon)

Pash B T 1970 *The Los Alamos Mission* (New York: Award)

Peierls R 1985 *Bird of Paradise: Recollections of a Physicist* (Princeton, NJ: Princeton University Press)

Piekalkiewicz J 1985 *The Air War: 1939–1945* (Poole: Blandford)

Poulsson J A 1982 *Aksjon Vemork: Vinterkrig på Hardangervidda* (Oslo: Gyldendal Norsk)

Powers T 1993 *Heisenberg's War: The Secret History of the German Bomb* (New York: Alfred Knopf)

Rae H K (ed) 1978 *Separation of Hydrogen Isotopes* (Washington, DC: American Chemical Society)

Ræder M 1946 'Leif Tronstad', Forhandlinger *Det Kongelige Norske Videnskabers Selskab* **26** 95–106

Reed T and Kramish A 1996 Trinity at Dubna *PT* **49** 30–5

Rhodes R 1986 *The Making of the Atomic Bomb* (New York: Simon and Schuster)

Rideal E K and Evans U R 1945 Prof. Leif Tronstad, O.B.E. *NA* **156** 74

Rozental S (ed) 1967 *Niels Bohr: His Life and Work as Seen by His Friends and Colleagues* (Amsterdam: North-Holland)

Rutherford E 1919 Collision of α particles with light atoms. IV. An anomalous effect in nitrogen *PM* **37** 581–7

—— 1920 Nuclear constitution of atoms *PRS* **97** 374–400

—— 1934 Discussion of heavy water *PRS* **144** 1–28

—— 1937 The search for the isotopes of hydrogen and helium of mass 3 *NA* **140** 303–5

Særland S 1923 Birkeland *Norsk Biografisk Leksikon* vol I (Kristiania: Aschehoug)

Sæter E and S 1995 *XU—i Hemmeleg Teneste 1940–1945* (Oslo: Det Norske Samlaget)

Schoenbrun D 1980 *Soldiers of the Night: The Story of the French Resistance* (New York: New American Library)

Segrè E 1970 *Enrico Fermi: Physicist* (Chicago: University of Chicago Press)

—— 1980 *From X-Rays to Quarks: Modern Physicists and Their Discoveries* (San Francisco: Freeman)

Shapley D 1978 Nuclear weapons history: Japan's wartime bomb projects revealed *Science* **199** 152–7

Shea W R (ed) 1983 *Otto Hahn and the Rise of Nuclear Physics* (Dordrecht: Reidel)

Sime R L 1996 *Lise Meitner: A Life in Physics* (Berkeley: University of California Press)

Soddy F 1906 The internal energy of elements *Journal of the Proceedings of the Institution of Electrical Engineers, Glasgow* **37** 7

—— 1915 Advances in the study of radio-active bodies *Engineering* **99** 604

Sørensen N A 1946 Minnetale over Professor Leif Tronstad *Årbok, Det Norske Videnskaps-Akademi* pp 113–34

Speer A 1962 *Inside the Third Reich* (New York: Macmillan)

Stokes R G 1988 *Divide and Prosper: The Heirs of I.G. Farben under Allied Authority* (Berkeley: University of California Press)

Stokker K 1991 In the footsteps of sabotage *The Norseman* **31** 14–7

Stuewer R H 1985 Bringing the news of fission to America *PT* **38** 49–56

Suess H 1968 Virus House: comments and reminiscences *Bulletin of the Atomic Scientists* **24** 36–9

Szilard L 1972 *The Collected Works: Scientific Papers* (Cambridge: MIT Press)

Trimble R F 1993 Odd Hassel *Nobel Laureates in Chemistry 1901–1992* ed L K James (Washington, DC: American Chemical Society and The Chemical Heritage Foundation) pp 514–9

Tronstad L 1934 Production of large quantities of heavy water *NA* **133** 872

Tronstad L and Brun J 1934 Über die Elektrolytische Anreicherung des Schweren Wasserstoffisotops im Wasser *Zeitschrift für Elektrochemie und angewandte physikalische Chemie* **40** 556–8

Urey H, Brickwedde F G and Murphy G M 1932 A hydrogen isotope of mass 2 *PR* **39** 164–5

—— 1932 A hydrogen isotope of mass 2 and its concentration *PR* **40** 1–15

Walker M 1989 *German National Socialism and the Quest for Nuclear Power 1939–1949* (Cambridge: Cambridge University Press)

—— 1990 Heisenberg, Goudsmit and the German Atomic Bomb *PT* **43** 10–8

Waltham C 1993 A brief history of heavy water *Physics in Canada* **49** 81–6 and 356–7

Washburn E W and Urey H C 1932 Concentration of the H^2 isotope of hydrogen by the fractional electrolysis of water *Proceedings of the National Academy of Sciences of the USA* **18** 496

Wattenberg A 1993 The birth of the nuclear age *PT* **46** 44–51

Weart S R 1976 Scientists with a secret *PT* **29** 23–30

—— 1979 *Scientists in Power* (Cambridge, MA: Harvard University Press)

—— 1983 The discovery of fission and a nuclear physics paradigm *Otto Hahn and the Rise of Nuclear Physics* ed W R Shea (Dordrecht: Reidel) pp 91–133

Weart S R and Szilard G W (ed) 1978 *Leo Szilard: His Version of the Facts* (Cambridge, MA: MIT Press)

Wells H G 1914 *The World Set Free: A Story of Mankind* (New York: Duntton)

Wheeler J A 1967 Mechanism of fission *PT* **20** 278–81

Whiting C and the Editors of Time-Life Books 1982 *The Home Front: Germany* (New York: Time-Life Books)

Wiggan R 1986 *Operation Freshman: The Rjukan Heavy Water Raid 1942* (London: William Kimber)

Wilcox R K 1995 *Japan's Secret War: Japan's Race Against Time to Build its Own Atomic Bomb* (New York: Marlowe)

Wilson D 1983 *Rutherford: Simple Genius* (Cambridge, MA: MIT Press)

Wilson J S 1945 *The 'Heavy Water' Operations in Norway, 1942–1945* (London: unpublished)

Winnacker K and Wirtz K 1979 *Nuclear Energy in Germany* Transl. D Goodman (La Grange Park, IL: American Nuclear Society)

Name index

Note: Years of birth and death (if known) are recorded after each name. Endnotes are indicated by a suffix 'n' and a note number.

Abrahamsen, Jennie Karoline (wife of Thor) [1899–1943], 216, n.10-73
Abrahamsen, Thor Nikolai [1899–1943], 168, 216–7, n.10-73
Akers, Sir Wallace Alan [1888–1954], 242, 243
Alikhanov, Abram Isaakovich [1904–1970], 278, 279
Allier, Lieutenant Jacques [1928–], 103
  on Aubert, 107
  and Bjarne Eriksen, 104
  and Cockcroft, 119
  and evacuation to Bordeaux, 127–8
  and Foley, 109
  London mission, 119–20
  meeting with Aubert, 106–7
  meeting with Dautry and Joliot, 104–5
  Oslo mission, 104, 105–9
  postwar activities, 288, 289
  and Reynaud, 124
  under German occupation, 145–6, n.9-57
Allison, Samuel King [1900–1965], 185
Alvarez, Luis Walter [1911–1989]
  confirms uranium fission, 82
  fails to find neutrons from fission, 87
Amaldi, Edoardo [1908–1989], 57, 60
  and *Alsos*, 248, n.16-5
  neutron bombardment studies, 60–3
  professor in Rome, 63, n.16-4
Anderson, Carl David [1905–1991]
  background, 50
  the positron, 50–1

Anderson, Herbert Lawrence [1914–1988], 88, 93
  and Chicago pile, 180
  and Columbia pile experiments, 173–4, n.11-12
  detects fission fragments, 81
Anderson, Sir John [1882–1958], 225, 239, 243
Appleton, Sir Edward Victor [1892–1965], 240
Ardenne, Manfred von [1907–], 271
Arnold, William Archibald [1904–], 80
Aston, Francis William [1877–1945]
  atomic weight of hydrogen, 22, 26
  on nomenclature for isotopic hydrogen, n.4-16
  fails to detect tritium, 134
  mass spectrograph, 134
Aubert, Axel [1873–1943], 46, 159
  Allier on, 107
  Bjarne Eriksen on, 106–7
  and Brun, 105, 155–6, 169, n.7-19
  on Brun's departure for England, 169
  death, 169
  on German uranium project, 150, 155–6, 217
  as Hydro's General Director, 46, 155–6, 212–3, 217
  meeting with Allier, 104, 106–7
  meeting with Mossé, 109
  perceived German sympathies, 106–7
  rejects German heavy-water request, 105
  on therapeutic uses of heavy water, 46

treatment by Germans, n.7-16
and tritium experiment, 134
on Vemork bombing, 217
Auger, Pierre Victor [1899–], 241–2

Bache-Wiig, Professor Jens
[1880–1965], 169
Bachke, Fredrik (Spec Mission Comd
Resist), 167–8
Bagge, Erich Rudolf [1912–], 191, 262
detained by *Alsos*, 262, 263, 271
and Diebner, 136–7
and Haigerloch pile, 256
and Heisenberg, 136–7
and Schumann conference of 1939,
137
Bahr-Berguis, Eva von [1874–1962], 70,
77
Baker, Henry [1698–1774], n.3-2
Barnes, T Cunliffe, 47
Barr, F T (Standard Oil), 179
Barth, Thomas Fredrik (Tom) Weiby
[1899–1971], 164
Barton, Sir Derek Harold Richard
[1918–], 42
Basche, H (Oberregierungsrat Dr), 137
Becker, Herbert, 15, 16
Becquerel, Antoine-Henri [1852–1908]:
radioactivity discovered by, 7
Behren, Major-General Karl Marie
von, n.12-31
Belitz, Herr (I G Farben) 153
Bell, Ronald Percy [1907–], n.4-17
Bender, Major, 145
Benedicks, Professor Carl Axel
Frederik [1875–1958], 32
Bennett, Major John M, 214–5, n.13-28
Berg, Morris (Moe) [1902–1972], n.16-5
'Berg, Mr' [alias name for Fredrik
Bachke], 167–8
Berkei, Friedrich, 163, 260
Bernal, John Desmond [1901–1971],
n.4-17
Bertrand-Vigne (French Military
Attaché), 114
Bethe, Hans Albrecht [1906–], 80
Bey, van der, 236
Bichelonne, Jean [?–1944], 104, 128,
n.8-25
Bielenberg, Christabel, 223
Biquard, Pierre, 55
Birge, Raymond Thayer [1887–1980]
failure to discover deuterium, n.4-12

suggests isotopes of hydrogen, 22,
23, 24, 26
Birkeland, Kristian Olaf Bernhard
[1867–1917], 33
background, 34
and Bjerknes, 34, 35
death, 34, 41
and Devik, 35
electromagnetic gun, 36
and Eyde, 36–7
arc furnace, 37
geophysics program, 34–6
and Nobel Committee, n.4-49
and Norsk Hydro, 38, 41
personal traits, 41
Terrella, 34–5
Bjerknes, Vilhelm Friman Koren
[1862–1951], 34, 35, n.4-49
Blackett, Patrick Maynard Stuart
[1897–1974], 53, 91, 255, n.2-6,
n.6-66
Bleakney, Walker [1901–], 26, 134
Blum, Léon [1872–1950], 102
Bock, General Feder von [1880–1945],
124
Bodenstein, Max [1871–1942], 132
Bohr, Erik [1920–1990], 79, n.6-34
Bohr-Nørlund, Margrethe [1890–1984],
69, n.6-30
Bohr, Niels Henrik David [1885–1962],
20, 21, 53, 279, n.6-34
brings news of fission to the US, 79,
81–2
and Fermi, 71
fission theory, 101, 120, 138, n.6-30,
n.6-36, n.6-77
and Frisch, 70, 78
and Hahn, 75, n.6-6
and Heisenberg's visit to
Copenhagen, n.10-57
and Jensen, 164
and liquid-drop model, 77
at Manchester, 12–3
and 'Maud Ray Kent', 119
and Meitner, 69, 74, 75, 118–9
and Meitner–Frisch fission
interpretation, 77–8, 79, 80
and occupation of Copenhagen, 110,
118
Boltwood, Bertram Borden
[1870–1927], 67
Boltzmann, Ludwig Edward
[1844–1906], 67
Bonhoeffer, Dietrich [1906–1945], 42

Bonhoeffer, Karl-Friedrich
    [1899–1957], 133
  and Harteck, 149
  and Norsk Hydro, 42–3
  and Taylor, 291
  and Urey, 291
Bopp, Friedrich (Fritz) Arnold
    [1909–], 223
Born, Max [1882–1970], 59
Bothe, Fritz (father of Walther), 15
Bothe, Walther Wilhelm Georg Franz
    [1891–1957], 13, 143, 188, 191,
    251, n.8-7
  and *Alsos*, 259, 263
  background, 15
  beryllium radiation studies, 15–6,
    18, 50, 138–9
  Bothe's 'mistake', 138, 139–40, 170,
    223, 297
  at first Schumann conference,
    137
  Geiger counter modifications, 15
  and Gentner, 140–1, 143–45, 146
  and Goudsmit, 259, 263
  and Harteck, 142
  and Heidelberg cyclotron, 139, 141,
    143, 144, 146, 259
  and Heisenberg, 140
  and Ingeborg Moerschner, 140–1
  and KWI for Medicine, 139, 259
  lackluster political attitude, 139
  and Lenard, 139
  neutron absorption measurements
    in carbon, 138, 139–40, 142, 147,
    170
  and Paris cyclotron, 145, 223
  and Planck, 15, 139
  polonium source, 15–6
  and Reich Research Council
    lectures, n.12-7
  Siberian internment, 15
  suggests carbon reflector for B-VII,
    223
Branch, Gerald Eyre Kirkwood
    [1886–?], n.4-19
Brauchitsch, Heinrich Alfred Hermann
    Walther von [1881–1948], 1
Brauer, Kurt (German minister), 111–2
Braune, Hermann [1886–?], n.6-2
Breit, Gregory [1899–1981], n.6-34
Bretscher, Egon [1901–], 123
Brickwedde, Ferdinand Graft [1903–],
    23, 25, 29, n.4-12

Briggs, Lyman James [1874–1963], 171,
    177, 181
  on Halban's Cambridge experiment,
    175
  on Urey's heavy-water project, 175
Broglie, Louis-César-Victor-Maurice
    Duc de [1875–1960], 53, n.3-18
Broglie, Louis-Victor [1892–1987], 53
Brown, Harrison Scott [1917–], 291
Brun, Jomar [1904–1993], 116, 157, 192,
    206, 209, 248, n.7-19, n.10-42
  and Abrahamsen, 168, 216–7,
    n.10-73
  and Aubert, 155–6, 169
  background, 43
  and Berg (F Bachke), 167–8
  'castor oil' sabotage, 158–9
  catalyst experiments, 162–3
  and Churchill, n.10-72
  on Cominco, 178
  electrolysis cells for Hassel, 42, 43
  on Falkenhorst, n.13-4
  flees Norway, 167–9, 297, n.10-73
  and *Freshman*, 169, 195
  and *Gerd*, 197
  on German interests in heavy water
    c. 1939, 105, 151, n.7-2
  and German requests for heavy
    water, 1933, 43–4
  and *Gunnerside*, 198, 204, 209, n.13-4
  heavy-water conference in Rjukan,
    1934, 45
  heavy-water experiments with
    Tronstad, 43–4, 295
  heavy water for Allier's party,
    107–8, n.7-17
  and heavy-water patents, n.4-77
  and heavy-water shipments to
    Germany, 189, 232
  and *Hydro* sinking, 228, n.14-34
  at ICI, 179, 228
  and Jensen, 155
  and Jones, R V, 168–9
  on Klaus Hansen's heavy-water
    demonstration, n.4-76
  at KWI for Physics, 155, 166, 169
  meeting with German Army
    Ordnance, Berlin, 153–5
  meeting with German scientists,
    Berlin, 153–5, 297
  meetings with German scientists,
    Rjukan, 150, 151, 153, 162–3,
    n.10-53, n.10-70

plans expansion of heavy-water
production at Hydro, 105, 150
plans heavy-water plant for UK,
166, 176–7, 228
platinum on charcoal, 179
postwar career, 288
provides London with German
plans for Vemork, 166–7, 168,
193
pseudonym, 169, n.10-72, n.19-1
recommends Hydro undertake
full-scale deuterium
production, 44–5
on smuggling heavy water to UK,
166–7, 168
and Suess, 162–4, 165, n.10-23
on Terboven, 163
tritium experiments, 133–4
and Tronstad, 150, 153, 156–7, 165–7,
168–9
and Welsh, 117, 168
and Wirtz, 115, 150–1, 155
Brun, Lorentz H B [1866–1944], 43
Brun, Tomy Johanne Svingjom
[1915–], 167, 168, 169
Bunsaku, Arakatsu (Kyoto Imperial
University), 280, 281
Burgess, Guy Francis de Moncy
[1911–1963], 277
Bush, Vannevar [1890–1974]
and *Alsos*, 249
and atomic bomb project
administration, 172, 242, 243
and Compton Committee, 173
and Hopkins, 172
and MAUD Report, 172
and OSRD, 180
and Roosevelt, 172, 173
at Trinity, 273
Butler, Nicholas Murray [1862–1947],
26
Bøhn, Per [1916–1960], n.14-34
Böhme, General Franz Friedrich
[1894–1947], 264
Bütefisch, Heinrich (Farben Director),
234, 237, 238

Cairncross, John, 277
Calvert, Major Horace K, 249, 251
Calvin, Melvin [1911–1997], 27
Carter, Baron (Belgian Ambassador in
Britain), 96

Chadwick, Sir James [1891–1974], 15,
16, 71, 138, 141, 240
α-bombardment experiments, 14, 53
background, 12–3
β-ray studies, n.3-7
in Bohr Institute parody, 21
British Mission to Washington,
244–5
and Canadian reactor project, 244–5
and Halban–Kowarski heavy-water
experiment at Cambridge,
175–6, n.11-22
Liverpool cyclotron, 79
and MAUD, 119, 123
neutron discovery, 14–5, 17, 18–20,
49–50, 53, 139, n.3-18
as Rutherford's lieutenant, 12, 14
at Trinity, 273
and wartime internment, 13–4
Chamberlain, (Arthur) Neville
[1869–1940], 73, 99
Chandrasekhar, Subrahmanyan
[1910–1995], 62
Cherwell, Lord. *See* Lindemann
Christian X (King of Denmark)
[1870–1947], 110
Chuikov, General Vasily I [1900–1982],
254
Churchill, Sir Winston Leonard
Spencer [1874–1965], 183, 243,
n.8-28
and Anderson, 249
Casablanca conference, 243
and *Gunnerside* mission, 209, and
Lindemann, 123
replaces Chamberlain, 113
and SOE, 156
Quebec conference, 294
Ciano, Galeazzo (Conte di Cortellazzo)
[1903–1944], 78
Clark, General Mark Wayne
[1896–1984], 247
Clusius, Klaus Alfred [1903–1963],
120, 121, n.12-5
Cockcroft, Sir John Douglas
[1897–1967], 56. 79, 135, n.11-22
and Frisch, 71, 122
heads Canadian heavy-water
reactor project, 244–5
and MAUD, 118–9, 122, 141
and Meitner, 71, 89, 122
and Norwegian heavy water, n.15-4
nuclear disintegration experiments,
53

and Rosbaud, 89, 248
at Solvay Conference of 1933, 53
and Tizard Mission, 174
Colette, Sidonie-Gabrielle [1873–1954], 65
Compton, Arthur Holly [1892–1962], 171, 242
advisory committee on uranium, 173, 176
on cosmic rays, 49
and Metallurgical Laboratory, 179
Conant, James Bryant [1893–1978]
and administration of atomic bomb project, 179, 242–3
and Anglo–American cooperation, 172, 177, 242–3
as British liaison, 172, 183
and Bush, 173, 242, 243, 274
and German uranium project, 193
and Halban, 177, 182–3
and Halban–Kowarski experiment at Cambridge, 182–3
on heavy water from Trail, 177, 178, 181, 243
and Mackenzie, 242, 243
and MAUD Report, 172
mistrust of Tube Alloys, 243
and Montreal project, 242–3
and Murphree, 181
promotes heavy water over graphite, 177
and S-1 project, 177, 182
and Top Policy Group, 173
at Trinity, 273, 274
and Urey, 175, 181–3
Corbino, Orso Mario [1876–1937], 63
as director of Rome physics laboratory, 59, 63
influence on Fermi, 59
prematurely announces transuranic elements, n.5-33
promotes cause of nuclear physics in Italy, 57
secures Fermi's appointment at the University of Rome, 59–60
suggests patenting radioisotope production, 63
Coster, Dirk [1889–1950], 66, 68–9, n.5-60
Crookes, Sir William [1832–1919]: and nitrogen fertilizers, 36–7
Curie, Eve Denise [1904–], 9, n.8-27, n.8-28

Curie, Marie Salomea Sklodowska [1867–1934], 9, 17, 52, 70, 75, 94, 126
and Frédéric Joliot-Curie, 8–9, 55
and Irène, 8–9
Nobel Prizes, 7–8, n.2-16
polonium discovery, 7
radium discovery, 7
and Radium Institute, 8, 94
Curie, Pierre [1859–1906], 7–8, 55, 95, n.2-15, n.2-16, n.2-21
Czulius, Werner, n.14-25

Dahl, Odd [1898–1994], 82, n.19-5
Daladier, Édouard [1884–1970], 88, 105, 106, 107, 124, 126
Dames, Wilhelm, 130, n.9-1
Darwin, Charles Galton [1887–1962], 12, n.2-4
α-particle scattering calculations, 3–4
Darwin, Charles Robert [1809–1882], 3
Dautry, Raoul [1880–1951], 127, 146
and Allier, 119, 127
and Allier Mission, 104–7
background, 101–2
and carbon procurement, 102
on heavy water, 104
and Joliot, 101–3, 106, 124, 125
Debye, Peter Joseph William [1884–1966], 60
and Meitner, 68
unseated from directorship of KWI for Physics, 137, 188, n.12-8
Dee, Peter Ivor [1904–], n.3-18
Devers, General Jacob Loucks [1887–1979], 261
Devik, Karl, 35
Devik, Olaf [1886–1986], n.4-45
Dickel, Gerhard [1913–], 120
Diebner, Kurt [1905–1964], 155, 156, 238
and Army Ordnance, 136–7, 142, 148, 153, 155, 186
background, 136
and Bagge, 136–7
character sketch, 136
detained by *Alsos*, 263, 271
and evacuation of heavy water from Vemork, 232, n.14-25
and Gerlach, 220, 254–6, 263
and Heisenberg, 136, 137, 148, 149, 221, n.9-27

and Joliot's laboratory, 143
and KWI for Physics, 137, 188, 209, 254
meeting with Brun, 153
meeting with Speer, 191
and pile research program, Gottow, 210, 219, 224, 257
and pile research program, Stadtilm, 224, 255–6, 260
and Schumann, 136
Dieseth, John, 229
Dill, Sir John Greer [1881–1944], 213
Dirac, Paul Adrien Maurice [1902–1984], 51, 53, n.5-24
Dodson, R W, 82
Droste, Gottfried von, 89
Drude, Paul Karl Ludwig [1863–1906], 32
Dunning, John Ray [1907–1975], 81
Döpel, Klara, 150, 174
Döpel, Robert [1895–?]: Leipzig pile experiments, 150–1, 174, 188–91, n.12-16

Eaker, Ira C [1896–1987], 194, 213–4
Earl of Suffolk & Berkshire, Charles Henry George Howard [1906–1941], 128–9
Ehrenfest, Paul [1880–1933], 59
Eide, Egil Hvoslef, 163
Einstein, Albert [1879–1955], 13, 52, 133, 171
Eisenhower, General Dwight David [1890–1969], 248, 271
Ellis, Sir Charles Drummond [1895–1980], 13, 14, 71
Enger, Arne [1882–?], 108
Eriksen, Bjarne Gotfred [1886–1976], 104
Eriksen, Colonel Birge Kristian [1875–1958], 111
Ermenc, Joseph John [1912–], 142
Esau, Professor Abraham [1884–1955], 130, 131, 135, 137, 141, 210–1, 219–20, 224, n.9-1
Eucken, Arnold Thomas [1884–1950], 133, n.8-13
Eyde, Samuel [1866–1940]
background, n.4-46
and Birkeland split up, 41
in diplomatic service, 41
meets Birkeland, 36–7

and need for nitrogen fertilizer, 36–7
Norsk Hydro founded, 37–8
and Norwegian hydropower, 37, n.4-46
partnership with Birkeland, 37
and Wallenbergs, 37
See also Birkeland; Birkeland–Eyde process; Norsk Hydro

Fajans, Kasimir [1887–1975], 24, 42
Falkenhorst, Nikolaus von [1885–1968], 1–2, 197, 207–8, 209, n.12-31, n.13-4
Farkas, Adalbert, 134
Farkas, Ladislaus [1904–1948], 134
Farrell, Thomas F, n.18-42
Fasting, Andreas, 158
Feather, Norman [1904–1978], 19, 123, n.3-18
Fehlis, Heinrich [1906–1945], n.12-31
Fenning, Frederick William [1919–1988], n.15-12
Fermi, Enrico [1901–1954], 64, 65, 90, 91, 142, 242, 276, n.6-34
absorption cross section for graphite, 174, n.11-12
adopts graphite as moderator, 170–1, 173–4, 177
and Amaldi, 57, 60–3 *passim*
background, 58–9
β-decay theory, 58
and Born, 59
and Cockcroft, 174
Compton on, 180
and Corbino, 59, 60, n.5-33
CP-1, first Chicago pile, 180, 189, 241, 279
decides to study artificial radioactivity with neutrons, 57–8, 60
discovers Fermi statistics, n.5-24
doctoral dissertation, 59
early chain-reaction studies, 87–8, 90–1, 97–8, 139, n.9-39
early pile studies at Columbia, 173–4
emigrates, 63, 71–2
fellowship abroad, 1923–24, 59
and heavy water as moderator, 173, 176, 181
and Ira Noddack, 61
and Joliot, 90

and Kowarski, 240
and Laura Fermi, 61, 63
learns of fission, 81
neutron bombardment studies, 60–3
Nobel Prize, 63, 71–2
and Persico, 58, 59, 60
and Placzek, 97
and Pontecorvo, 61, 63
professor at Columbia, 63, 79
and Rasetti, 59, 60, 61
and Rutherford, 58, 61
at Scuola Normale Superiore, 59
and Segrè, 60–3 *passim*
slow neutron discovery, 61–3, 295
at Solvay Conference, 57
supposed transuranic elements, 61,
    72
and Szilard, 90–1, n.6-53
temporary post in Florence, 59
and Trabacchi, 60
transfers Columbia work to
    Chicago, 177, 179–80
at Trinity, 273
use of term 'pile', n.11-1
wins chair at University of Rome,
    59–60
Fermi, Laura Capon [1907–], 61, 63,
    74
Feuerlein, Fritz (Gestapo), n.12-31
Finaly, Horace (Director General
    Banque de Paris), n.10-1
Fischer, Emil Hermann [1852–1919],
    67
Fischer, Erich Horst [1910–], 223
Fisher, Major Russell (*Alsos*), 263
Fleischmann, Rudolf, 251
Fleming, John Adam [1877–1956],
    n.6-34
Flerov, Georgii Nikolaevich
    [1913–1990], n.18-16
Flügge, Siegfried [1912–], 89, 97, 137
Foley, Major Francis (Frank) Edward
    [1884–1958]
  and Allier Mission, 109
  and battle for Rjukan, 114
  and Bertrand-Vigne, 114
  and British legation in Oslo, 109,
      112–3
  and *Broompark* evacuation, 128
  and General Ruge, 113
  and German attack on Norway, 113,
      114
  and heavy water, 109, 113
  and Rosbaud, n.5-58

Foss, Antonius, 42
Fowler, R D, 82
Fowler, Professor Ralph Howard
    [1889–1944], 171
Fox, alias for Tronstad
Frank, F Charles (assistant to Jones),
    168
Freiss, alias for Allier, 104, 106
Freundlich, Herbert Max Finley
    [1880–1941], 32
Friesen, Sten von [1907–], 71
Frisch, Auguste (Gusti) Meitner
    [1877–1951], 74
Frisch, Jutz, 74, 75
Frisch, Otto Robert [1904–1979], 84,
    89, 123, 141, 172
  on atomic bomb 121–2
  background, 69–70, n.3-8
  in Birmingham, 120–1
  on chain reaction, 83, 121
  and Cockcroft, 71, 122
  in Copenhagen, 70, 75, 78, 79, 87
  critical mass calculation, 121, 248
  fission cross-section estimate, 121
  fission recoil experiment, 80
  and Frisch–Peierls memorandum,
      120, 121–2, 296, n.8-9, n.11-5
  and Hahn, 122
  interprets Hahn–Strassmann results
      with Meitner, 77–8, 79, 80, 81–2,
      295, n.6-16, n.6-30
  on Irène Curie, n.6-23
  on measuring neutrons from fission,
      87
  meeting with Bohr, 79, n.6-19
  and Meitner–Frisch letter to *Nature*,
      80–1, n.6-16, n.6-30
  and Møller, 83
  and neutron absorption
      measurements in deuterium,
      102, n.11-12
  and Oliphant, 120–1, 122
  and Peierls, 120, 121–2
  and Placzek, 80
  and Przibram, 69, n.3-8
  and term 'fission', 80
  on uranium separation, 120–1
Fromm, General Friedrich [1888–1945],
    191
Fuchs, Klaus Emil Julius [1911–1969],
    277
Furman, Major Robert R, 249

Gamow, George Anthony [1904–1969], 79

Garben, Chr W, 104, 107, 108

Gasperi, Major Mario, 247

Gaulle, Charles-André-Marie-Joseph de [1890–1970], 127, n.8-28

Geib, Karl-Hermann [1908–1949], 234–5, 271

Geiger, Hans Johannes Wilhelm [1882–1945], 12, 138
  α-particle scattering, 3
  and Bothe, 15
  and Chadwick, 13
  Geiger counter, 15
  and uranium project, 136, 137, n.12-7
  in World War I, 13–4

Gentner, Wolfgang [1906–1980], 140
  and Bothe, 143, 145
  and discovery of artificial radioactivity, 54–5, 143
  and Goudsmit, 259
  and Joliot-Curie, 143–6, n.9-52
  and Langevin, 146

Gerlach, Walther [1889–1979], 146, 222, 249, 262
  background, 219
  detained by *Alsos*, 263, 271
  and Diebner, 220, 263
  and Esau, 219–20
  on Haigerloch pile, 259
  and heavy water, 224, 255–6
  and Hechingen move, 253, 254–6
  and Heisenberg, 219–20, 253, 254–6
  ideological views, 220
  as plenipotentiary for nuclear physics, 219–20, 223, 254, 255–6
  and Rosbaud, 220, 254–5, 259

Giauque, William Francis [1895–1982], 22, 27, 28, 29, n.4-12

Goering (Göring), Hermann [1893–1946], 1–2, 219, n.9-24

Goldschmidt, Bertrand [1912–], 241–2

Goldschmidt, Victor [1888–1947], 42, 236

Goudsmit, Samuel Abraham [1902–1978]
  and *Alsos*, 135, 250, 251, 253, 259–60, 262–3, 272
  background, 249–50
  and Bothe, 259–63
  electron spin, 249
  and Furman, 250
  and Gentner, 259–60

and Groves, 253
and Hahn, 262–3
and Harteck, 263
and Joliot, 251
and Pash, 262
and Uhlenbeck, 249

Gough, H J, 119

Graue, Georg, n.5-57

Green, George Kenneth [1911–1977], 82

Grieg, Johan Nordahl Brun [1902–1943], n.7-25

Griggs, David Tressel [1911–], 260

Groth, Wilhelm [1904–], 131–2, n.9-4

Groves, General Leslie Richard [1896–1970], 24
  and administration of atomic bomb project, 180, 183
  and *Alsos*, 247, 249, 250, 252–3
  and Anglo–American cooperation, 243, 244, 245
  and Auer plant, 252–3
  and Chadwick, 245
  and Chalk River project, 245
  and Chicago physicists, 184
  and Farm Hall transcripts, 273
  and *Freshman*, 213
  and Goudsmit, 253
  and Oppenheimer, 245, 274
  and Pash, n.16-2
  recommends bombing Vemork, 213
  and Rosbaud, 290
  takes charge of Manhattan Project, 180
  and Trail heavy-water plant, 180, 183–4
  and Trinity, 274

Gubbins, Major General Colin McVean [1896–1976], 156

Görnnert, Fritz, n.14-56

Haakon VII, King of Norway [1872–1957], 110, 111, 112–3

Haber, Fritz [1868–1934], 42, 133, n.8-7

Hafstad, Lawrence Randolf [1904–?], 82

Hagen, Sverre (pseudonym for Jomar Brun), 169, n.19-1

Hahn, Otto [1879–1968], 81, 89, 122, 130, 137
  actinium–lanthanum experiment, 65–6, 73–5
  and *Alsos*, 262–3

anti-Nazi views, 65, 66
background, 67–8
barium conclusion, 75–7
and Cockcroft, 89
correspondence with Meitner, 71, 74, 75–7, 78
and Curie–Savitch's radiochemical studies, 65–6, 73–4
director of KWI for Physical Chemistry, 65–6
discovers radiothorium, 67
early collaboration with Meitner, 68
and Farm Hall, 271, 274–5
in Fischer's Institute, 68
and German uranium project, 89, 191, 219, n.12-7
Hahn–Strassmann paper to *Naturwissenschaften*, 76, 78, 80, 295
on Irène Curie's chemistry, 66
and Joliot, 65, 66
KWI for Chemistry bombed, 221–2
KWI for Chemistry evacuated to Tailfingen, 254
meets Meitner, 66
meets Meitner in Copenhagen, 75, n.6-6
and Meitner's escape from Germany, 66, 69, 71
and Meitner's relatives, 74–5
and Ramsay, 67
and Rosbaud, 76
and Rutherford, 67
Rutherford on, 67, 68
Strassmann's collaboration with, 73–6, n.6-2
in World War I, 68
See also Meitner, Lise; nuclear fission
Halban, Hans von [1908–1964], 64, 81, 91, 96, 100, 139, n.15-7
after 1944, 289
during Allier's mission, 106, 109
background, 83–4
at Bohr's Institute, 83–4
bringing heavy water to England, 123, 125–9 *passim*, 296
and Canadian authorities, 244
chain-reaction studies at Cambridge, 141–2, 174–6, 181, 189, 296, n.11-23
chain-reaction studies at Ivry, 98–9, 100–1, 125, 174

chain-reaction studies in Paris, 84–8, 93–4, 98, 275, 296
at Collège de France, 84
and Francis Perrin, 241
French citizen, 99, 106
and Goldschmidt, 241–2, n.15-2
and Kowarski, 84, 99, 102, 106, 240, 245
and MAUD Committee, 141, 174
and Montreal team, 239, 240–4 *passim*, 298
and need for heavy-water moderator, 102
neutron absorption measurements, 102, 174, n.11-12
and Placzek, 87, 89, 97, 101, 242
at Radium Institute, 83–4
on reactors versus bombs, 93
and Tube Alloys Directorate, 176–7, 239
and Urey, 175, 177, 182–3
and US authorities, 176–7, 182–3, 239, 244
and Weisskopf's cable, 91
Hambro, Sir Charles [1897–1963], 261
Hanle, Wilhelm [1901–], 89
and graphite as a moderator, 139–41
Hansen, Klaus: heavy-water demonstration, 47, n.4-76
Hansteen, Major General Wilhelm von Tangen [1896–1980], 198
Hanstein, H B, 88, 93
Harkins, William Draper [1873–1951], 14
Harris, Air Chief Marshal Sir Arthur [1892–1984], 221, 222
Harrison, General Eugene (Chief of Intelligence, 6th Army), 261
Harteck, Paul [1902–1985], 147, 179, 191, 222, 264
advocates German heavy-water program, 148–9
background, 132–3
and Bonhoeffer, 133, 149
and Bothe's graphite experiment, 142
and Bütefisch, 234, 237–8
carbon–hydrogen as moderator, 224
catalytic exchange process, 151–2, n.10-22
declines leadership of German fission project, 137
demonstrates heavy-water properties, 135, n.4-17, n.9-20

detained by *Alsos*, 263, 271, 272,
   n.12-41, n.17-10
dry ice pile experiment, 142–3, 150,
   188, 296
dual-temperature exchange process,
   233–5, 238, n.14-44
and Esau, 224
fusion experiment at Cambridge,
   133–4, 295, n.9-9
and Geib, 234–5
and Gerlach, 237–8
and Goudsmit, 263
and Groth, 131–2, 135
and Hassel, 235–6
and Heisenberg, 142–3, 148–9
and I G Farben, 149, 233–5, 237–8
Irving on, 132
and Italian heavy-water plants, 211
and Jensen, 164
letter to Schumann, 132, 136, 296,
   n.9-4
and Lewis, 31, 135
low-pressure distillation column,
   236–7
meeting with Brun in Berlin, 153,
   155
and Norsk Hydro compensation,
   235
prepares heavy water at Cambridge,
   31, 134–5, 294
professorship at RPI, 142, 290
professorship at University of
   Hamburg, 135, 290
and residual heavy water at
   Vemork, 224, 236, 254
retrofitting Vemork plant, 153, 155,
   162–3
reviews heavy-water processes,
   236–7
and Rosbaud, 235–6
and Rutherford, n.9-8, n.9-17
and Schumann, 135–6
and Schumann conference, 1939, 137
and Stern, 131
and Suess, 149, 152, 222
Suess on, 290
on uranium separation, n.12-19
visits Norsk Hydro, 151, 254, 297
and Wirtz, 151
Hassel, Odd [1897–1981]
   background, 41–2
   and Harteck, 235–6
   and heavy water, 42–3
   imprisoned, 236

and Rosbaud, 235–6
Hauge, Jens Christian [1915–], 265
Haugland, Knut Magne [1917–], 195,
   200, n.12-26
Haukelid, Knut [1911–1994], 198
Hautecloque, Captain Viscount
      Philippe de. *See* Leclerc
Haworth, William, 123
Heisenberg, Elisabeth [1914–], 263
Heisenberg, Werner Carl [1901–1976],
      6, 59, 140, 255, 260
   and *Alsos*, 251, 252, 262, 263, 271,
      n.16-5
   and Bagge, 136
   and chain-reaction experiments,
      142–3, 151
   and chain-reaction theory, 138, 148,
      210, 220–1, n.12-7
   commitment to Germany, 165
   on critical mass of U-235, 275
   and destruction of Hahn's Institute,
      221–2
   and destruction of Leipzig facilities,
      222
   and Diebner, 136, 148, 188, 209, 221,
      n.9-27
   and Esau, 210–1
   and family, 254, 256, 263
   at Farm Hall, 272, 274–5
   and Gerlach, 219–20, 253, 256
   on German atomic bomb, 164, 191,
      275
   and German heavy-water stocks,
      210, 219, 221, 254–5, 282
   and Harteck, 142–3, 148–9
   heavy-water estimate for sustaining
      chain reaction, 189
   and heavy-water reactors, 148–9,
      189–91, 221, 222–3, 254, 256–8,
      299
   and Jensen, 164
   on nuclear structure, 53
   popular lecture on uranium fission,
      n.12-7
   postwar years, 289–90
   and production of heavy water,
      148–9
   relocation to Hechingen, 253–4, 256
   S-matrix theory, n.14-4
   and Speer conference on German
      program, 191–2
   succeeds Debye as head of KWI for
      Physics, 188, 209, n.16-5

uranium report to Army Ordnance,
148, 296
visit to Copenhagen, n.10-57
and Vögler, 223
and Wick, n.16-4, n.16-5
and Wirtz, 223
Helberg, Claus Urbye [1919–] 199,
200, 201, 208–9, n.12-26
Herold, Paul, 142, 234
Hertz, Gustav Ludwig [1887–1975],
271
Hertz, Heinrich Rudolph [1857–1894],
34
Hess, Kurt Joseph [1888–1961], 66, 69
Hess, Victor Franz (Francis)
[1883–1964], 49
Hevesy, György Charles von
[1885–1966], 12
Hiby, Julius, n.16-16
Hill, Archibald Vivian [1886–1977],
171
Himmler, Heinrich [1900–1945], 187,
n.14-25
Hitler, Adolph [1889–1945], 1, 88, 92,
127, 131, 148, 186, 253, n.6-39,
n.8-13, n.10-40
and Führerbefehl on saboteurs,
196–7, n.12-31
and Goering, 2, n.9-24
Reichstag speech, 1 September 1939,
99
and Speer, 254
and Speer on uranium project, 192
Hochheim, Ernst, 43
Hoffman, Gerhard [1880–1945], n.9-1
Hoffmann, Eric, n.12-31
Hole, Njål [1914–], 165, 248
Holmes, Sherlock, 157
Hopkins, Harry Lloyd [1890–1946],
172
Houtermans, Fritz Georg [1903–], 193
Howe, Clarence Decator (DC)
[1886–1960], 239
Hugo, Victor-Marie [1802–1885], 65
Hund, Friedrich [1896–], n.16-4
Hutchins, Robert Maynard
[1899–1977], 24
Hutton, Robert Salmon [1876–1970],
130
Hylleraas, Egil Anderson [1898–1965],
164
Höcker, Karl-Heinz [1915–], 210
Hörlein, Heinrich, 66

Idland, Kasper [1918–1968], 198, 199,
200
Ingebretsen, Olav, 206
Ioffe, Abram Fedorovich [1880–1960],
277

Jeans, Sir James Hopwood
[1877–1946], 52
Jensen, Johannes Hans Daniel
[1907–1973]
Bohr, meeting with, 164
Brun, meeting with, 153, 155
information leak on German
program, 164
Norwegian scientists, meeting with,
164
Jensen, Peter Herbert [1913–1955], 139
Jentschke, Willibald Karl [1911–], 89
Johansen, Gustav (night watchman,
NH), 206, 208
Johnston, Herrick, 22, n.4-12
Joliot, Hélène [1927–], 9, 49, n.2-21
Joliot, Pierre [1932–], 9, 17, 49, n.2-35
Joliot-Curie, Frédéric [1900–1958], 57,
60, 61, 63, 68, 119, 141, 170, 174,
241
and Allier, 104, 105, 127, 289
Allier mission planned, 105, 108
α bombardment, radiation from, 17,
53
and *Alsos*, 250–1
artificial radioactivity, 54–5, 56, 94,
295, n.5-11, n.5-12
background, 8
and Dautry, 101–3, 104, 105–6, 124–5
and Earl of Suffolk, 128
fission fragments recorded, 82
and Francis Perrin, 93
and Gentner, 143–6
and Hahn, 65, 66
and Halban and Kowarski, 83–4,
106
heavy-water moderator, need for,
102–3, 119, n.6-94
and Irène Curie, 8–9, 55, 63, 65, 68,
289
laboratory equipment, 10, 17, 52, 54
and Marie Curie, 8, 9
and Meitner, 53
and neutron discovery, 21, 49,
n.3-18
neutron mass, 52
Nobel Prize, 55, 83

Norwegian heavy-water escapade
   in France, 104–9, 124–7
and nuclear bombs, 93–6, 124–5
nuclear chain-reaction studies, 83,
   88, 89–90, 93–4, 96, 100, 120
nuclear chain-reaction system,
   patents on, 94–5
and nuclear fission discovery, 81
in occupied France, 143–6, 250–1
and Paris cyclotron, 144–5, 146
and positron discovery, 49–50
postwar career, 288–9
predicts nuclear chain reaction, 83
reactor studies, 94, 95, 98–9, 100–2,
   124–5, 295–6
recoil protons from neutral
   'beryllium radiation', 17–8, 50
relocation from Paris to Ivry, 98–9
and Schumann, 143–4
secondary neutrons from fission,
   82–3, 84–7, 88, 89–93, 296,
   n.6-75
at Solvay Conference of 1933, 52–4
and Szilard, 89–93
and Union Minière, 95–6
wartime resistance activities, 250–1
Joliot-Curie, Irène [1897–1956], 51, 63,
   64, 65, 68, 93, 100, 126, 128, 129,
   145, 146, 250, n.8-28
   α-ray studies, 9
   artificial radioactivity, 54–5, 94
   background, 8
   and Hahn, 68
   Hahn on, 66
   health, 51, 64, 128, 129
   and Joliot, 8–9, 63–4, 289, n.2-19
   and Marie Curie, 8–9
   and Meitner, 68
   and neutron discovery, 21, 49
   neutron mass, 52
   Nobel Prize, 55
   and nuclear fission discovery, 81
   and politics, 64
   and polonium source, 54
   and positron discovery, 50
   and Radium Institute, 8, 9, 65
   recoil protons from neutral
     'beryllium radiation', 17–8, 50
   and Strassmann, 73
   and thorium investigation, 64–5,
     73–4
   and uranium investigation, 64
Jones, Reginald V [1911–], 168, 249,
   252, 261, 272, n.12-36

Joos, Jakob Christoph Georg
   [1894–1959], 89, n.9-1
Jordan, Pascual [1902–1980], 59
Joseph, Emperor Franz, 83

Kaftanov, Sergei, 279
Kamerlingh Onnes, Heike [1853–1926],
   52
Kapitsa, Pyotr Leonidovich
   [1894–1984], 277
Kay, William [1879–1961], 4
Kayser, Fredrik [1918–], 198, 199, 200,
   204
Keitel, Field-Marshal Wilhelm
   [1882–1946], 187
Kendall, RAF Wing Commander
   Douglas, 252
Khariton, Iulii Borisovich [1904–1996]
   chain-reaction studies, 276, 277
   and the first bomb test, 279
   and thermonuclear weapons, 291
Khlopin, Vitalii Grigor'evich
   [1890–1950], 276
King, William Lyon Mackenzie
   [1874–1950], 239
King-Salter, Lieutenant-Colonel E J C,
   113
Kinsey, Bernard, 79
Kipling, Rudyard [1865–1936], 65
Kirsch, Gerhard Theodor [1890–1956],
   19
Kittelsen, Theodor Severin
   [1857–1914], 39–40
Kjelstrup, Arne [1913–1995], 200, 204,
   n.12-26
Knall-Demars, Jehan, 106, 108–9
Knudsen, Gunnar [1848–1928], 36, 37
Koch, Jørgen [1909–1971], 102, n.11-12
Koht, Halvdan [1873–1965], 112
Korsching, Horst (German scientist),
   262, 263
Kowarski, Lew [1907–1979] , 81, 91,
   93, 96, 100, 139, 176
   background, 84
   bringing heavy water to England,
     123, 127–9 *passim*, 296
   and CERN, 289
   and chain-reaction problem, 98,
     n.6-75
   chain-reaction studies at Cambridge,
     141–2, 174–6, n.11-23
   chain-reaction studies at Ivry, 98–9,
     100–1, 125, 174

chain-reaction studies in Paris, 84–8,
    93–4, 98
  at Collège de France, 84
  during Allier mission, 106, 109
  and Fermi's Chicago team, 240
  French citizen, 99, 106
  and Halban, 84, 240, 245
  Halban on, 84
  on Joliot's team, n.6-46, n.6-53
  and MAUD Committee, 141, 174
  and Montreal project, 240–1, 245–6
  on need for heavy-water moderator,
    102, 240, n.6-94
  neutron absorption measurements,
    102, 174
  on neutrons per fission, n.6-75
  at Radium Institute, 84
  and ZEEP, 245–6, 289
Kramers, Hendrik Antonie
    [1894–1952], n.19-5
Kummetz, Rear Admiral, 111
Kurchatov, Igor Vasil'evich
    [1903–1960], 275, 291
  argues for feasibility of chain
    reaction, 276, 277
  and graphite-pile project, 278–9
  put in charge of atomic project,
    277–8, 279

Lamb, Arthur Becket [1880–1952], 24
Lamb, Willis Eugene [1913–], 81
Landsdale, Colonel John, 262
Landsverk, Jon, 268, 269
Landsverk, Sveinung, 268
Langevin, Paul [1872–1946], 53, 55,
    146, 241, n.2-21
  and Joliot, 8
Langmuir, Irving [1881–1957], 24
Larsen, Alf Henry [1911–], 163, 206
Larsen, E J, 47
Laslett, Laurence Jackson [1913–1993],
    70
Laue, Max Theodor Felix von
    [1879–1960], 222, 254, 262, 271
Laugier, Henri [1888–?], 95, 124
Laurence, George Craig [1905–], 244,
    n.15-1
Lauritzen, Charles Christian
    [1892–1968], 30
Lawrence, Ernest Orlando [1901–1958],
    27, 57, 70, 79, 91, 95, 244, 273
  and Advisory Committee on
    Uranium, 175

deuteron experiments, 57, n.6-31
  and Gentner, 141–2, 144
  misses artificial radioactivity, 56–7
  neptunium and plutonium
    identified, 171
  and neutron stability, 56
  and Nishina's 60-inch cyclotron, 279
  at Solvay Conference of 1933, 53, 56
  thermal neutron capture cross
    section for deuterium, 175, 181
  and Tuve, n.6-37
  and Urey, 29
  on Urey's temper, n.4-16
Lechien, Gustav, 95, 96
Leclerc, Captain Viscount Philippe de
    [1902–1947], 250
Leeb, General Emil (Chief, Army
    Ordnance), 186
Leen, Richard E, 24
Lenard, Geheimrat Phillip Eduard
    Anton von [1862–1947], 139
Léopold, III (King of Belgium)
    [1901–1983], 124
Leprince-Ringuet, Louis [1901–],
    n.3-18
Levina, S, 276
Lewis, Gilbert Newton [1875–1946],
    23, 41, 43, 57, n.4-76
  background, 28
  Berkeley tenure, 27, 28
  compared to Langmuir, 24
  death, 291
  deuterium nomenclature, 29–30
  deuterium researches, 28–9, 291
  doctoral dissertation, 28
  and Harteck, 31
  mouse-feeding experiment, 30, 47,
    n.4-76
  and Rutherford's Cambridge team,
    30–1, 135
  school of chemistry, 27, 291, n.4-19
  scientific researches generally, 28,
    291
  and Urey, 23, 24, 28–9, n.4-16
  in World War I, 28
Libby, Willard Frank [1908–1980], 27,
    291
Lier-Hansen, Knut [1918–], 227,
    229–31
Lindemann, Frederick Alexander
    (Lord Cherwell) [1886–1957],
    71, n.5-63, n.8-13
Lindemann, Thv, 33

Linge, Captain Martin [1894–1941],
n.12-37
Livingood, John Jacob [1903–], 57
Lognvik, Johans, 268, n.17-28
Lognvik, Torgeir, 267–8, n.17-28
Lorentz, Hendrik Antoon [1853–1928],
52
Loritz (Werkschutzführer bei SS und
Polizeiführer Nord), 267
Lo Surdo, Antonino [1880–1949], 60
Lowzow, Major, 114–5
Lund, Roscher [?–1998], 193
Lüttke, Georg, n.5-58

Macdonald, Ronald T, 28, 29, 30, 31
Mackenzie, Chalmers Jack, 239
Marsden, Sir Ernest [1889–1970], 12
α-particle scattering, 3–4
H-particles observed, 4
nuclear disintegration experiments,
4
Marshall, General George Catlett
[1880–1959], 173
Masa, Takeuchi, 281
Masatoschi, Okochi, 279
'Master' [code name for Brun], 165–7
Mattauch, Josef [1895–1976], 130, 137,
n.9-1
Maxwell, Robert [1923–1991], 290
May, Alan Nunn [1911–], 123
Mayer, Maria Goeppert [1906–1972],
164
Meinich-Olsen, Kjell, 265
Meitner, Lise [1878–1968], 13, 16, 73,
80, 82
background, 66–7
and Bohr, 69, 74–5, 118
and Chadwick, 71
and Cockcroft, 71, 89, 118–9, 122
contributes wartime intelligence,
165
and Coster, 68–9, n.5-60
criticizes Joliot-Curie neutron
experiment, 53–4
and Curie–Savitch radiochemistry
studies, 65–6
declines to work on atomic bomb,
165
early collaboration with Hahn, 68
and Ellis, 71
escape from Copenhagen, 118
escape from Germany, 66, 68–9

and Frisch in Copenhagen, 69–70,
74–5
and Frisch in Sweden, 77–8
and Hole, 165
interprets Hahn–Strassmann results
with Frisch, 77–8, 79, 295,
n.6-16, n.6-30
joins Hahn, 66–8
and Jutz Frisch, 74–5, n.6-6
and Kurt Hess, 66, 69
and KWI for Chemistry, 53
Laura Fermi on, 74
and Maud Ray Kent telegram, 118–9
meets Hahn in Copenhagen, 75
and Mentzel, 66, 219
personality traits, 68
role in fission discovery, Meitner's
claim to, n.6-18
and Rosbaud, 69, 165
and Siegbahn's Institute, 69, 71
and Strassmann, 75, n.6-2
in Sweden, 70–1
in World War I, 68
See also Hahn, Otto; nuclear fission
Mentzel, Rudolf [1900–], 66, 219–20,
222
Menzel, Donald Howard [1901–1976],
22, 23, 24
Menzies, Sir Stewart Graham
[1890–1968], 272
Meyer, Robert C, 82, n.6-34
Meyer, Samuel L, n.4-29
Meyer, Stefan [1872–1949], 67
'Mikkel' [code name for Tronstad],
165–7
Milch, Field Marshal Erhard
[1892–1972], 191
Millikan, Robert Andrews [1868–1953],
49–50
Mitchell, Margaret [1900–1949], 65
Mittag-Leffler, Magnus Gustav (Gösta)
[1846–1927], n.4-49
Moerschner, Ingeborg, 140–1
Montel, Paul, 119
Moon, Philip Burton [1907–], 97
Morshead, Owen, 129
Moseley, Henry Gwyn Jeffreys
[1887–1915], 12
Mossé, Lieutenant Fernand, 106, 108,
109
Mountbatten, Admiral Lord Louis (1st
Earl) [1900–1979], 194
Moureu, Henri [1899–], 125, 126
Muller, Captain Jean, 106, 109

Müller, Walther, 15
Munch, Peter [1870–1948], 110
Murphree, Eger V, 179, 181, 182
Murphy, George Moseley [1903–1969], 24, 25, 29
Mussolini, Benito [1883–1945], 73
Møller, Christian [1904–1980], 83, 164

Nernst, Hermann Walther [1864–1941], 14, n.4-19, n.8-13
  and Solvay, 52
Neumann, John Ludwig von [1903–1957], 79
Nielsen, Kjell [1915–], 265, n.13-16
  and *Hydro* sinking, 225–7, 231
Nier, Alfred Otto Carl [1911–], 171
Nilssen, Bjarne [1879–1956], 43, 107, 151, 168, 206
Nishina, Yoshio [1890–1951], 279–81
Noddack, Ida [1896–1979], 61
Noddack, Walter [1893–1960], 61
Noguchi, Jun [1873–?], 281
Nordhagen, John Olav, 284
Nunn May, Alan. *See* May, Alan Nunn

Occhialini, Guiseppe Paolo Stanislao [1907–], 53
Oliphant, Marcus (Mark) Laurence Elwin [1901–], 119, 120, 121, 122, 133, 134, 135, n.3-5
Oppenheimer, (Julius) Robert [1904–1967], 245
Oster, Hans [1888–1945], 123, n.10-1
Ottar, Brunulf, 164, 248, n.10-59

Palmstrøm, Colonel Rolf, 157
Paneth, Friedrich (Fritz) Adolph [1887–1958], 242
Pash, Lieutenant Colonel Boris T
  and *Alsos I*, 247–8, n.16-5
  and *Alsos II*, 250–1, 253, 259, 260–4 *passim*
  background, n.16-2
  and capture of German scientists, 262–3
  and capture of Haigerloch cave, 261–2
  and Oppenheimer, n.16-2
Pauli, Wolfgang [1900–1958], n.8-7
  and neutrino, 21, n.3-7, n.3-22
Pegram, George Braxton [1879–1958], 91, 176, n.9-39

Peierls, Heinrich (father of Rudolf), n.8-7
Peierls, Rudolf Ernest [1907–1995]
  and atomic bomb, 121–2
  background, 121, n.8-7
  critical mass calculations, 120–1
  on fast-neutron fission, n.6-77
  and Frisch, 120, 121–2
  and Frisch–Peierls memorandum, 120, 121–2, 123, n.8-9
Perrin, Francis [1901–1992], 53, 99, 100, 241
  critical mass calculations, 93–4, 121
Perrin, Jean-Baptiste [1870–1942], 6, 7, 56, 65, 83, 241
Perrin, Sir Michael (Wilcox) [1905–1988], 248, 273
Persico, Enrico [1900–1969], 58, 60
Pervukhin, Mikhail Georgievich [1903–1978], 278
Petterssen, Lieutenant Colonel Sverre [1898–1974], n.12-30
Pettersson, Hans [1888–1966], 19
Placzek, George [1905–1955]
  at Bohr Institute, 80
  as experimentalist, 80
  and Fermi, 97–8
  and Frisch, 80
  and Halban, 87, 89, 97, 242
  and Montreal project, 242
  and neutron capture in uranium, 98
  personality, 80
  suggests looking for uranium fission fragments, 80, n.6-23
Planck, Max Karl Ernst Ludwig [1858–1947], 15, 52, 133, 221, 222, n.8-7
Pleijel, Professor H, 72
Poincaré, Jules Henri [1854–1912], 34
Polyani, Michael [1891–1976], 133
Pontecorvo, Bruno M [1913–], 61, 63, 242
Portal, Sir Charles, 214
Poulsson, Jens Anton [1918–], 194, 199, 200, n.12-26
Powell, Cecil Frank [1903–1969], 123
Poynting, John Henry [1852–1914], 34
Prankl, Friederich (Vienna), 89
Preiswerk, Peter [1907–], 64
Prout, William [1785–1850], n.2-4
Przibram, Karl [1878–1973], 69, n.3-8
Pynten, Per [1933–], 288
Pétain, Marshal Philippe [1856–1951], 126, n.8-25

Quisling, Major Vidkun (Abraham
    Lauritz-Johnson) [1887–1945],
    112, n.7-38

Raeder, Grand Admiral Erich
    [1876–1960], 1, 187, n.7-24
Ramsay, Sir William [1852–1916], 6, 67
Randers, Gunnar [1914–1992], n.19-5
Rasetti, Franco Rama Dino [1901–],
    59, 60–1, 63
Ray, Maud, 118–9
Rediess, Wilhelm [1900–1945], n.12-34
Reid, Margaret, 112
Reimers, Jan Herman [1914–], n.7-39
Rendulic, Lothar [1887–1971], 264
Renthe-Fink, Cecil von, 110
Reynaud, Paul [1879–1966], 105, 124,
    126, 127
Ribbentrop, Joachim von [1893–1946],
    78
Richardson, Owen William
    [1879–1959], 118
Rideal, Eric Keightely [1890–1974]
    and Harteck, n.9-22
    on ortho–para hydrogen system,
        n.4-17, n.9-22
    and Royal Society discussion on
        heavy hydrogen, 31, 41, n.4-17
    C P Snow on, n.4-39
    and Tronstad, 31, 41, 123
Riezler, Wolfgang Heinrich Sigmund
    [1905–], 146
Rittner, Major T H, 273, 274
Roane, Owen D ('Cowboy'), 214
Roberts, Richard Brooke [1910–1980],
    n.6-34
    confirms uranium fission, 82
    observes neutrons from fission, 92
Robinson, Harold Robert [1889–1955],
    n.3-7
Roosevelt, Franklin Delano
    [1882–1945], 70, 78, 171, 172–3,
    183, 243, n.8-28
Rosbaud, Paul Wentzel Matteus
    [1896–1963]
    after 1945, 290
    background, n.5-58
    and Blackett, 255
    and Cockcroft, 89, 130
    and Colonel Wilson, 290
    and Gerlach, 259–60
    and Hahn, 76, n.9-2
    and Hahn–Strassmann paper, 76, 78

and Haigerloch experiment, 259–60
and Harteck, 235–6
and Hassel, 235–6
and Mattauch, 130
and Meitner, 69, 165
undercover role, 89, 130–1, 248, 252,
    255, 290, n.5-58
and 'Uranium Club', 130–1
Rosenblum, Salomon, 53
Rosenfeld, Léon [1904–1974], 79, 81,
    n.6-34
Ruark, Arthur E [1899–], 23
Rubens, Heinrich [1865–1922], 13–4
Rue, Warren De La [1815–1889], 34
Ruge, Major General Otto [1882–1961],
    113, 114
Rundstedt, General Karl Rudolf Gerd
    von [1875–1953], 124
Russinov, Lev, n.18-16
Rust, Bernhard [1883–1945], 130
Rutherford, Sir Ernest (Lord
    Rutherford of Nelson)
    [1871–1937], 2, 18, 30, 52, 58, 61,
    66, 70, 119, 132, 171, 279, n.6-37
    α-particle scattering, 4
    Bakerian Lecture, 11, 21
    and Bohr, 12–3
    and Chadwick, 12–3, 14–5, 18–9
    chairs 'discussion on heavy
        hydrogen', 27, 30–1
    deuteron predicted by, 11–2, 22
    and discovery of artificial
        radioactivity, 55
    and Hahn, 67
    and heavy-water researches, 30–1
    helium-3 nuclei first observed, 133
    hydrogen-3 nuclei first observed,
        133
    and Marie Curie, 7
    on 'moonshine' talk, 6, n.2-10
    neutron predicted by, 11–2, 18, 21,
        22
    on nomenclature for H^2 nuclei, 30,
        133, 135
    nuclear atom, 3
    nuclear disintegration experiments,
        4–5
    on nuclear energy, 6, 83
    nuclear transformation experiments,
        14, 135
    proton identified, 5
    triton predicted by, n.3-5
    and tritium-concentration
        experiment, 1935–36, 133–4

working style, 135
in World War I, 4, 5–6
Rutherford, Mary (Lady Rutherford of
Nelson) [?–1954], 7
Röntgen, Wilhelm Conrad [1845–1923],
7
Rønneberg, Joachim [1919–], 198, 199
Rørholt, Sergeant Bjørn, 116, n.7-38

Sarasin, Edward (with De La Rue), 34
Sas, Colonel Jacob, 123
Savitch, Pavel, 64, 65, 73, 74, 81
Schacht, Hjalmar [1877–1970], 78
Scherrer, Paul [1890–1969], 248
Schmidt, Gerhard Carl Nathaniel
[1865–1949], 7
Schoepke, Consul Erhard, 151, 153,
159, n.10-18
Schrödinger, Erwin [1887–1961], 133
Schumann, Erich, 145, n.9-53
and Diebner, 136
and Esau, 135
and Joliot, 143–4
uranium project administration,
135–7, 186–7, 191
Schweigaard, Major Anton, 282
Seaborg, Glenn T [1912–1999], 27, 242
Seeling, Stabsarzt Werner Fritz,
n.12-31
Segrè, Emilio Gino [1905–1989], 60, 63
neutron bombardment studies, 60–3
Selborne, 3rd Earl of, Roundell Cecil
Palmer [1887–1971], 156
Sengier, Edgar
and Joliot, 95–6
and Tizard, 96
Shepherd, Dr L D'Arcy, 166
Sidgwick, Nevil Vincent [1873–1952],
n.4-17
Siegbahn, Karl Manne Georg
[1886–1978/79], 69, 71
Simon, Franz (Francis) Eugen
[1893–1956], 122, 123, n.8-13
Skinnarland, Einar [1918–], 158, 165,
169, 192, 211
background, 157
and *Freshman*, 195
and *Galtesund* affair, 157–8
and *Grouse*, 195
and *Gunnerside*, 198, 200, 205, 208
and *Hydro* sinking, 224–5, 226–8,
231
and *Sunshine*, 265, 268–9

Skinnarland, Torstein, 169, 195, 198,
n.12-33
Skobel'tzyn, Dmitry, 17, 20
Smith, Edgar R, n.4-29
Smith, General Walter Bedell
[1895–1961], 261
Snow, Charles Percy [1905–1980],
n.4-39
Soddy, Frederick [1877–1956]
discovery of isotopy, 26
disintegration theory of
radioactivity, 6
social and economic concerns, 6,
n.2-12
views on deuterium, 26
Solvay, Ernest [1838–1922]: and
Solvay Conferences, 52
Sommerfeld, Arnold Johannes
Wilhelm [1868–1951], n.8-7
Spaatz, General Carl [1891–1974], 260
Speer, Albert [1905–1981]
and Esau, 219–20
and German uranium project,
191–2, 193, 221, 253, n.12-19
and Hitler, 192, 254
Spiller, Captain Ederhard, 112
Stalin, Joseph [1879–1953], 252
Starheim, Captain Odd Kjell
[1916–1943], 158
Starke, Kurt [1911–], 187
Stephansen, Nicolai [1889–?], 105, 150,
151, 162, 163, 217, n.7-2
Stern, Otto [1888–1969], 60, 70, 131,
219
Stetter, Georg, 137
Stimson, Henry Lewis [1867–1950],
173
Stonehaven, Lord, 96
Storhaug, Hans [1915–1995], 198, 199
Strassmann, Friedrich (Fritz) Wilhelm
[1902–1980], 81
background, n.6-2
and barium discovery, 75
and confirmation of fission, 75–6,
295, n.6-18
early uranium experiments, n.6-17
and Hahn, 73, 75–6, 78, n.6-2
Hahn–Strassmann paper, 76–7, 78,
80–1
and Meitner, 75, 78, n.6-17
and thorium investigation, 65–6,
73–4
Strømsheim, Birger, 198, 199, 200, 204,
229

Süffert, Fritz, 76
Suess, Hans Eduard [1909–], 234
  and Brun, 162, 163–4, n.10-53
  catalyst tests at Vemork, 162–3
  catalytic-exchange process, 148–9,
    152, 187, 235, 237, n.10-22,
    n.10-23
  on Harteck, 290
  information leaks on German
    program, 163–4, 192
  offered asylum in Britain, 165
  suggests heavy water as moderator,
    148–9
Syverstad, Gunnar Bryde [1910–1945]
  and heavy-water contamination, 158
  and *Hydro* sinking, 225, 226, 228,
    231
  and *Sunshine*, 267–70
Szilard, Leo [1898–1964], 6, 79, 101,
    133, 174, 276
  background, 70
  and Britain's Academic Assistance
    Council, 70
  concern over nuclear energy, 83,
    90–1
  and Einstein, 171
  and Halban, 89, 91
  on heavy water, 173
  and Joliot-Curie, 89–93
  patents on chain-reaction concepts,
    n.6-39
  slow neutron experiment, 89–91
Sørensen, Captain Erling, 231
Sørensen, N A, 47, 269, n.10-36
Sørlie, Rolf, 200

Takeo, Yasuda, 279
Tamm, Igor Evgen'evich [1895–1971],
    291
Tate, John Torrence [1889–1950], 90
Taylor, Hugh Stott [1890–1974]
  background, 174–5
  and Bonhoeffer, 291
  catalytic exchange process, n.11-35
  and Cominco, 178, 181
  heavy-water catalyst, 176, 178–9
  on Lewis, 291
  and Urey, 174–5, 176, 276
Teller, Edward [1908–], 79, 90, n.6-34
Terboven, Josef [1898–1945], 148, 163,
    197, 264
  and *Gunnerside*, 207–8, 209
  and Herøya raid, 212–3

Thomson, Sir George Paget
    [1892–1975], 96, 171
  chain-reaction experiments of 1939,
    97, 120, 122
  and MAUD (initially Thomson)
    Committee, 119, 122, 141
  and MAUD Report, 172
  memorandum to Tizard, 97
Thomson, Sir Joseph John [1856–1940],
    3, 11, 97
Thornton, Robert Lyster [1908–], 57
Tizard, Sir Henry Thomas [1885–1959]
  and Britain's radar network, 80
  and Frisch–Peierls Memorandum,
    122
  and G P Thomson, 97
  meeting with Sengier, 96
  skeptical of nuclear energy, 120,
    171,
  and Thomson Memorandum, 97
Tizard Mission, 171–2
Tojo, Hideki [1884–1948], 279
Torp, Oscar [1893–1958], 227
Trabacchi, Professor Giulio Cesare
    [1884–1959], 60
Tronstad, Edla Obel [1912–1997], 32,
    116, 265
Tronstad, Hans Larsen [?–1903], 32
Tronstad, Leif Hans Larsen
    [1903–1945], 31–2, 33, 63, 65,
    134, 264, 288
  academic training, 32
  on Aubert, 169
  background, 32
  and Brun, 150, 165–9, 177, 193, 195
  Cambridge University stipend, 33,
    41
  and Churchill, n.10-72
  deuterium researches, 47
  doctoral dissertation, 33
  earliest researches, 32
  eulogy of, 47–8, 269
  flees Norway, 116, 153, 156, 297,
    n.7-39
  and *Freshman*, 167, 169, 193, 195
  on German interests in heavy water
    c. 1939, 150
  and *Gunnerside*, 167, 197–200, 205,
    209, 211
  and Hauge, 265
  heavy-water conference, Rjukan
    1934, 45
  and Herøya raid, 212
  and *Hydro* sinking, 227–8

and invasion of Norway, 115–6
killed in action, 268–9, 299
laboratory notes on heavy water,
44–5, n.4-68
marriage, 32
as 'Mikkel', 165
Nature, letter of 1934, 45
as Norsk Hydro consultant, 41, 43,
45, 47
with Norwegian High Command,
London, 156–7, 165–7, 189, 193,
211–3, 217, 224–5, 248, 264–5
personal attitude, n.17-16
posthumous tributes, 270, n.17-30
preliminary heavy-water
experiments with Brun, 44–5,
295
professorship at NTH, 33
proposes Hydro undertake
heavy-water production, 43–4
resistance network, Trondheim, 116,
153, n.7-38, n.7-39
as sportsman, 33
students, 115, 116, 165, 197, n.10-42
and Suess, 165
and *Sunshine*, 264–70
as teacher, 33
and Vemork bombing, 213, 217
wartime journals, 116, 205, 265,
n.10-72
and Welsh, 156, 225, n.10-72
and Wilson, 156, 198
Zeitschrift für Elektrochemie, paper of
1934, 45
Turner, Louis Alexander [1898–1977],
n.6-75
Tuve, Merle Antony [1901–1982], 79,
82
on confidality of fission research,
91–2
fission demonstration, 82, n.6-31,
n.6-34
nuclear structure studies at DTM, 82
Tzschirner, Lieutenant Colonel (Chief,
Wehrwirtschaftstab Norwegen),
153

Uhlenbeck, George Eugene
[1900–1988], 249
Urey, Harold Clayton [1893–1981], 27,
41, 46, 187, 241, 276
Atoms, Molecules, and Quanta, 23, 24
background, 23–4

on Birge, n.4-12
and Birge and Menzel's isotopic
proposal, 22–3, 24, 26
on Bohr, 24
at Bohr's Institute, 24
and Bonhoeffer, 291
at Columbia, 24–6
and Conant, 175, 182–3
deuterium announced, 26
deuterium discovery, fortuitous
nature of, 26
deuterium first identified
spectroscopically, 24–5
deuterium first liquefied, 25–6
deuterium nomenclature, 29–30,
n.4-16
deuterium spectrum confirmed,
25–6
doctoral dissertation, 24
as editor of *Journal of Chemical
Physics*, 29
at Fermi Institute, 291–2
and Halban, 176–7
and Halban–Kowarski heavy-water
experiment, 174, 176–7, 182–3,
n.11-23
Hutchins on, 24
at La Jolla, 292
and Lewis, 28–9
Lunar theory, 292, n.19-42
postwar career, 291–2
and space science program, 292
and Taylor, 174–5
and tritium-concentration
experiment, 133, 134
and wartime heavy-water project,
175–6, 178, 181–3, 236–7

Viten, Thor (Viggo) [1910–1987], 265
Volkoff, George M, n.15-1
Vögler, Albert [1877–1945], 186, 219

Wallace, Henry Agard [1888–1965],
172
Wallenberg, Knut Agaton [1853–1930],
37
Wallenberg, Marcus (Marc)
[1864–1943], 37, n.7-12
Walton, Ernest Thomas Sinton
[1903–1995], 135
Warburg, Emil Gabriel [1846–1931], 14
Wardenburg, Frederic, 260

Washburn, Edward Wight [1881–1934], 28, n.4-29
Watson-Watt, Robert Alexander [1892–1973], 79
Webster, H C, 19
Weil, George Leon [1907–], 180
Weizsäker, Carl-Friedrich von [1912–], 188, 191
Weisskopf, Victor [1908–], 91
Wells, Herbert George [1866–1946], 6, 83
Welle-Strand, Erik, 116
Welsh, Lieutenant Commander Eric ('Theodor') [1897–1949]
  and *Alsos*, 249, 261, 262, 272, 273
  background, 117
  and Brun, 117, 168
  death, 291
  and *Freshman*, n.12-36
  and heavy water, 117, 225
  and Jones, n.12-36
  and Meitner, 71
  personality, 156
  and Rosbaud, 193, 290
  and SIS, 156, 248, 249, 291, n.12-36
  and Tronstad, 116–7, 156, 225
  and Wilson, 156
  and XU network, 193
Wergeland, Harald Nicolai Storm [1912–], 164–5, 193, 248
Wheeler, John Archibald [1911–], 101, 184–5
Wick, Gian Carlo [1909–1992], 248, n.16-4
Wigner, Eugene Paul [1902–], 79
Wilkinson, A B (Sqn Ldr RAF), 196
Wilson, Colonel John (Jack) Skinner [1888–1969]
  and SOE, 156, 158, 198, 209, 211, 227, 264, 290
  and Tronstad, 156, 198, 264–5
  and Welsh, 156
Wilson, Charles Thomson Rees [1869–1959], 135

Wirtz, Karl (German authority on deuterium) [1910–], 115, 191
  and American science, 290
  background, 147
  and Bonhoeffer, 147, 148
  and Brun, 150, 151, 153–5
  detained by *Alsos*, 262, 271
  and Diebner, 148
  on German brutalities in Norway, 164–5
  and Harteck, 151, 290
  on heavy-water production in Norway, 187
  and Heisenberg, 188, 223, 256
  and Hiby, n.16-16
  on *Hydro* sinking, 232
  at Kernforschungszentrum Karlsruhe, 290
  meeting with Brun in Berlin, 153–5
  postwar career, 290
  and uranium pile experiments, 150–1, 188, 189, 222–4, 254, 256–8
  and 'Virus House', 150, 188
  visits Norsk Hydro, 149–50, 151, 153, 163, 164–5
  and Wergeland, 164–5
Witzeil, Admiral Carl, 191
Wood, Robert Williams [1868–1955], 26

Zeeman, Pieter [1865–1943], 249
Zel'dovich, Iakov Borisovich [1914–1987], 276, 277, 278
Zhukov, Marshal Georgy Konstantinovich [1896–1974], 254
Zicgner, Ottoni von (fiancée of Wirtz), 149–50
Zincke, Ernst Carl Theodor [1843–1928], 67
Zinn, Walter Henry [1906–], 88, 92, 93
  and CP-1, 180
  and CP-3, 185, 245–6

Subject index

Note: Endnotes are indicated by a suffix 'n' and a note number.

Abwehr, 108, 109, 123, 137
Academic Assistance Council, 70
Academy of Sciences, Paris, 82, 100, n.8-18
Academy of Sciences, Stockholm, 71, 81
Accademia dei Lincei, n.5-33
Actinides, 74
Activated charcoal, 179
Advisory Committee on Uranium, US President's, 174
Alamogordo Bombing Range, 273
Allied Force Headquarters (AFHQ), 247–8
Allier Mission. *See (in name index)* Allier, J.
α–hydrogen scattering, 3–5
α–particles/rays
  composition of, in terms of electrons and protons, 11
  as nuclear projectiles, 6
  range of, 9, 54
  Rutherford's scattering experiment (1906), 3
Alsos, 249
  in Brussels, 251
  German scientists round-up, 262–3, 272
  Goudsmit and, 249–51, 253
  in Haigerloch, 261–3
  Heisenberg and, 259, 263
  in Italy, 247–8
  origin of name, 247
  in Paris, 250–1
  in Stadtilm, 260
  in Strassburg, 251–2

American Institute of Physics: Tate Medal, 290
American Nuclear Society, 290
American Physical Society
  Christmas meeting (1931), 26
  discussion on deuterium (1933), 27, n.4-15, n.4-16
American Products Company (Niagara Falls), 37
Ammonia, 267, 281
Ammonia (ferry), 232
Amplifier
  linear, 19
  proportional, 80
Angular momentum (atomic), quantization of, 11
Anschluss proclaimed (1938), 66
Antony (Joliot's home), 146
Arc furnaces (Norsk Hydro), 37
Argentina, 285
Army Ordnance (German), 132, 135–7, 143–5, 149, 151, 163, 179, 186, 209, 224, 225, 235, n.12-2, n.12-7, n.12-19
Arzamas-16 (Soviet weapons facility), 291
Atomic bomb
  Farm Hall revelation, 274–5
  first Soviet (RDS-1), 279
  French calculations on, 93–4
  Frisch–Peierls memorandum on, 121–2
  Heisenberg on, 191
  Japanese conclusions on feasibility of, 280
  Kazakhstan test, 279
  MAUD Report on, 172, 173

mechanism, 122
predetonation problem, 122
Sahara test, 96
strategic issues associated with, 122
Trinity test, 273–4
yield of Trinity 'gadget', 274
Atomic energy: from radioactivity, 6, n.2-12
Atomic Energy of Canada Limited (AECL), 286
Atomic Energy Commission (French), 288, 289
Atomic nuclei: constituents of, 23
Atomic weights
of radioactive elements, 26
two systems, 22
Atom smashing
Perrin on, 6
Rutherford on, 6
Atoms, Molecules, and Quanta, 23
Auer-Gessellschaft, 142, 220–1, 252, n.18-4
Aurora borealis: Birkeland's researches, 34–6
'Ausonium', 72
Aviemore, Norwegian depot at, 198
Åndalsnes, 115

Badische Anilin- und Sodafabrik: partnership with Norsk Hydro, 44
Badische process, 44
Balmer spectrum: of hydrogen isotopes, 24–6
Baltic States, 1
Bamag-Meguin Company, 237, 238
Banque de France, 125, 126
Banque de Paris et des Pays-Bas, 38, 104
Barium, 7
in discovery of nuclear fission, 75, 76, 78, n.6-17
Bergen, Norway, 110, 114, 117, n.7-24
Berlin
battle of, 221–2, 223, 238, 254
Free University of, n.18-4
KWI for Physics evacuated from, 255
University of, n.8-7, n.8-13
Berlin-Dahlem, n.18-4
Beryllium
penetrating radiation from, 16–20, 50

properties of, 16
β-decay: theory of, 58
β-rays: continuous spectrum, 14, n.3-7
Birkeland–Eyde process (nitrogen fixation), 37–8, 39–40, 44, n.4-53
Bismuth, 7
Blücher, sinking of, 111
Bletchley Park: codebreakers, 248
Bohr Institute for Theoretical Physics
Frisch at, 70, 80, 83, 84
under German occupation, 118
Heisenberg's visit (1941), n.10-57
Jensen's visit (1942), 164
Meitner's visit (1940), 118
parody on Goethe's *Faust*, 21
Placzek and, 80–1
Urey's fellowship at, 24
Bordeaux, 127–9, 146
B-org, 265, n.17-13
Boy Scouts Association, 156
Bradley–Lovejoy process (nitrogen fixation), 37
Briggs Uranium Committee, 171, 172, 173, 175, 176, 177
British Admiralty: and patents for chain reactors, n.6-39
British Airborne Division, First, 194, n.12-31
British Association (for the Advancement of Science), n.3-18
British Ministry of Economic Warfare, 214
Bromma Airfield, 168
Broompark, and heavy-water evacuation, 128–9
Brownian motion, 93
Bruce, Ontario, 284, 285
Bubble chambers, 289

Cadmium (neutron absorber), 94, 105, 180, 185, 258
'Cambridge five', 277
Canada: reactor work in, 239–40, 242–6, 284
CANDU reactor program, 284, 289
Carbon
as moderator, 101, 138, 139–40, 141–2, 170, 173–4, 177
neutron absorption cross section, 102, 138, 139, n.9-39
Carlsberg House of Honor (Copenhagen), 69

Carnegie Institution of Washington, DC, 79, 172
Carrier, in radiochemical procedures, 64
Casablanca Conference (1943), 243
Cavendish Laboratory
 Chadwick's administrative role, 14
 fusion experiment, 133, 134–5
 Halban–Kowarski experiments, 141–2, 174, 175–6
 Harteck's heavy-water project, 31, 134–5
 and Montreal project, 240–1
Cellophane, 17
CERN (European Organization for Nuclear Research), 289
Chaffinch, operation, 228
Chain reaction, 82–3
 divergent, 88, 93–4, 175–6
 multiplication factor, 100, n.6-91
 secrecy and, 89–93
 theoretical work in France, 93–4, 98, 100–1
 theoretical work in other countries, 97–8, 120–1, 242, 276–7
 See also nuclear energy; nuclear fission; nuclear reactors
Chalk River Laboratory: establishment of, 245
Chemisch-Technische Reichsanstalt, 210
Chirchik, n.18-21
Chirchik Nitrogen Combine, 276, 278
Chosin Reservoir, 281
Clarendon Laboratory, 122, 123, n.8-13
Clermont-Ferrand, 125–7
Clinton Engineer Works, 184
CNRS (Chaisse/Centre Nationale de la Recherche Scientifique), 94, 101
Cockcroft–Walton accelerator, 279
Collège de France, 55, 64, 65, 109, 126, 143, 144, 146, 174, 250, 251, 289
Collège Sévigné, 8
Columbia University, 72
 Fermi's appointment to, 79
 Szilard and, 79
Combined Policy Committee (CPC), 244–6 *passim*
Cominco (Consolidated Mining and Smelting Company of Canada, Ltd)
 Brun on, 178
 electrolytic hydrogen plant, 178

heavy-water plant, 178–9, 180–1, 183–4
 origins of, 178
Compound 808 (plastic explosive), 228
Comptes Rendus, 9, 87
Congress of Chemistry, Tenth International, 65
Copenhagen
 German Cultural Institute, n.10-57
 meeting between Hahn and Meitner, n.6-6
Cosmic rays
 Compton's explanation, 49
 Millikan's views on, 49, 50
 nature of, 49
Cosmic-ray studies: positron discovered during, 50–1
Crimean War, 111
Cross section, n.6-85
Culmann et Compagnie, 144
'Curiosium', 66
Cyclotrons
 Berkeley, 95, 279
 Copenhagen, 70, 118, n.7-20, n.9-54
 Heidelberg, 139, 146, 259, 262, n.9-54
 Leningrad, 275
 Liverpool, 79, n.9-54
 Miersdorf, 144
 Paris, 56, 144–5, 235, n.9-54
 Stockholm, 71, 118, n.7-20, n.9-54
 Tokyo, 262, 279, 280

Defense Industries Limited, 245
Degussa (Deutsche Gold und Silber Scheideanstalt), 210, 251, 255
DEN (Deutschland-England-Norwegen)-Group, 113
Denmark
 invasion of, 110
 nonaggression pact with Germany, 110
DENOFA, 42
Department of Terrestrial Magnetism (DTM), Carnegie Institution, 82
Deuterium
 announcement of, 26
 discovery of, 24–6
 first identified spectroscopically, 24–6
 first liquefied, 25
 fortuitous nature of discovery, 26
 isotopic abundance, 26, 134

Lewis's work on, 28–30
as moderator, 101, 102–3, 125, 173,
    176, 181–2
in neutrino detection, 286
neutron absorption cross section,
    102, 148, 175, 181
nomenclature for, 27, 29–30, n.4-16
ortho–para system of, n.4-17, n.9-22
prediction of, 12, n.4-12
properties of, 28, 147, n.4-17
Royal Society discussion on, 27, 41,
    135, n.4-17
separation of, 25–6
Soddy on, 26, n.4-17
tritium versus, 133, 134
See also heavy water; deuteron
Deuterium-hydride, 164
Deuterium oxide. *See* heavy water
Deuteron
discovery of, 24–6
neutrino, interaction with, 286
as nuclear projectile, 30, 133, 135
prediction of, 12, 22
Deutsches Museum: and fission
    discovery, n.6-18
Dimona reactor, n.19-18
Dovre, 115, 116
Du Pont Company
heavy-water plants, 184
and Manhattan Project, 183–4
Du Pont ordnance works, 184
'Dustbin', as internment center, 271
Dysprosium, 85

École d'Electricité Industrielle Charlat,
    8
École de Physique et de Chimie
    Industriélle de la ville de Paris
    (EPCI), 7, 8, 241, 251
Egersund, 196, n.12-31, n.12-32
Egypt, 34, 36, 114
Eidanger Salpeter Works, 212
Eighth Air Force, US, 212, 213–5, 221,
    237, 249, 260
Eldorado Gold Mines, Ltd, 97
Electrolyte: need for, in electrolysis, 44
Electromagnetic gun: Birkeland's,
    33–4, 36, 37
Electron–positron pair production, 51,
    53
Electron–proton model of matter, 11,
    23
Electron spin, 249

Elverum, 112
Eureka homing device, 192, 195
European Nuclear Energy Agency, 289
Examen artium, 32
Exponential experiments: origin of
    name, 180

Fabrique d'Elektrolyseurs
    Hydroxygene SA, 44
Fall Gelb, 123
Farm Hall
commando training school, 199, 272
detainees, 271
German scientists' internment, 272,
    274–5, n.12-41
microphone installations, 272
transcripts from, 272–3
Fermi–Dirac statistics, n.5-24
Fermi Institute for Nuclear Studies,
    24, 291
'Festung Norwegen', 264
Finland, 1, 124, 264
Finnmark, 34, 35, 264, n.17-12
Fornebu Airport, 1–2, 8, 108–9, 110,
    112
Fort Eben Emael, 124
F-Project (Japan), 280, 281
Fractional crystallization, 75
France
armistice and surrender, 127
fall of Daladier government, 124
invasion of, 124
Popular Front, 102
resistance movement, 250–1
Reynaud named Premier, 124
Secret Service (Deuxième Bureau),
    104
Free French Forces, 250
Freshman
advance party. *See Grouse*
German response, 196–7, n.12-31
Groves and, 213
mission fails, 195–7
objectives, 194
planning for, 167–9, 193–4, 195
postwar military trials, n.12-31
Front National, 250
Führerbefehl, Hitler's, 197, n.12-31
Fusen Reservoir, 281
Fusion, thermonuclear, n.9-9
d–d reaction, 133
d–p reaction, 133

first experimental demonstration, 133, 134

Galtesund affair, 157–8
Gebirgsarmee, German 20th, 264
Geiger counters, 10, 13, 15, 21, 54, 60
Geochemical researches, Urey's, 291
Geophysical researches, Birkeland's, 34–6
Gerd, 197
German High Command, Oslo, 207
German Post Office, Miersdorf, 144
Gestapo, 111, 156, 197, 213, 231, 267, n.12-31, n.12-32
Girton College (Cambridge), 71
Glace Bay, 284
Glenmore training camp, 198
Glider operations
    and fall of Fort Eben Emael, 124
    first Allied, 194
    See also Freshman
Glomfjord works, 282, 283
Glucinium: early name for beryllium, 16
Goethe's *Faust*, 21
Gottow laboratory, 209–10, 224, 225
Graphite. *See* carbon
Grating spectrograph, 24
Grini Concentration Camp, 196, 209
Grouse, 192, 194–5, 197, 198, n.12-26
Guano, Chilean: for fertilizers, 36
Gudbrandsdal, 113, 114, 116
Gunnerside
    anniversary observances, 288
    Churchill on, 209
    damage caused by, 206–7
    decorations bestowed, 209
    German reaction, 206–8, 209
    Grove and, 213
    Helberg's narrow escape, 208–9
    objectives, 198
    preparations on Hardanger Plateau, 200–2
    retracing of route (1990), 288
    strike on Vemork, 202–5
    training for mission, 198–9
Göttingen, University of, 290
Göttingen Physics Institute, 59

Haber–Bosch process (ammonia), 42, 106, 281
Hafnium, n.5-60

Haigerloch
    Alsos reaches, 261–3
    pile laboratory, 255, 256–8, 259–60, 261–2, 282
Haldde Observatory, 35
Halden reactor, n.10-65, n.19-6
Hamar, 112
Hamburg: fire raids, 222
Hanford: plutonium-production reactors, 184–5, 245
Hardanger Plateau, 115, 194, 202, 231
    and *Grouse/Swallow*, 192, 194–5, 197, 199, n.12-26
    and *Gunnerside*, 200, 205
    Helberg's chase, 208–9
    Lake Møs, 192, 194, 208, 209, 265, 267, 268
    Lake Såheimsbotn, 167
    Skoland Marsh, 192, 195
    and *Sunshine*, 265, 267–9, n.17-15
    Ugleflåt, 265
Hardangervidda. *See* Hardanger Plateau
Harnack Haus, 187, 191, 254
Heavy hydrogen. *See* deuterium
Heavy ice, 210, n.4-17
Heavy water, 2, 8, 17, 33, 113, 119, 144, 156, 260, 288, 292
    abundance of, 43, 134
    Allied commando attacks, 169, 193, 194–205, 206–8, 225–33, n.12-39, n.12-43
    Allied concerns over German plans, 156–7, 165–7, 193–4, 197, 211, 213–4, 224–5, 247–9
    and Allier mission, 104, 105–9
    American bomber strike, 212, 214–7
    Berlin discussions (Diebner's office) on, 153–5
    Bertrand-Vigne on, 114
    in biological systems, 27, 30, 46–7, n.4-29, n.4-76, n.7-2
    Cambridge experiments on, 141–2, 174–6, 277
    'castor oil' sabotage, 150, 158–9
    code names for, 115, 126, 149, 157, 165–6, 184
    cost of, 30–1, 47, 103, 155, 285
    and Eidanger salpeter works, 212
    evacuation from France, 123, 125–9, 145–6, 149, 185
    evacuation from Norway, 105–9, 145, 153, 185

Farm Hall discussions on, 275,
n.10-22
Fermi on, 173, 176
frozen, 210
German interests in, *c.* 1939, 104–5,
107, 114
graphite versus, 138, 147, 173, 176,
177, 180, 240, 278
Hansen's experiment, 47, n.4-76
ICI and, 31, 179, 228, 242, 243, 244,
n.11-33
I G Farben and, 105, 106–7, 149, 153,
233–5, 237–8, 247, 278, 283
industrial production (Canada),
284–5, n.14-44
industrial production (Cominco),
178–9, 180–1, 183–4, 242, 286,
n.11-31
industrial production (Du Pont),
183–4, 286, n.11-31
industrial production (Germany),
151, 224, 233–5, 236–8, 286
industrial production (postwar, by
Norsk Hydro), 282–3, 284
industrial production (prewar, by
Norsk Hydro), 45–8, 105
industrial production (Soviet
Union), 276, 278–9
industrial production (Standard Oil
Company), 176
industrial production (wartime, by
Norsk Hydro), 147, 149–51, 153,
158–62, 167, 189, 193, 211–2,
224–5, 235, 236, 286, n.11-31
industrial production (worldwide, *c.*
1977), 284, 285
industrial production first proposed
(Hassel), 42, 43
international control of, 285
Italian sources for, 211, n.13-11
Ivry experiments on, 125
and Japanese atomic project, 275,
281
Jensen on, 164
lost through sabotage, 207, 211–2,
232, 286
methods of shipment to Germany,
189
as moderator, 101, 102–3, 125, 138,
148–9, 150–1, 173, 176, 177,
188–90, 240, 276, 278, 281
mouse experiment, 30, 47, n.4-76
in neutrino detection, 286
neutron diffusion length in, 151

patents for, n.4-77
preparation of (Harteck), 31, 134–5
preparation of (Hassel), 42
preparation of (Lewis), 28–9, 31, 41
preparation of (Taylor), 174–5, 178–9
prewar demands for, 45–6, 114
production, methods for, 41, 44,
45, 148–9, 151–2, 159–62, 175,
178–9, 187, 233–4, 236–7,
n.10-17, n.10-22, n.12-5
properties of, 43, 135, 293, n.4-17,
n.9-20
reactors, 184–5, 239, 245–6, 279, 283,
284, 289, 291
Rutherford on, 30–1
and Soviet atomic project, 275–9
subcritical pile experiments, 141–2,
174–6, 188–90, 210–1, 220–1,
223, 239, 256–8, 259–60
Suess on, 163–4
and tritium preparation, 133–4
Urey versus Conant on, 181–3
wartime stocks in Germany, 210–2,
219, 222, 224, 232–3, 256, 261–2,
282, 286
Welsh on, 116–7
See also Norsk Hydro
Hechingen
Alsos visits, 262
French occupation, 260, 262
KWI for Physics relocates to, 224,
252, 253–4, n.16-16
reactor project, 252–3
Helium
impractical as a moderator, 98
from radium decay, 6
Helium-3, 133
Helouan, Egypt, 34
Helsingør, 110
Herøya, 214, 218, 226, 228, 271, n.13-15
ammonia plant, 283
Eidanger Salpeterfabrikker, 212–3
Nordisk Lettmetall, 148, 212, n.13-16
'Hesperium', 72
'High Velocity Corrosion' (Canadian
Tube Alloys Project), 240
Hird (Quisling troops in uniform),
208
Hiroshima bombing: announcement
of, 274
Horsa gliders, 192, 194, 195–6
H-particles, 4
'Humbug', operation, 253
Hydro (ferry). *See* Lake Tinn operation

Hydroelectric power
  for ammonia production, 36–9, 178
  for light-metal industry, 148
  versus oil power, 283
Hydrogen
  atomic weight, 22, 26
  heavy. *See* deuterium
  isotopes of, 22, 25–6
  liquid, 25, 187, n.4-12
  as moderator, 62–3, 98
  ortho–para system, n.4-17
Hydrogen bomb, Soviet, 291

I G Chemie, 44, 113
I G Farbenindustrie, 43, 105, 142,
    148, 149, 233–8, 247, 259, 278,
    n.11-33
  enters into partnership with Norsk
    Hydro, 44, 113
  dissolution after WWII, 283, n.4-67
Imperial Chemical Industries (ICI), 31,
    46, 177, 179, 228, 242, 243, 244,
    n.11-33
'Independent particle' model, of
    nucleus, 77
India, 283, n.19-18
Inert gases, discovery of, 67
Institute for Advanced Study
    (Princeton), 79
Institute for Chemical Physics
    (Leningrad), 276
Institute de Chimie Nucléaire, 143
Institute of Physical Chemistry
    (Moscow), 278
Institute for Theoretical Physics
    (Leipzig), 222
Internasjonal Farvefabrikk A S, 117
Ionization chamber, 17, 19, 20, 21, 80
Isotopes, first proposed, 24
Isotopy, Soddy's definition of, 26–7,
    n.4-17
Italy, 63, 247–8, n.16-4, n.16-5

Japan
  air raids on, 280, 281
  Army Aviation Technology Research
    Institute, 274
  Committee on Research in the
    Application of Nuclear Physics,
    280
  Navy Technology Research Institute,
    280

JEEP (nuclear reactor). *See* Kjeller
    reactor
Joachimstahl mine, 7
Jornada del Muerto, 273
Journal of Chemical Physics, 29
'Juice' (code name for heavy water),
    157
Jungfraujoch Research Station, 49

Kaiser Wilhelm Institute for
    Chemistry, 53, 68, 221–2, 254,
    262
Kaiser Wilhelm Institute for Medicine,
    138, 139, 259
Kaiser Wilhelm Institute for
    Physical Chemistry and
    Electrochemistry, 32
Kaiser Wilhelm Institute for Physics,
    137, 147, 155, 162, 187, 188, 209,
    210, 220, 234, 237, 252
  bunker laboratory, 221, 222, 254,
    272, n.18-4
  evacuation of, 254–6, 262
Kaiser Wilhelm Society, 186–8, 191
Karlsruhe, 290
Kellogg Company, 175
Kelly Hospital (Baltimore), 19
Kernforschungszentrum Karlsruhe,
    290
Kjeller Aerodrome, 214, 215
Kjeller reactor, 283, n.19-5
Klystron, n.11-3
Knaben molybdenum mines, 214, 215
Konan, 281
Kongsberg, 114
Kon-Tiki, 288, n.12-26
Korean Hydro Electric Company, 281
Kriegsmarine, 264
Kristiania. *See* Oslo
Krupp Company, 36
Königsberg, n.7-24
Kurchatov Institute, 278
Kyoto Imperial University, 280

Laboratoire Curie, 143
Lake Såheimsbotn, 167
Lake Tinn, 115, 226, 288. *See also* Lake
    Tinn operation
Lake Tinn operation
  lives lost, 232, n.14-34
  planning for, 226–9
  plastic charges placed, 230–1
  postwar disputes regarding, n.14-35

results of, 232, 286
  sinking of *Hydro*, 231–2
'Lamplight', n.17-15
L'Arcouest, 9, 17, 51, 64, 99, 128
Légion d'Honneur, 28, 146
Lend-Lease, 278
Leningrad Physico-Technical Institute,
  275
Lensmann, 265
Light-metal industry, Norwegian, 148,
  212, n.13-16
Lillehammer, 112
Linde Eismachinen Fabrik, 233, 235,
  237
Linge Company, 158, 198, 209, 264,
  n.12-37
Liquid-drop model of nucleus, 77
London
  arrival of Halban and Kowarski,
    123, 129
  arrival of Tronstad, 116
  center for British secret operations,
    157
  Manhattan Project Office, 249
  Royal Albert Hall rally, 70
  as war breaks out (1939), 99
  Wormwood Scrubs Prison, 129
Los Alamos, 273
Low Countries, invasion of, 123–4
Luftwaffe, 2, 110, 112, 124, n.14-54

Magnetron, cavity, 172
Malmø, 149
Manchester University, 3, 4, 12
Manhattan District of the US Corps
    of Engineers. *See* Manhattan
    Project
Manhattan Project
  Atomic Bomb Mission to Japan, 281
  Farm Hall detainees learn of, 274–5
  Goudsmit briefed on, 250
  London office, 249
  Los Alamos Laboratory, 273
  start of, 180
Manifesto della Razza, 71
Mass defect, nuclear, 77–8
Mass spectrograph, Aston's, 22
'Masurium', n.6-16
MAUD Committee, 118–9, 123, 141,
    171, 174, 176, 177
  Report of, 172, 173, 176, n.11-5
Max Planck Haus, 137
Max Planck Institute for Physics, 289

Maxwell's equations: Birkeland's
    solution of, 34
Me-163 (German jet fighter), n.14-54
Metallurgical Laboratory, 179, 181, 193
Merano, 211
Merseburg, 233, 237, 278
Metallwirtschaft, n.5-58
Metropole, Hotel (Brussels), 52
MI6. *See* Secret Intelligence Service
Milorg Forces, 116, 157, 248, 264,
    265, 268, 269. *See also* XU
    organization
Moderators, 62–3, 94, 98, 101, 138, 170,
    240
  dry ice, 142, 255
  paraffin, 62, 97
  water, 63, 97
  See also carbon; heavy water;
    hydrogen
Montecatini electrolysis plants, 211
Montreal team, 240–2, 245
Monts Dore, 126, 127
Moon, origin of, 292
'Moonlight', n.17-15
'Moonshine', in Rutherford's remark,
    6, n.2-10
Moscow: Canadian Embassy, 234
Moscow Electrode Factory, 278
Munich, 222, 254, 263, 290
  heavy-water black-market affair,
    282
Mæl, 226, 228, 229–33 *passim*

Narvik, 110, 114, n.7-24
Nasjonal Samling (NS), 268
National Bank of Norway, 112
National Bureau of Standards, 25,
    n.4-12
National Defense Research Committee,
    172
Nature, 58
Nazi–Soviet Pact, 124, 127
Neptunium, 138, 171, 274
Neutrino
  detection, 286
  Pauli's proposal, n.3-7, n.3-22
Neutron bombardment studies
  in Berlin, 75–6
  at Columbia, 88
  Fermi's, 60–3
  Irène Curie's, 64
  in Paris, 84–7

Neutron density-distribution curves,
    85–7, 139
Neutrons
    in artificial radioactivity, 58, 60–1
    Chadwick's search for, 14–5
    discovery of, 18–20, n.3-18
    fast, 88
    and human population behavior,
        n.12-7
    magnetic properties of, 77
    number per fission, 82–3, 87–8, 89,
        n.6-75
    and oil logging, 242
    origin of name, n.3-22
    in parody on *Faust*, 21
    Rutherford's 'prediction', 11, 21, 22
    search for, in cosmic rays, 50
    slow, 62–3
    thermal, n.5-38
New Mexico, 273–4
Nickel catalysts, 152, 155, 162–3, 179,
    238
NI Project (Japan), 280
Nitro-cellulose (plastic explosive),
    228–9
Nitrogen, 254
    fertilizers from, 36–7, 225
    induced nuclear transmutation of,
        4–5, n.2-6
    Norsk Hydro exports of, 114, 235
    production under deuteron
        bombardment of carbon, 57
Nobel Committee: Birkeland and,
    n.4-49
Nobel Foundation, 69
Nobel Institute for Experimental
    Physics (Siegbahn Institute), 69,
    71
Nobel Prize
    Barton, 42
    Becquerel, 7
    Calvin, 27
    Curies, 7
    Fermi, 63
    Giaque, 27
    Hassel, 41, 42
    Jensen, 164
    Joliot-Curies, 55
    Libby, 27
    Mayer, 164
    Seaborg, 27
    Siegbahn, 69
    Urey, 26, 27
Norden bomb sight, 171–2, 215

Norges Tekniske Høyskole (NTH)
    Brun's professorship, 288
    Tronstad's professorship, 33, 164
    Tronstad's training at, 32
Norsk Hydro-Elektrisk
        Kvælstofaktieselskab ('Norsk
        Hydro'), 33, 41, 102, 103, 141,
        146, 147, 148, 149, 157, 177, 178,
        254, 281, 288, 289, n.4-77
    and Allier mission, 104, 105–9
    American bomber attacks, 212,
        214–8
    ammonia production processes,
        37–8, 283
    anniversary of, n.4-50
    Badische affair, 44, 106
    and battle for Rjukan, 113–5
    Birkeland and, 36–7, 40–1
    Birkeland–Eyde arc process, n.4-53
    Board of Directors, 155, 213, 289,
        n.10-1
    commando attacks, 169, 192,
        193–205, 206–8, 225–33, n.12-39,
        n.12-43
    compensation for wartime
        operations and damage, 235
    and Dimona reactor, n.19-18
    Eyde and, 36–7, 40–1
    and first French reactor, 289
    founding, 37–8
    under German control, 115, 147–8,
        149–50, 158–9, 197, 206–8,
        211–3, 224–5, 235, 264, n.7-16,
        n.13-4
    German interest in heavy water,
        104–5, 107, 114, 155–6, 193, 248
    Glomfjord works, 282, 283
    heavy-water production
        discontinued, 284
    heavy-water production (postwar),
        282–3
    heavy-water production (prewar),
        45–8, 105
    heavy-water production (wartime),
        149–50, 151, 153, 158, 161–2,
        189, 236, 286
    Herøya works, 115, 148, 212–3, 283,
        n.13-16
    and I G Farben, 44, 105, 106, 107,
        113, 148, 149, 153, 235, 236, 283
    internal plant sabotage, 158–9
    and Kjeller reactor, 283
    and Nordisk Lettmetall AS, 148
    Notodden works, 38, 153, 159

and occupied France, 148, n.10-1
petrochemical initiatives, 283
possible German industrial
  sabotage, 264, 266–7
postwar startup strategies, 271
Rjukan works, 38, 212–3, 215–7
Swedish connection, 37–8
Såheim works, 153, 159, 224, 236,
  238
tritium preparation attempt, 133–4
Tronstad and, 41, 43–5, 150, 153,
  165–7, 177, 189, 193, 197–200,
  211–3, 224–5, 264–5
Vemork works, 43–4, 45, 150, 151,
  158–63, 165–6, 200–5, 206–8,
  209, 211–8, 224–6, 228, 236, 238,
  283–4
wartime heavy-water plant
  expansion, 150–5, 159–62, 178
See also Brun; deuterium; heavy
  water; Tronstad
Norsk Industriarbeidermuseum. *See*
  Norwegian Industrial Workers
  Museum
Norsk Teknisk Museum: Birkeland
  archives, 35
North Korea, 281
Norway
  Allied and Home Forces near end
    of WWII, n.17-12
  evacuation of, 112–3
  German forces in, near end of
    WWII, 264
  invasion of, 110–3, 114–5, n.7-24
Norwegian Army, mobilization of, 115
Norwegian BBC transmissions, 266
Norwegian Brigade, Scotland, n.17-12
Norwegian Engineering Society, 110
Norwegian gold bullion, evacuation
  of, 112, n.7-25
Norwegian High Command, London,
  156, 165–6, 168, 212, 213, 214,
  225, 227, 248, 264–7 *passim*, 269,
  n.10-59
Norwegian Home Forces. *See* Milorg
  Forces
Norwegian Independent Company
  No. One. *See* Linge Company
Norwegian Industrial Workers
  Museum, 232, 284, 288, n.13-2
Norwegian Light Metal Works, 148,
  212, n.13-16
Norwegian Resistance. *See* Milorg
  Forces

Norwegian Resistance Museum, 288,
  n.13-2
Norwegian War College, 288
Notgemeinschaft der Deutschen
  Wissenschaft, n.6-2
Notodden works, 38, 107, 115, 157,
  159, n.4-54
Novaya Zemlya, 35
NRX (reactor), 245, 246
Nuclear disintegrations: experiments
  of Cockcroft and Walton, 53
Nuclear energy: from fission, 77–8, 83,
  121
Nuclear fission
  Bohr's reaction to discovery, 79
  delayed neutrons from, 125
  energy release from, 77–8, 83, 121
  Hahn–Strassmann paper, 76–7
  Joliot-Curie's reaction to discovery,
    81
  liquid-drop model of, 77–8
  Meitner–Frisch interpretation, 77–8
  neutron multiplication, 82–3, 85–8,
    89–91
  origin of name, 80
  recoil experiments (Frisch), 80
  verification (Joliot), 81
  verification (in US), 81–2
Nuclear reactors, 93–4, 96, 102, 119,
  130, 138, 147, 163, 170, 179, 186,
  187, 191–2, 239, 252, 275, 281,
  287, 290, n.11-1
  and element 94, 138, 182, 274, 286
  fission product poisoning, 184–5
  graphite-moderated, 180, 278–9,
    n.15-20
  Harteck on, 224, n.12-19
  heavy-water moderated, 148, 278,
    279, 286
  heavy-water moderated (CANDU),
    284
  heavy-water moderated (CP-3),
    184–5, 239, 245–6, 298
  heavy-water moderated (Kjeller),
    283, n.19-5
  heavy-water moderated (NRX), 245,
    246, 284, n.15-20
  heavy-water moderated (Soviet),
    279, 299
  heavy-water moderated (ZEEP),
    245–6, 289, 299
  heavy-water moderated (Zoé),
    289

lattice designs, 98, 101, 125, 148, 210, 220, 222–3, 277
onset of reactivity, n.16-23
patents on, 94–5, 241
plutonium-producing, 184, 185, 243, n.15-20
for submarine propulsion, 102, 186, 287
theory of, 93–4, 97–8, 100, 102, 120, 121, 276–7, n.6-76, n.6-91
See also moderators; subcritical pile experiments
Nuclear transformations
under α-bombardment, 53
artificially induced, 4–5, 14
energy release during, 6

Oerlikon Machienfabrik, 145
Office of Scientific Research and Development, 180
Office for Strategic Services (OSS), 236, 248, n.16-5
Oil-well logging, 242
Optical polarization: in thin-film measurements, 32, 33
Oranienburg, 252, n.18-4
Oscarsborg, 111
Oslo, 32, 35, 41, 42, 47, 116, 149, 157, 165, 168, 196, 206, 209, 229, 231, n.7-39, n.12-26
and Allier mission, 1, 104, 105–9 *passim*, 119, 143
Bohr's 1940 visit, 110, 118
British legation, 112
escape of Cabinet and Parliament, 112
escape of Royal Family, 112
German attack on, 110–2, 115
German military liaison office, 151, 153, 236
German Minister in, 111–2
Gestapo headquarters, n.12-31
Norsk Hydro arc furnaces, 38
Norsk Hydro offices, 115, 224, 265
Reichskommissariat, 213, 224
Resistance Museum, 288
Oslo, University of, 34, 236
Birkeland demonstration, 33–4, 36
closed by Germans, 236
and Hassel, 42, 235–6
Wergeland's colloquium, 164, 193
Oslofjord, 1, 108, 109, 111–2
Oxygen, isotopes of, 22, n.4-12

Paraffin, 17
as moderator, 61–3
Paris
Alsos entry, 250–1
Bjarne Eriksen's visit, 104
during German occupation, 250–1
evacuation of (1940), 125–6
heavy-water evacuation from, 125
life in September 1939, 99–100
Préfecture of Police, 251
University of, 8
uprising (1944), 251
Pasteur Institute, 8
Pathfinder Mosquito aircraft, 221
Pechkranz electrolyzers, 44, 45, 47, 150, 159–61, 211
Peenemünde, 193, 269, n.10-36
Pergamon Press, 290
Petrochemicals: and Norsk Hydro, 283
Philips Company, Eindhoven, 251
'Phony War', 1, 114
Photon: creation/annihilation of, 51
Physikalische Berichte, 248
Physikalische Zeitschrift, 248
Physikalisch-Technische Reichsanstalt (PTR), 69, 130, 210, 220, n.8-13
Bothe at, 13, 15
Chadwick at, 13
'Pile': Fermi's use of term, n.11-1
'Pinnochio', 241
Pitchblende, 7, 95, 280
Platinum catalysts, 179, 181
Plutonium, 125, 138, 171, 184, 242, 243, 274, 277–8, 286, n.9-36
Poland
military campaign against, 1, 99
polonium named after, 7
Soviet attack on, 253, 254
Polonium, 9, 17, 52
α-emission from, 10, 19
discovery of, 7
naming of, 7
preparation of, 10
sources of, 9–10, 16, 19
toxicity, 10
Port Hawkesbury, 284, 285
Positive electron, 50–1
Positron
discovery of, 50–1, 53
prediction of, 51
term coined, 51
theory of, 51, 53
Potash lye, 44, 45, 225–6, 232

Poynting vector: Birkeland's
    expression for, 34
'Preparation 38'. *See* uranium oxide
Princeton University: Physics Journal
    Club, 81
'Product Z', 126
Project Y, 273
Protactinium (element 91), 68
Proton
    compound structure of, 52
    first established as nuclear
        constituent, 5
    recoils, from γ-ray impacts, 17
    in slowing down neutrons, 62–3
    term coined, n.2-4
Proximity fuse, 219
Pumice stone, 152

Quantum theory, 52, 59
Quebec Conference (1943), 244
Quisling's Party. *See* Nasjonal
    Samling

Radar research, British, 120
Radiation Laboratory, MIT, 249
Radioactive tracers, 65
Radioactivity, 5
    apathy towards discovery, 7
    artificial, 54–6, 57–8, 61
    atomic energy from, 6
    as an atomic property, 7
    Curies' researches on, 7–8
    discovery of, 7
    of tritium, 134
Radiochemistry: laboratory
    procedures, 64–5, 74–5
Radiothorium, 67
Radium
    Chadwick's student project on, 12
    Curie's gift, 9
    discovery of, 7
    'isotopes' of, in Hahn–Strassmann
        investigation, 74–6
    world's leading producer, 95
Radium Institute (Leningrad), 275, 276
Radium Institute (Paris), 95, 145
    founding of, 8
    Joliot-Curies at, 8, 9, 55–6, 64–5, 98,
        289
Radon, 60, 88
Rauland, 209, 267, n.17-15
Rebecca receiving device, 195

Red Army, 186, 253, 254, 271, 277,
    n.17-12
Reich Education Ministry, 66, 71, 130
Reich Research Council, 66, 135, 146,
    186–8 *passim*, 224, 235, 251, 254,
    260
Rensselaer Polytechnic Institute, 142,
    290
Reproduction factor, n.6-91
Research Corporation, 95
Restoration Branch, US Military
    Government, 282
Rhodesia, 36
Rhodium foil, 88
Ricerca Scientifica, 58
Riken Institute, 279
    Nuclear Research Laboratory,
        279–80
Riom, 126, 127
Rjukan, 33, 37, 49, 105, 107, 115, 150,
    157, 158, 166, 167, 168, 192,
    194, 197, 238, 282, 288, n.4-54,
    n.7-17, n.12-26
    Allier visit, 107
    battle for (1940), 114–5
    bombing raid, 214–7, 236
    geographic setting, 38–9, 113, 200–1
    growth of town, 40
    and *Gunnerside* attack, 200–5 *passim*
    and Herøya raid, 212–3
    and *Hydro* sinking, 226–7, 232–3
    Norsk Hydro first established at, 38
    and operation *Sunshine*, 265, 267–8,
        n.17-15
    origin of name, 38
    waterfalls, 37, 38–9
Rome
    and *Alsos*, 247–8, n.16-5
    fall of (1944), n.16-5
Rome, University of
    experimental physics, *c.* 1927, 60
    Fermi appointed to chair in physics,
        60
Rothschild Foundation, 8
Royal Air Force: Bomber Command,
    221, 228
Royal Society
    discussion on heavy hydrogen
        (1933), 27, 30–1, 41, 46, n.4-17
    Rutherford on, 61
Royan, Bay of, 129, 143, 149
Ruhleben camp: Chadwick's
    internment at, 13–4
Ruhr Dams, 209

Sahara: French nuclear test in, 96
Saltpeter, 37
San Francisco Chronicle, 82
Scapa Flow, 110, 113
Schering Quinine Factory, Leipzig, 189
Schotten Gymnasium, Vienna, 132
Science News Letters, 51
Scrubbing procedure, in
     Harteck–Suess process, 152
Scuola Normale Superiore, 59
Secret Intelligence Service (SIS), 156,
     165, 168, 193, 248, 291, n.12-36
Sevastopol, 163
'Shell model' of atomic nucleus, 164
'Shetland Bus', 157
Siemens Company, 139, 141
Skagerrak, 2
Skitten Airdrome, 192, 195, 196
Skoland Marsh, 192, 195
Skylark, 116
S-matrix theory, Heisenberg's, n.14-4
Société des Terres Rares, 252
Sodium carbonate, 52
Sola Airfield, 110
Solvay Conferences, 52
   first (1911), 52
   seventh (1933), 52–3, 56, 57
Sorbonne, the, 8, 9, 55
Soviet All-Union Conference on
     Isotopes (1940), 276
Soviet All-Union Conference on
     Nuclear Physics (1940), 276
Soviet intelligence mission (Germany,
     1945), 271–2
Space physics: Birkeland's
     contributions to, 34–6
Spain, 114
Special Forces. *See* Special Operations
     Executive
Special Operations Executive (SOE),
     157, 165, 169, 193, 194, 195, 199,
     211, 248
   creation of, 156
   Norwegian Section of, 156, 198, 205,
     227, 228, 264, 290
   and SIS, 156, 193
Special Training School Number 17,
     198–9, 267
Special Training School Number 61
     (Farm Hall), 199, 272
Springer Verlag, 69, 89, 254, 290,
     n.5-58
Stadtilm 'laboratory', 224, 255–6, 260
Stagg Field, Chicago, 180, 279

Stalingrad, 275, 277
Stalin Organ, 233, 234, 237
Standard Oil Company
   and Cominco, 179, 181
   pilot heavy-water plant, 176
'Starlight', n.17-15
Stern–Gerlach experiment, 70, 219
Stockholm
   Academy of Sciences, 71
   Norwegian military office, n.10-59
   US Embassy, 236
Storting (Norwegian Parliament), 37,
     112
Strasbourg, University of, 251–2
Stuart Oxygen Co., 179
Subcritical pile experiments, 276
   in Berlin, 150 (B-I); 188 (B-II); 189
     (B-III); 222–3 (B-VI); 223 (B-VII);
     254, 256–8, 259, 260–2 (B-VIII)
   at Cambridge, 141–2, 174, 175–6, 189
   at Chicago, 180, 181
   at Columbia, 97–8, 170–1, 173–4, 287
   in Gottow, 210 (G-I); 210–1, 219
     (G-II); 219, 221 (G-III)
   in Hamburg, 142
   at Ivry, 98–9, 100–1, 125
   at Leipzig, 150–1, 188 (L-I); 189
     (L-II); 189 (L-III); 189–91, 210–1,
     297, n.12-16 (L-IV)
   in Montreal, n.15-12
   Thomson's, 97, 120
Sudbury, Ontario, 286
Sudbury Neutrino Observatory:
     heavy-water application, 286
Sunshine, 264–7, 269
   aftermath, n.17-28
   objectives, 264, 266–7
   operational areas, n.17-15
   Tronstad's demise, 268–9
Supernovae, 286
Supreme Headquarters, Allied
     Expeditionary Forces (SHAEF),
     214
Svalbard, 35
Swallow, 198–200
Sweden: Norwegian police troops in,
     n.17-12
Såheim works, 159, 224, 232, 238,
     n.10-55

Tailfingen, 222, 254, 262
Technetium, n.6-16

Telemark, 38–40, 49, 114, 115, 162, 163, 264
  Milorg forces, 264–5, 268, 269
  Mount Gausta, 38
  Vestfjord Valley, 38, 108, 113, 200, 201, 204, 209, 215, 284
  See also Hardanger Plateau; Lake Tinn; Rjukan; Vemork
Telemark Regiment, 114
'Terrella', Birkeland's, 34–5
The Shape of Things to Come, 83
The Times of London, 27, 83
The World Set Free, 6, 83
Thomson Committee, 118–9
Thorium
  commercial uses, 252
  radioactivity of, 7, 14
Thorium investigation
  by Hahn and Strassmann, 73–4
  by Irène Curie, 64–5, 73–4
Thorium X (Th-X), 67
Thorpe Abbotts, 214
Tinnoset, 226
Tinnsjøen. *See* Lake Tinn
Tizard Mission, 171–2, 174
Tobacco seeds, germination of, 30
Todt Organization, 148
Tokyo, 34
  Imperial University of, 280
  See also Riken Institute
Top Policy Group, 172–3
Toulouse: uranium stocks at, n.9-53
Trail, British Columbia, 178. *See also* Cominco
Trandum, 196, n.12-31
Transuranic elements, 72, 73, 78, 138
  Fermi's search for, 61–2
  See also actinides; neptunium; plutonium
'Trennrohr', (Clusius–Dickel tube), 120–1
Trinity College, Cambridge, 141
Trinity test, 273–4
'Triterium', 133
Tritium, 133
  Cambridge analysis of Norsk Hydro sample, 134
  natural abundance of, 134
  radioactivity of, 134
  stability of, 133
Tritium oxide: Norsk Hydro's attempt to isolate, 133
Troms County, Norway, 115

Trondheim, 110, 112, 115–6, 150, 153, n.7-38, n.7-39, n.10-42, n.17-30
'Trondheim cables', 116
Troyes, 146
Tube Alloys, Directorate, 239, 240, 242, 248, 249
Tungsten, 286
Tübingen, 253

Ugleflåt, 265
U-boats
  German campaign, 240
  German mission to Japan, 280
  nuclear reactors for, 102, 186, 287
Union Minière du Haut-Katanga, 95, 96, 251, 252
University of California: Department of Chemistry and G N Lewis, 27, 28, 291
Uranbrenner. See nuclear reactors
Uranium
  activity of, 7
  barium formation, 81
  first available in metallic form for German pile experiments, 189
  Hahn's investigation, 66, 74–6, 77–8
  hexafluoride, 280
  industrial demands for, 96
  Irène Curie's investigation, 64–5
  nitrate, 87
  oxide, 93, 96–7, 120, 125, 135, 141–4, 174, 188–9, 210, 276, 280, n.19-5
  sources of, 95, n.18-4
Uranium-235
  critical mass of, 120, 121, 275
  fission of, confirmed, 120, 171, 275
  fission of, predicted, 101, 138, 275
  geometric cross section of, 121
  as nuclear explosive, 172, n.9-36
  versus plutonium, 277–8
  separation of, 120–1, 122, 280
Uranium-238
  resonance capture of neutrons, 97–8, 125
  uranium-239 from, 101, 138
'Uranium Club': history of, 89, 130, 248
Uranyl nitrate, 93
Urfeld, Bavaria, 222, 253, 254, 256, 263

V-1, V-2 flying bombs, 192, 269, n.10-36
Van de Graaff accelerators, 82, 279

Vemork, 39, 105, 107, 117, 157, 178, 189, 288
  bombing raid on, 214–8, 219, 236, 286
  electrolysis plant, 42–4, 113, 283–4
  Freshman operation and, 194–7
  geographical setting, 167, 194
  Gunnerside attack on, 198–205, 206–8, 209, 288
  heavy-water plant, 33, 39, 43–5, 47–8, 147, 150–3, 155, 156, 158–63, 165–8, 177, 211–2, 224–5, 236, 238, 282, 286, n.10-42, n.12-34
  and Herøya raid, 212–3
  and *Hydro* sinking, 226–7, 229, 231, 232
  power station, 157, 236, 284, 285
  See also Norwegian Industrial Workers Museum
Versailles, Treaty of, 3
Vichy regime, 261
Vidda. *See* Hardanger Plateau
Vilnius, 84
'Virus House', 150, 155, 188
Voss, 116
Våer hamlet, 201, 202
Vaagsøy, commando raid on, n.12-37

Wallenberg Foundation, 37, 69
Washington, DC: British Mission to, 245
Washington Conferences on Theoretical Physics, 79, 81–2
'Washington Cup', 195

Water: discrepancy in density measurements, 24
Wehrwirtschaftstab Norwegen, Schoepke's role in, n.10-18
Weser Day, 110
Weserübung, 1–2, 110
Wick, Scotland, 192, 196
Wilson cloud chamber, 10, 50–1, 279
Windsor Castle: heavy water and, 129
'World room', Birkeland's discharge vessel, 34–5
World War I, 6
  Chadwick's experience in, 13–4
  end of, 3
  Geiger's experience in, 14
  Lewis's experience in, 28
  Norway during, 44
  Rutherford's experience in, 4, 5–6
  Urey's experience in, 24

Xenon, 185
X-ray diffraction, 59
X-rays, discovery of, 7
XU organization, 164, 165, 193, 248, n.10-59

Yalta Conference (1944), 252
Yeast cultures, growth of, 30
Yugoslavia, 264

Z, neutron propagation factor, 223
ZEEP (reactor), 245–6, 289
Zoé (reactor), 289
Zones of occupation, Allied, 252–3